T0156001

Computer Communications and Networks

The **Computer Communications and Networks** series is a range of textbooks, monographs and handbooks. It sets out to provide students, researchers, and non-specialists alike with a sure grounding in current knowledge, together with comprehensible access to the latest developments in computer communications and networking.

Emphasis is placed on clear and explanatory styles that support a tutorial approach, so that even the most complex of topics is presented in a lucid and intelligible manner.

More information about this series at http://www.springer.com/series/4198

Reinhard Gotzhein

Real-time Communication Protocols for Multi-hop Ad-hoc Networks

Wireless Networking in Production and Control Systems

Reinhard Gotzhein
Chair for Networked Systems
Technische Universität Kaiserslautern
Kaiserslautern, Rheinland-Pfalz, Germany

ISSN 1617-7975 ISSN 2197-8433 (electronic)
Computer Communications and Networks
ISBN 978-3-030-33321-8 ISBN 978-3-030-33319-5 (eBook)
https://doi.org/10.1007/978-3-030-33319-5

This Springer imprint is published by the registered company Springer Nature Switzerland AG
The registered company address is: Gewerbestrasse 11, 6330 Cham, Switzerland

I dedicate this monograph to my mentors Lothar Heinhold, Gregor von Bochmann, and Jürgen Nehmer, whom I thank sincerely for their enduring support and advice in different phases of my scientific career.

Preface

The use of wireless real-time communication technologies for the flexible networking of sensors, actuators, and controllers is a crucial building block for future production and control systems. With WirelessHART and ISA 100.11a, two technologies that have been conceived for industrial use are currently available. Nevertheless, our research on wireless real-time communication systems has shown further potential, resulting in more flexible, versatile, and robust solutions that can be implemented on today's low-cost and resource-constrained hardware platforms.

This monograph presents fundamentals, results, and trends of communication protocol research that enable flexible real-time multi-hop networking in production and control systems, using wireless communication technologies. Core functionalities such as time synchronization, global medium slotting, automatic topology detection, medium access schemes, deterministic arbitration, duty cycling, dynamic quality of service multicast routing with mobility support, network clustering, and application-specific middleware are identified and described, and suitable protocols are presented. For real-time support, deterministic protocols, i.e., protocols with predictable behavior, are required; this concerns, e.g., precise bounds for synchronization accuracy and transfer delays, exclusiveness of medium access, uniqueness of the winner of network-wide arbitration, and timely and reliable delivery of packets.

While state-of-the-art and state-of-the-practice in the area are surveyed and assessed, the focus of this monograph is on protocols that have been developed at the Chair for Networked Systems of the University of Kaiserslautern. Each protocol is presented in detail, and it is shown how these protocols are integrated into the wireless real-time multi-hop communication system ProNet (Production Network) 4.0. Due to its extensive functionality, its real-time support, its application-specific middleware, and its high degree of flexibility, ProNet 4.0 is particularly suitable for use in future networked production and control systems. ProNet 4.0 has been prototypically implemented on the Imote2 hardware platform equipped with the Chipcon CC2420 radio transceiver and has been installed and operated in the SmartFactory^{KL} at the German Research Center for Artificial Intelligence (DFKI) in Kaiserslautern.

The book focuses on core functionalities for wireless real-time multi-hop networking with time-division multiple access (TDMA) and their integration into a flexible, versatile, fully operational, and self-contained communication system. The protocols presented in the book are located on medium access control (MAC) layer and above and build on the PHYsical (PHY) layer of standard wireless communication technologies. This is different from existing books on wireless communication, which mainly treat functionalities located on PHY and MAC layer, and which do not focus on real-time multi-hop networks and self-contained real-time protocol stacks.

The book is directed toward graduate students, Ph.D. students, researchers, and technical professionals in the disciplines of computer science, control engineering, electrical engineering, mechanical engineering, and production engineering. The chapters of the book are largely self-contained and can be studied in any order. However, it is recommended to read the introductory Chap. 1 first, followed by other chapters of particular interest, preferably in the order of the book. Chapters 2–10 present results that are put into a larger context. They all start with a subchapter on foundations, followed by subchapters on results of the author and his research group, related work, and conclusions. Chapter 11 is on implementation aspects, Chap. 12 presents case studies, and Chap. 13 summarizes results and gives an outlook.

Kaiserslautern, Germany Reinhard Gotzhein

Acknowledgements

This monograph summarizes and extends research work that has been done at my Chair for Networked Systems at the Computer Science Department of the University of Kaiserslautern in Germany over the past 12 years. It all started with ideas about deterministic tick synchronization, resulting in the protocol Black Burst Synchronization (BBS) first published by Thomas Kuhn and myself in 2008. We soon realized that this is a core functionality of wireless real-time communication systems, and subsequently developed further ideas and deterministic protocols, integrated them into a protocol stack later called Production Network (ProNet) 4.0, realized a prototypical implementation, and showed feasibility in several case studies. Involved in this work have been my Ph.D. students (in alphabetical order) Philipp Becker, Tobias Braun, Dennis Christmann, Markus Engel, Ingmar Fliege, Alexander Geraldy, Anuschka Igel, Marc Krämer, Christopher Kramer, and Thomas Kuhn, my colleagues Thomas Jaitner, Lothar Litz, Stefan Siegmund, and Fabian Wirth, and many Master students. I am indebted to all of them for their dedicated teamwork and their individual contributions to this large topic. I am also grateful for valuable feedback and encouragement from my colleagues during the early stage of writing this monograph, especially from Ferhat Khendek, Steven Liu, Parya Memar, Arnd Poetzsch-Heffter, and Jens Schmitt, and for constructive input from my Ph.D. students toward completion.

Contents

About the Author

Reinhard Gotzhein is holding the Chair for Networked Systems at the University of Kaiserslautern. He received his Ph.D. from the University of Erlangen-Nuremberg in 1985 and his habilitation from the University of Hamburg in 1992. During this period, he spent two one-year research stays (Postdoc, Invited Professor) at the University of Montreal in Canada. In 1993, he joined the University of Kaiserslautern, where he is currently holding a full professor position. His areas of research include communication systems, ad hoc networks, protocol engineering, and formal description techniques. Reinhard Gotzhein has published about 150 reviewed scientific papers. He has chaired and co-chaired several international and national symposia and workshops, has served on 80+ programme committees, on the editorial board and as guest editor of several scientific journals, and as speaker and vice-speaker of coordinated research projects.

Abbreviations

ABT	Assisted Bicycle Trainer
ACE	Algorithm for Cluster Establishment
ACK	Acknowledgement
ACTP	Arbitrating and Cooperative Transfer Protocol
AMRA	Asset Monitoring for Rail Applications
AODV	Ad hoc On-demand Distance Vector
ASN	Absolute Slot Number
ATDP	Automatic Topology Discovery Protocol
BAS	BiPS application scheduler
BBC	Black Burst Contention
BBQR	Black Burst-based QoS Routing
BBS	Black Burst Synchronization
BGP	Border Gateway Protocol
BiPS	Black burst integrated Protocol Stack
BRTS	BiPS real-time scheduler
CAN	Controller Area Network
CAP	Contention Access Period
CBT	Channel Busy Time
CCA	Clear Channel Assessment
CFP	Contention Free Period
CIFS	Contention Inter Frame Space
CORBA	Common Object Request Broker Architecture
CPU	Central Processing Unit
CSMA	Carrier Sense Multiple Access
CTS	Clear-To-Send
DCE	Distributed Computing Environment
DCF	Distributed Coordination Function
DCOM	Distributed Component Object Model
DIFS	DCF Inter Frame Space
DMA	Direct Memory Access

DNS	Domain Name System
DSH	Decentralized Sync Header
DSRP	Distributed Slots Reservation Protocol
EDCA	Enhanced Distributed Channel Access
EDF	Earliest Deadline First
ERP	Enterprise Resource Planning
FCFS	First Come, First Served
FDMA	Frequency-Division Multiple Access
FF	First Fit
FFD	Full Function Device
FIFO	First-In, First-Out
FSF	Fewest Slots First
GDB	GNU Debugger
GPIO	General-Purpose Input/Output
GPS	Global Positioning System
GSM	Global System for Mobile Communications
GSR	Global State Routing
GTS	Guaranteed Time Slot
HART	Highway Addressable Remote Transducer
HCF	Hybrid Coordination Function
HNC	Heterogeneous Network Clustering
IoT	Internet of Things
KPI	Key Performance Indicator
LAN	Local Area Network
LBT	Listen Before Talk
LCA	Linked Clustering Algorithm
LCF	Least Conflicts First
LDC	Local Data Collector
LLC	Logical Link Control
LWB	Low-power Wireless Bus
MAC	Medium Access Control
MACAW	Multiple Access with Collision Avoidance for Wireless
MAD	Maximal Allowable Delay
MATI	Maximal Allowable Transfer Interval
MEF	Maximum-Error-First
MEP	Message Exchange Pattern
MMU	Memory Management Unit
MP	Master Present
MRF	Most Reuse First
MSC	Message Sequence Chart
NAV	Network Allocation Vector
NCS	Networked Control System
NSF	Nearest Slot First
NTP	Network Time Protocol
OLSR	Optimized Link State Routing

OMG	Object Management Group
OOK	On–Off Keying
OPC	Open Platform Communications
OS	Operating System
OSF	Open Software Foundation
OSI	Open Systems Interconnection
OSPF	Open Shortest Path First
PAN	Personal Area Network
PCF	Point Coordination Function
PHY	PHYsical
PIFS	PCF Inter Frame Space
ProMid	Production Middleware
ProNet	Production Network
QMR	QoS Multicast Routing
QoS	Quality of Service
RBS	Reference Broadcast Synchronization
RCRM	Remote Clock Reading Method
RFD	Reduced Function Device
RFF	Rotating First Fit
RID	Radio Interference Detection
RMAC	Routing-enhanced MAC
RMI	Remote Message Invocation
RREP	Route REPly
RREQ	Route REQuest
RSS	Received Signal Strength
RSSI	Received Signal Strength Indicator
RTS	Request-To-Send
SDMA	Space-Division Multiple Access
SFD	Start of Frame Delimiter
SIFS	Short Inter Frame Space
S-MAC	Sensor-MAC
SMTX	Smallest Maximum number of Transmissions
SNR	Signal-to-Noise Ratio
SOA	Software-Oriented Architecture
SOAP	Simple Object Access Protocol
SOF	Start of Frame
SOTF	Start of Time Frame
TCP	Transmission Control Protocol
TDMA	Time-Division Multiple Access
TEP	Transmission End Point
TO	Transmission Opportunity
TOD	Try-Once-Discard
TPSN	Time-sync Protocol for Sensor Networks
TSN	Time-Sensitive Networks
TSP	Transmission Start Point

TTP	Time-Triggered Protocol
UA	Unified Architecture
UART	Universal Asynchronous Receiver–Transmitter
UDDI	Universal Description, Discovery and Integration
UDP	User Datagram Protocol
UWBR	Unusable Wasted Bandwidth Ratio
VI-P	Vertical Integration of Production processes
VoIP	Voice over IP
W3C	World Wide Web Consortium
WAN	Wide Area Network
WCET	Worst-Case Execution Time
WCF	Windows Communication Foundation
WiDom	Wireless Dominance
WLAN	Wireless Local Area Network
WNCS	Wireless Networked Control System
WSDL	Web Services Description Language
WSN	Wireless Sensor Network
ZDO	ZigBee Device Object

Chapter 1
Introduction

This chapter establishes the context and scope of this book and gives a survey of its topics. We discuss the notion of real-time system, address control systems as an important application area, identify technological challenges of using wireless communication technologies, and briefly survey the state-of-the-practice.

1.1 Context and Scope

"A *real-time computer system* is a computer system in which the correctness of the system behavior depends not only on the logical results of the computation, but also on the physical instant at which these results are produced. … [It] is always part of a larger system … called *real-time system* [38]." Typical real-time systems are control systems [33] regulating the operation of dynamic technical systems, e.g., air conditioning systems, antilock braking systems, or automated production systems.

A real-time computer system may be distributed, consisting of a set of nodes connected via a real-time communication system. These systems exhibit specific requirements regarding performance, reliability, guarantee, and synchronicity. Performance addresses efficiency aspects concerning resources and timeliness, e.g., processing/transmission rates and delays. Reliability refers to the availability of system operation, for instance, downtimes or packet loss rates. Guarantee defines the degree of commitment, e.g., best effort (no commitment) or deterministic guarantee (hard commitment). Synchronicity is another type of requirement, referring to the coordination accuracy of concurrent activities.

To connect the nodes of a distributed real-time computer system, wired technologies are in general preferable, as the quality of links is highly predictable. On the downside is the need for cables, which reduces flexibility in terms of node placement and topology changes due to node mobility or replacement. Here, wireless communication technologies can play an important role. To increase flexibility, wireless networks should be self-organizing, as in mobile ad hoc

R. Gotzhein, *Real-time Communication Protocols for Multi-hop Ad-hoc Networks*, Computer Communications and Networks,
https://doi.org/10.1007/978-3-030-33319-5_1

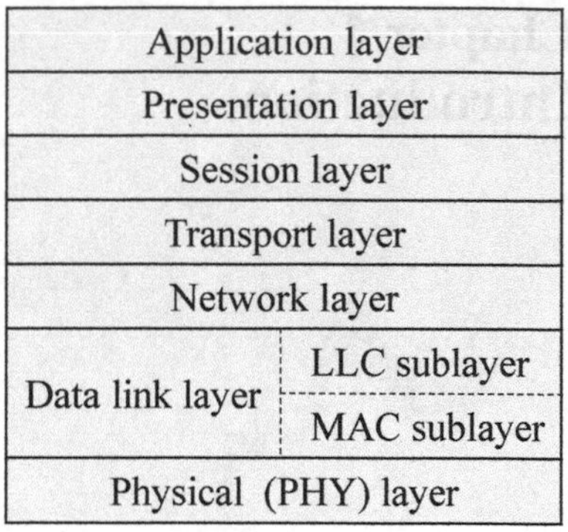

Fig. 1.1 Extended Open Systems Interconnection (OSI) reference model [32]

Application layer
Presentation layer
Session layer
Transport layer
Network layer
Data link layer: LLC sublayer, MAC sublayer
Physical (PHY) layer

networks. This raises the question whether wireless networks can be used in real-time systems, with strong requirements on predictability of operation.

This book focuses on wireless communication in distributed real-time computer systems. More specifically, core functionalities for wireless real-time multi-hop networking with time-division multiple access (TDMA) and network self-organization, and their integration into a flexible, versatile, fully operational, and self-contained communication system are treated. The communication protocols presented in the book are located on Medium Access Control (MAC) layer and above and built on the PHYsical (PHY) layer of standard wireless communication technologies (see Fig. 1.1). This is different from the existing books on wireless communication, which mainly treat functionalities located on PHY and MAC layer, and which do not focus on real-time multi-hop networks and self-contained real-time protocol stacks.

1.2 Wireless Networked Control Systems

A specific type of real-time system is the control systems regulating the operation of dynamic technical systems, with the objective to achieve optimal system behavior. Modern control systems learn and memorize dynamic behavior and variable targets, to provide stable and robust operation in the sense that defined margins of operation are reached and maintained, despite dynamic disturbances. For instance, the stabilization criterion of an air conditioning system is to reach and maintain target temperature and moisture, in the presence of disturbances caused, for instance, by persons and/or air exchange with the environment. Here, the quality of control can be characterized by the ability to compensate for disturbances, stabilization delay—the duration until a stable system state is (re-)established—oscillation behavior during stabilization resulting in deviations from stable system states, and predictability of the stabilization process.

To achieve its objective, a control system is composed of three types of devices. Sensors sample the state of the technical system, e.g., temperature or velocity.

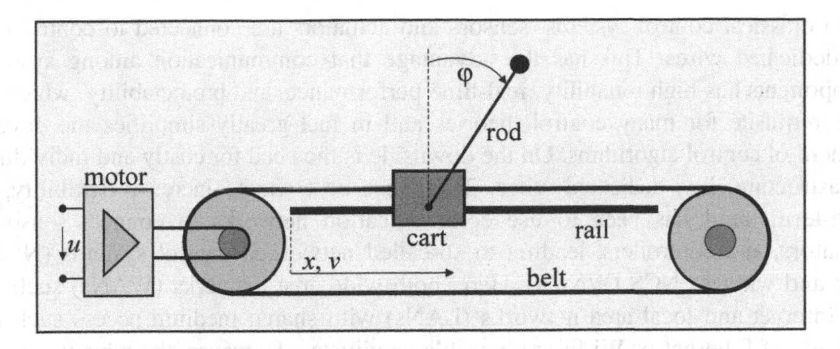

Fig. 1.2 Inverted pendulum (see [18], Fig. 2)

Actuators apply steering values to stabilize the technical system. Controllers determine the current state of the technical system by gathering sensor values, aggregating and processing these values, calculating appropriate steering values, and applying these values through actuators.

A well-known example of a control system is the inverted pendulum shown in Fig. 1.2. Here, the technical system consists of a cart that is mounted to a belt. The belt can be (de-)accelerated by a motor in both directions such that the cart moves back and forth, with changing velocity. Attached to the cart is a rod (pendulum) with a weight at the end that can swing around one axis. Control objective is to stabilize the rod in an upright position, by moving the cart. Disturbances are caused, for instance, by frictions of the rail, reaction delays of the motor, and pushes to the rod. The state of this technical system is characterized by position x and velocity v of the cart, and by the angle φ of the rod, and is sampled periodically, e.g., every 36 ms, by sensors. Here, it is essential that sampling of measured values occurs synchronously. From the history of samplings, a voltage u applied to the motor to (de-)accelerate the cart is calculated by the controller and applied. Thus, the cart is moved to swing the rod into an upright position and keep it there. Obviously, the ability to stabilize the rod depends on various factors, such as precision of measurements, length of the rail, geometry of the pendulum, strength and reaction time of the motor, sampling period, and sophistication of the control algorithm.

The inverted pendulum is an example of a *closed-loop* (or feedback) control system, which acquires dynamic feedback about the system status to determine and apply proper steering values to achieve defined stabilization objectives. In *open-loop* control systems, there is no direct dynamic feedback from the system itself; control action is independent of the system state, and, for instance, triggered by a clock or a human operator. An example of this type of control system is the collection and aggregation of production parameter values, e.g., key performance indicators (KPIs) such as the ratio of broken and produced pieces. KPIs may then be monitored, for instance, via an enterprise resource planning (ERP) system, without immediate feedback to the production line.

In classical control systems, sensors and actuators are connected to controllers by dedicated wires. This has the advantage that communication among system components has high reliability, real-time performance, and predictability, which is a prerequisite for many control theories and in fact greatly simplifies the development of control algorithms. On the downside is the need for costly and individual infrastructure, i.e., dedicated wires. To reduce cost and to increase flexibility, a long-term trend has been to use communication networks to connect sensors, actuators, and controllers, leading to so-called networked control systems (NCS) [34] and wireless NCS (WNCS). Here, both wide area networks (WANs) such as the Internet and local area networks (LANs) with shared medium access such as variants of Ethernet or Wi-Fi are possible candidates. However, the advantages of dedicated wires, namely high reliability, real-time performance, and predictability, are typically lost when general-purpose technologies are applied. This calls for the conception of more elaborate control theories, e.g., predictive controllers, constraints on the use of communication systems, e.g., traffic control, and customized communication technologies.

In Fig. 1.3, the inverted pendulum is revisited. Sensor values x_k (position of the cart), v_k (velocity of the cart), and φ_k (angle of the rod), which are sampled periodically and synchronously at points in time t_k, are now communicated over a wireless network to the controller, which calculates steering values u_k (voltage to be applied to the motor) and sends them to the actuator. To determine steering values, the controller may take the entire history of sensor values at discrete points in time t_1, …, t_k into account. Due to the nature of the (wireless) network and the traffic load, sensor values as well as steering values may be delayed, corrupted, or lost. Furthermore, synchronicity of samplings may be reduced, as synchronization is to be achieved by message exchange, too. Obviously, this will have an impact on the achievable quality of control. To cope with this situation and to achieve stability of the inverted pendulum, a more advanced control algorithm and a customized communication system are required. Yet, if delays and/or losses exceed certain thresholds, stability will suffer or be out of reach.

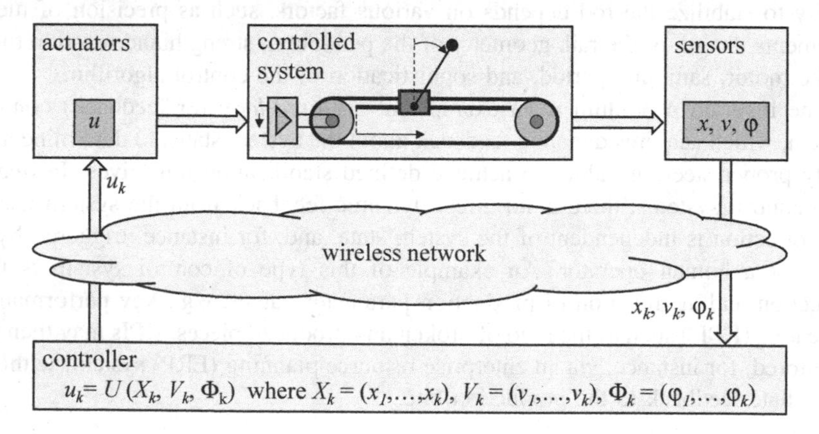

Fig. 1.3 Wireless networked control system "inverted pendulum"

1.3 Technological Challenges

Real-time computer systems exhibit specific requirements regarding performance, reliability, and guarantee. Matching these requirements is a technological challenge. Execution of real-time computer systems is usually decomposed into tasks, which are triggered by the occurrence of events (event-triggered) or at determined points in time (time-triggered). To meet hard deadlines, the worst-case execution times (WCETs) of tasks must be known, a problem that is aggravated by measures for efficiency improvements such as memory caches and multi-core architectures. Furthermore, to guarantee schedulability of tasks, their occurrence in time has to be sufficiently predictable. In distributed real-time systems such as networked control systems, this also applies to message exchange, i.e., communication tasks. In summary, to realize hard real-time requirements, the worst-case timing behavior of a real-time computer system has to be entirely predictable. This calls for deterministic operating system functionalities, and for deterministic communication protocols.[1]

Another technological challenge is the fact that real-time computer systems often operate under strong resource limitations, e.g., regarding computation, data storage, signal conversion, communication, and energy. These limitations are the result of the pressure to reduce size, weight, and cost of hardware and operation. Furthermore, there is a strong trend toward infrastructureless systems, i.e., systems with battery-driven nodes and wireless transceivers, which demands, in particular, energy-efficient hardware and operation.

The trend toward wireless real-time computer systems, where nodes communicate via radio, is a further technological challenge. Connectivity in wireless networks depends on channel quality, which can vary due to noise, interference, terrain, and weather conditions. Furthermore, connectivity can vary due to node mobility, node density, adaptive and heterogeneous transmission power, reflections, and the operational mode (e.g., idle, receive, transmit, power down) of nodes.

Operation of distributed real-time computer systems over wireless communication technologies requires, among other things, predictable communication behavior regarding packet delay and packet loss. To achieve such behavior, we adopt three complementary measures. First, we place an assumption on network connectivity, called "single-network property" (defined in the next paragraph). Second, we devise and implement deterministic communication protocols that provide reliable message exchange under this assumption (content of this book). Third, we assume that the coupled system can cope with situations where network connectivity is impaired, e.g., by applying predictive algorithms or by moving to a fail-safe state (not further addressed in this book).

[1]Though it is feasible in practice to develop such systems, it is often sufficient to relax hard real-time requirements to stochastic requirements, where real-time constraints are satisfied on average or with a specified probability.

We define the *single-network property* as follows: Let V be a set of nodes, and let E be a set of directed labeled edges between subsets $V \times V$ of pairs of nodes. An edge (v, v') is labeled by c, if there is a communication link from v to v', i.e., if in the absence of interference by other nodes, v' can receive messages sent by v reliably[2]. It is labeled by i, if there is an interference link, i.e., v can distort receptions of v', and labeled by s, if there is a sensing link, i.e., v' can detect transmission energy of v. We assume that a communication link is also an interference link, and that an interference link is a sensing link too. The single-network property holds if and only if there are paths of communication links between all pairs of nodes of V, and no nodes running a different protocol and using the same radio channel are in interference or sensing range of any node in V.

Certainly, in wireless networks, the single-network property is a strong assumption that needs justification. It is obvious that the property is a prerequisite to achieve predictable communication behavior, as required for distributed real-time systems. A straightforward solution is the use of licensed radio channels, where license owners have complete access control (dimension "frequency"). A possible threat is the additional use of such frequencies by cognitive radio networks with non-perfect secondary users. Another solution is the operation of distributed real-time computer systems on protected space—private sites with controlled access, e.g., production facilities (dimension "space"). Here, a possible threat is the operation of foreign transceivers in sensing range. The property can also be satisfied by the use of protected time intervals[3] (dimension "time"), with the same threat, i.e., foreign transceivers in sensing range. Finally, the three approaches can be combined. If the mentioned possible threats cannot be ruled out, the third assumption that the distributed real-time system can cope with situations where network connectivity is impaired is needed.

1.4 State-of-the-Practice

For several years, there is an increasing drive to use wireless communication technologies in networked control systems. With WirelessHART [40] and ISA 100.11a [37], there exist two wireless technologies that have been specifically devised for control purposes in industrial plants. Both technologies support time-division multiple access (TDMA) with reservations for exclusive medium

[2]In graph-based network models, links are assumed to be reliable. Other types of models are physical models, which are based on propagation properties and stochastic models defining link reliabilities.

[3]As supported by network slicing in the context of 5G networks.

access and multi-hop networks. Therefore, maximal transmission delays, freedom of frame collision[4], and synchronicity can be guaranteed in principle. Nevertheless, there are several drawbacks as follows:

- Both WirelessHART and ISA 100.11a are based on the 2450 MHz DSSS physical layer of IEEE 802.15.4, which has a low gross data rate of only 250 Kbps per channel. Although both technologies can use multiple channels, this strongly limits achievable net data rates.
- For the operation of wireless networks, the knowledge about communication and interference topologies is crucial to ensure interference-free time slot reservations. Neither WirelessHART nor ISA 100.11a supports the automatic detection of these topologies.
- To meet real-time requirements, it is crucial that there are reliable paths of communication links between all pairs of network nodes, and that there are no nodes in interference range running a different protocol and using the same radio channel(s). Neither WirelessHART nor ISA 100.11a checks this single-network property (see Sect. 1.3). In fact, practical experiments with WirelessHART have shown a strong increase of frame losses when WLAN nodes are operated simultaneously [39]. ISA 100.11a devices are basically capable of detecting medium occupancy, however, at the cost of undefined transmission delays.
- There is currently only a small number of devices compatible with WirelessHART or ISA 100.11a.

Widely spread, available, and low cost are the digital radio technologies Wi-Fi [35], Bluetooth [31], and ZigBee [36]. Neither of these technologies has been developed for distributed real-time systems, therefore, their suitability is strongly constrained:

- Using shared medium access with contention (e.g., Wi-Fi distributed coordination function (DCF), Wi-Fi enhanced distributed channel access (EDCA), ZigBee contention access period (CAP)), upper bounds for transmission delays cannot be guaranteed. Furthermore, due to non-exclusive medium access, destructive frame collisions resulting in frame loss can occur.
- Controlled medium access (e.g., Wi-Fi point coordination function (PCF), Wi-Fi hybrid coordination function (HCF), Bluetooth, and ZigBee contention free period (CFP)) would in principle be suitable to achieve maximal transmission delays, however, is constrained to single-hop networks.

In the scientific literature, a variety of measures to improve real-time capabilities of wireless networks has been proposed, addressing all kinds of communication functionalities. Due to the large number of publications, it is very difficult to provide a reasonably concise survey. Therefore, we will provide insights into functionalities we consider as relevant for wireless real-time communication systems and will refer to short lists of related work only.

---

[4]Transmissions overlapping in time, frequency and space are called *collisions*.

1.5 Survey of the Book

The book focuses on core functionalities for wireless real-time multi-hop networking with TDMA and their integration into a flexible, versatile, fully operational, and self-contained communication system called Production Network (ProNet) 4.0 [1]. The number 4.0 indicates that ProNet has been developed and deployed in the context of Industry 4.0, a future project paving the way to the intelligent factory and referred as the Fourth Industrial Revolution. We will now give a survey of the book, which is structured according to the core functionalities, based on the architecture of ProNet 4.0 as shown in Fig. 1.4.

Chapter 2. Tick and time Synchronization—ProSync

Tick synchronization establishes network-wide reference points in time. This is needed, in particular, to structure time by creating global time slots, which can be reserved exclusively for communication among nodes, thereby avoiding collisions. Time synchronization, in addition, adjusts local clocks to a common time. This is needed to timestamp measurements of sensors and to synchronize data sampling. In this chapter, we present *Black Burst Synchronization (BBS)* [2–5], a deterministic protocol with precise upper bounds for clock offset and convergence delay, implemented in protocol module ProSync.

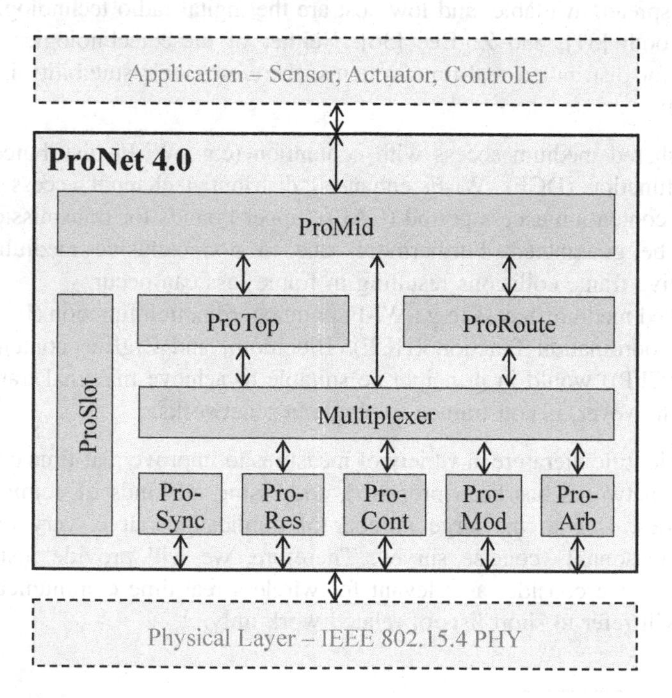

Fig. 1.4 Conceptual architecture of ProNet 4.0 (see [1])

Chapter 3. *Global Time Slotting—ProSlot*

Global time slotting is crucial for effectively using the wireless medium by networked production and control systems. Here, physical and virtual time slotting are distinguished. Physical time slots provide a strictly periodic time structure and are formed by micro slots and super slots. Micro slots are then combined into virtual time regions of flexible length, purpose, and placement. In this chapter, we study and compare the objectives and solutions of *global time slotting* [19, 20] (see ProSlot).

Chapter 4. *Automatic Topology Detection—ProTop*

In wireless networks, information about network topology is of great importance. Knowledge about the communication and interference topologies enables more efficient and reliable use of the wireless medium, and some protocols even require knowledge about nodes in sensing range. Manually measuring and configuring topology information are very cumbersome. Rather, it is crucial to detect topology automatically at system startup and to maintain and update the topology status during operation. In this chapter, we present *Automatic Topology Discovery Protocol (ATDP)* [6], which is capable of detecting the aforementioned topology types in TDMA-based wireless multi-hop networks and has been implemented in protocol module ProTop.

Chapter 5. *Medium Access Schemes—ProRes, ProCont, ProMod*

In TDMA-based wireless networks, time slots can be reserved exclusively, or medium arbitration is used to resolve contention. In the wireless domain, a variety of contention schemes is known, such as (passive) random backoff or (active) busy tones. In addition, statistical and strict frame priorities can be enforced. In this chapter, we study and compare different medium access schemes implemented in protocol modules ProRes and ProCont. In addition, we present a novel access scheme for restricted deterministic contention called *mode medium access* [7–9, 29], implemented in ProMod.

Chapter 6. *Deterministic Arbitration—ProArb*

An advanced functionality of wireless real-time communication systems is deterministic arbitration. Here, an arbitrary number of nodes contend, for instance, to determine a unique master node (leader election). Contention can be deterministically resolved, e.g., by using busy tone or binary-countdown protocols. In this chapter, we present *Arbitrating and Cooperative Transfer Protocol (ACTP)* [10, 11], which classifies as a deterministic binary-countdown protocol for wireless multi-hop networks and has been implemented in protocol module ProArb.

Chapter 7. *Duty Cycling*

Duty cycling is concerned with the task to switch hardware components (e.g., CPU, memory, AD converter, transceiver) to active mode when needed, and to sleep or idle mode when not needed, in order to reduce energy consumption and thereby extend node and thus network lifetime. In this chapter, we present a flexible and energy-efficient *duty cycling protocol* for wireless networks [13, 14] supporting weakly periodic activity periods.

Chapter 8. *Quality of Service Multicast Routing with Mobility Support—ProRoute*

In multi-hop real-time networks, quality of service (QoS) routing support is needed. In networked control systems, where services of a single sensor may be used by several controllers, multicast routing can save network resources. Furthermore, in a production environment, there may also be mobile nodes such as autonomous robots, which calls for mobility support. In this chapter, we study these aspects further and present the *QoS multicast routing (QMR)* protocol [16], a specialized routing protocol for partially mobile wireless TDMA networks implemented in ProRoute (see also [12]).

Chapter 9. *Network Clustering—ProMid*

Clustering is a well-known approach to improve scalability in wireless networks by reducing the complexity of the topology using overlay structures. Networked control systems are typically heterogeneous, consisting of nodes with different capabilities. This calls for specialized clustering algorithms that take constraints regarding functionality and energy into account. In this chapter, we present *heterogeneous network clustering (HNC)* [17], which we have devised for control applications in networked production and control environments, and implemented in production middleware (ProMid) (see also [15]).

Chapter 10. *Middleware for Networked Production and Control Systems—ProMid*

For the application developer, a middleware should provide suitable abstractions, hiding, e.g., the distributed system nature, and should offer application-specific views, thereby making the system more intuitive to use. In the context of networked production and control systems, this means that periodic and event-triggered real-time services offered by sensor and actuator nodes and used by controllers are to be provided. Furthermore, the middleware should operate a distributed service registry, where application services can be dynamically registered and looked up [18]. In this chapter, we study these functionalities in more detail and present our production- and control-specific service-oriented middleware called *production middleware (ProMid)*.

Chapter 11. *Implementation Aspects of ProNet 4.0—Multiplexer, BiPS*

Distributed real-time systems present a particular challenge, as stringent timing constraints are to be observed. In particular, this concerns the implementation of protocols for wireless multi-hop networked production and control systems. In this chapter, we study these issues further and outline a real-time-capable implementation framework [19, 20] specifically devised for the protocols presented in previous chapters called *Black burst integrated Protocol Stack (BiPS)* incorporating a component *multiplexer*. With this framework, time-critical protocols can be isolated from less preferential applications running on the same node. Finally, we show the integration of protocols into ProNet 4.0 using this framework.

Chapter 12. *ProNet 4.0 Case Studies*

The communication system ProNet 4.0 devised in this book has been prototypically implemented and deployed in an industrial context. In this chapter, we present three case studies. The first case study [21] addresses the wireless networking of different types of control systems, exploiting novel functionalities provided by ProNet 4.0. The second case study [22, 23] is about the realization of Try-Once-Discard, a protocol from the control systems domain, in a wireless multi-hop network, using advanced functionalities of ProNet 4.0. In the third case study [24], ProNet 4.0 forms part of a setting for remote maintenance in a production plant. The case study ranges from field level to management level, comprises a production facility, a wireless sensor network, an autonomous robot, a multimedia system, and an enterprise resource planning (ERP) system, and exploits various communication functionalities of ProNet 4.0 on all protocol layers. Taken together, the case studies provide evidence that ProNet 4.0 offers rich functionalities that cover a large variety of real-time communication requirements.

Chapter 13. *Conclusions and Future Research*

In this chapter, we summarize the results reported in this book, present conclusions, and report on our future research in the area of wireless real-time multi-hop communication systems [25–28, 30].

Literature

Chair for Networked Systems

1. Gotzhein R (2014) ProNet 4.0—a wireless real-time communication system for Industry 4.0. White Paper, Networked Systems Group, Department of Computer Science, University of Kaiserslautern. http://vs.informatik.uni-kl.de/publications/2014/Go14/whitePaperEN-ProNet4.0.pdf. Last Accessed 27 Aug 2019
2. Gotzhein R, Kuhn T (2008) Decentralized tick synchronization for multi-hop medium slotting in wireless ad hoc networks using Black Bursts. In: Proceedings of the 5th annual IEEE communications society conference on Sensor, Mesh, and Ad Hoc Communications and Networks (SECON 2008), San Francisco, USA, pp 422–431, 16–20 June 2008
3. Gotzhein R, Kuhn T (2011) Black Burst Synchronization (BBS)—a protocol for deterministic tick and time synchronization in wireless networks. Comput Netw 55(13):3015–3031
4. Engel M, Christmann D, Gotzhein R (2014) Implementation and experimental validation of timing constraints of BBS. In: Krishnamachari B, Murphy AL, Trigoni N (eds) 11th European Conference on Wireless Sensor Networks (EWSN 2014), vol 8354. Springer LNCS, Oxford, pp 84–99, 17–19 Feb 2014

5. Gotzhein R, Kuhn T (2011) Method, computer program product and system for the tick synchronization of nodes in a wireless multi-hop network. European Patent Office, Az EP 2195949, 31 Aug 2011 (date of granting)
6. Kramer C, Christmann D, Gotzhein R (2015) Automatic topology discovery in TDMA-based ad hoc networks. In: Proceedings of the 11th International Wireless Communications & Mobile Computing Conference (IWCMC 2015), Dubrovnik, Croatia, pp 634-639, 24–28 Aug 2015
7. Braun T, Gotzhein R, Kuhn T (2014) Mode-based scheduling with fast mode-signaling—a method for efficient usage of network time slots. In: Proceedings of the 6th International Conference on Computer Science and Information Technology (ICCSIT 2013), Paris, France, 20–21 Dec 2013; Journal of Advances in Computer Networks (JACN), vol 2, issue no. 1, pp 48–57
8. Gotzhein R, Kuhn T (2012) Zeit- und Prioritäts-gesteuerter Sende/Empfangsknoten, Deutsches Patent- und Markenamt, Az 10 2010 039 488, 6 June 2012 (date of granting)
9. Braun T (2016) Reliable mode-based communication and virtual prototyping in the development of distributed real-time systems. Ph.D. Thesis (in German), Computer Science Department, University of Kaiserslautern
10. Christmann D, Gotzhein R, Rohr S (2012) The Arbitrating Value Transfer Protocol (AVTP) —deterministic binary countdown in wireless multi-hop networks. In: Proceedings of the 21st International Conference on Computer Communication Networks (ICCCN 2012), Munich, Germany, pp 1–9, July 30–Aug 2, 2012
11. Christmann D (2015) On the development of a wireless binary countdown protocol and the applicability of SDL to such time-critical systems. Ph.D. Thesis, Computer Science Department, University of Kaiserslautern
12. Becker P, Birtel M, Christmann D, Gotzhein R (2011) Black Burst-based Quality-of-Service Routing (BBQR) for wireless ad-hoc networks. In: Proceedings of the 11th international conference on New Technologies in Distributed Systems (NOTERE 2011), Paris, France, pp 1–8
13. Christmann D, Gotzhein R, Krämer M, Winkler M (2010) Flexible and energy-efficient duty cycling in wireless networks with MacZ. In: Proceedings 10th annual international conference on New Technologies of Distributed Systems (NOTERE 2010), Tozeur, Tunisia, pp 121–128, May 31–June 2, 2010; Journal on Concurrency and Computation: Practice & Experience, vol 25, Issue no. 2, February 2013, pp 218–233
14. Fliege I, Geraldy A, Gotzhein R, Jaitner T, Kuhn T, Webel C (2006) An ambient intelligence system to assist team training and competition in cyclin. In: Moritz EF, Haake S (eds) Developments in sports, vol 1: the engineering of sports 6. Springer Science and Business Media, New York, pp 103–108
15. Sefati H, Gotzhein R, Kramer C, Schloesser S, Weiss M (2018) Dynamic overlay line topology establishment and repair in wireless networks. In: IEEE Wireless Communications and Networking Conference (WCNC 2018), Barcelona, Spain, 15–18 Apr 2018
16. Gebhardt J, Gotzhein R, Igel A, Kramer C (2015) QoS multicast routing in partially mobile wireless TDMA networks. In: Proceedings of IEEE Global Communications Conference (Globecom 2015), San Diego, USA, 6–10 Dec 2015
17. Kramer C, Christmann D, Gotzhein R (2016) A clustering algorithm for distributed service registries in heterogeneous wireless networks. In: Proceedings of wireless days 2016, Toulouse, France, 23–25 Mar 2016
18. Chamaken A, Litz L, Krämer M, Gotzhein R (2009) Cross-layer design of wireless networked control systems with energy limitations. In: European Control Conference (ECC 2009), Budapest, Hungary
19. Christmann D, Braun T, Engel M, Gotzhein R (2016) BiPS—a real-time-capable protocol framework for wireless sensor networks. In: Proceedings of the 6th international conference on Pervasive and Embedded Computing (PEC 2016), Lisbon, Portugal, pp 17–27, 25–27 Jul 2016

20. Engel M, Kramer C, Braun T, Christmann D, Gotzhein R (2019) BiPS—a real-time-capable protocol framework for wireless networked control systems and its application. In: Obaidat M, Cabello E (eds) E-Business and Telecommunications (ICETE 2017), Communications in Computer and Information Science, vol 990. Springer, pp 313–336
21. Haupt A, Gotzhein R et al. (2014) Control and communication co-design. In: Lunze J (ed) Control theory of digitally networked dynamic systems. Springer, pp 328–348
22. Christmann D, Gotzhein R, Siegmund S, Wirth F (2014) Realization of try-once-discard in wireless multi-hop networks. IEEE Trans Industr Inf 10(1):17–26
23. Duc LH, Christmann D, Gotzhein R, Siegmund S, Wirth F (2015) The stability of try-once-discard for stochastic communication channels—theory and validation. In: 54th conference on Decision and Control, Osaka, Japan, 15–18 Dec 2015
24. Berg M, Gotzhein R et al (2015) Vertical integration and adaptive services in networked production environments. In: Proceedings of ERP Future 2015, vol 245. Springer LNBIP, Munich, Germany, pp 1–16, 16 Nov 2015
25. Mathews K, Kramer C, Gotzhein R (2017) Token bucket based traffic shaping and monitoring for WLAN-based control systems. In: 28th IEEE annual international symposium on Personal, Indoor, and Mobile Radio Communications (PIMRC 2017), Montreal, Canada, 8–13 Oct 2017
26. Kohlstruck C, Mathews K, Gotzhein R (2019) r_{min}-routing—discovery and operation of routes in wireless ad-hoc networks with specified statistical minimum reliabilities. In: 44th IEEE conference on Local Computer Networks (LCN 2019), Osnabrück, Germany, 14–17 Oct 2019
27. Engel M, Gotzhein R (2018) Dynamic computation and adjustment of channel hopping sequences for cognitive radio networks based on quality metrics. In: International conference on Embedded Wireless Systems and Networks (EWSN 2018), Madrid, Spain, 14–16 Feb 2018
28. Aragao P, Engel M, Gotzhein R (2018) A three-dimensional stabilization protocol for time-slotted multi-hop cognitive radio networks with channel hopping. In: The 32nd IEEE international conference on Advanced Information Networking and Applications (AINA 2018), Cracow, Poland, pp 32–39, 16–18 May 2018
29. Braun T, Gotzhein R (2017) Zeit- und Prioritäts-gesteuerter Sende/Empfangsknoten – Technische Realisierungen mit FlexRay und LIN, Deutsches Patent- und Markenamt, Az 10 2012 200 475, Oct 17, 2013 (date of granting); European Patent Register, EP2614996 B1, Nov 11, 2017 (date of granting)
30. Kramer C, Mathews K, Gotzhein R (2019) Cooperative fair bandwidth scaling in contention-based wireless networks using time token bucket. In: 38th IEEE International Performance Computing and Communications Conference (IPCCC 2019), London, UK, 29–31 Oct 2019

Further References

31. Bluetooth SIG, Inc. Bluetooth Specification Version 4.0, Dec 2009
32. Day JD, Zimmermann H (1983) The OSI reference model. In: Proceedings of the IEEE 71 (12):1334–1340
33. Dorf RC, Bishop RH (2016) Modern control systems, 13th edn. Pearson Education
34. Gupta RA, Chow M-Y (2008) Overview of networked control systems. In: Wang F-Y, Liu D (eds) Networked control systems—theory and applications. Springer, pp 1–23
35. Institute of Electrical and Electronics Engineers. IEEE Standard 802 Part 11: Wireless LAN Medium Access Control (MAC) and Physical Layer (PHY) Specifications. IEEE Computer Society, New York

36. Institute of Electrical and Electronics Engineers (2011) IEEE Standard 802 Part 15.4: Low-Rate Wireless Personal Area Networks (LR-WPANs). IEEE Computer Society, New York
37. International Electrotechnical Commission (2012) Industrial Communication Networks—Wireless Communication Network and Communication Profiles—ISA 100.11a (IEC 62734 ed 1.0), Geneva, Switzerland
38. Kopetz H (2011) Real-time systems—design principles for distributed embedded applications, 2nd edn. Springer
39. Petersen S, Carlsen S (2009) Performance evaluation of WirelessHART for factory automation. In: Proceedings of 12th IEEE international conference on Emerging Technologies and Factory Automation, ETFA 2009, Palma de Mallorca, Spain, pp 1–9, 22–25 Sep 2008
40. International Electrotechnical Commission (IEC) (2010) Industrial Communication Networks—Wireless Communication Network and Communication Profiles—WirelessHART (IEC 62591 ed 1.0), Geneva, Switzerland

Chapter 2
Tick and Time Synchronization

Synchronization is a core functionality of distributed real-time computer systems. In this chapter, we explain foundations of time synchronization and of the weaker notion of tick synchronization, present our protocol BBS (*Black Burst Synchronization*) [2–5], survey and assess related work, and draw conclusions.

2.1 Foundations

In this chapter, we establish the context of tick and time synchronization, explain concepts and requirements, and address areas of operation.

2.1.1 Context

Time synchronization (also called *clock synchronization*) is essential for the operation of distributed real-time computer systems. Nodes use physical clocks, which may deviate in clock value, speed, and speed change. The objective of time synchronization is to keep the deviation of clock values of all network nodes within tight bounds, by aligning them from time to time. This is needed, for instance, for data fusion, where data of multiple sources are timestamped locally and combined by another node [7], or for sampling data at specified points in time [12].

Tick synchronization (also called *heartbeat synchronization*) is weaker than time synchronization, in that no clock values are exchanged. The purpose of tick synchronization is rather to establish network-wide reference points in time. This is

The original version of this chapter was revised: The terms "$d_{\text{round m}}$" and "$d_{\text{maxTickOffset m}}$" have been replaced with "$d_{\text{round\_}m}$" and "$d_{\text{maxTickOffset\_}m}$," respectively. The correction to this chapter is available at https://doi.org/10.1007/978-3-030-33319-5_14

R. Gotzhein, *Real-time Communication Protocols for Multi-hop Ad-hoc Networks*, Computer Communications and Networks,
https://doi.org/10.1007/978-3-030-33319-5_2

sufficient, e.g., for network-wide medium slotting as basis for exclusive reservation of time slots [17], and for duty cycling, where nodes change between energy modes in a synchronized way [29].

In distributed systems, synchronization requires the signaling of reference points in time. This can be achieved by messages of an external reference clock (*external synchronization*), as provided, for instance, by Global Positioning System (GPS) [16] or DCF77 [11]. If clocks of nodes forming a networked system are to be synchronized among themselves, for instance, if external synchronization is not an option, synchronization can only be achieved by message exchange among these nodes (*internal synchronization*). This calls for synchronization protocols defining synchronization message formats, rules for their exchange, and an algorithm to determine reference points in time locally and to adjust clocks. A prominent example is Network Time Protocol (NTP) [22], which has become the de facto standard for internal synchronization in the Internet and is used, among other things, for the timely treatment of stock market orders and transaction logging of distributed databases. Another well-known example is the highly accurate internal synchronization of the 24 satellites of GPS [16], which is used, e.g., for position control of vehicles and legal land surveys.

2.1.2 Concepts and Requirements

In the literature (e.g., see [27]), it is common to use the following synchronization concepts (see Fig. 2.1). *Real time* denotes the actual time, with values from a domain *Time*. Access to real time is by means of devices measuring time, called *clocks*, which are modeled as functions c_x: *Time* → *Time*. We can conceive a perfect (reference) clock c_p that always returns the current value of real time, i.e., for all $t \in Time$, $c_p(t) = t$. In practice, there are physical clocks, which measure local time deviating from the values of a perfect clock. By *local time* of a node v, we refer to the value $c_v(t)$ reported by the physical clock c_v of node v at real time t. Figure 2.1 shows values reported by a perfect clock c_p, a clock c_v that runs fast, and a clock $c_{v'}$ that loses time.

Synchronization accuracy of node v at time t is expressed by the *time offset* (*clock offset*) $d_{\text{time Offset}}(t) = c_v(t) - t$, i.e., the difference between local time reported by clock c_v and real time (see Fig. 2.1), or relative to the clock $c_{v'}$ of some

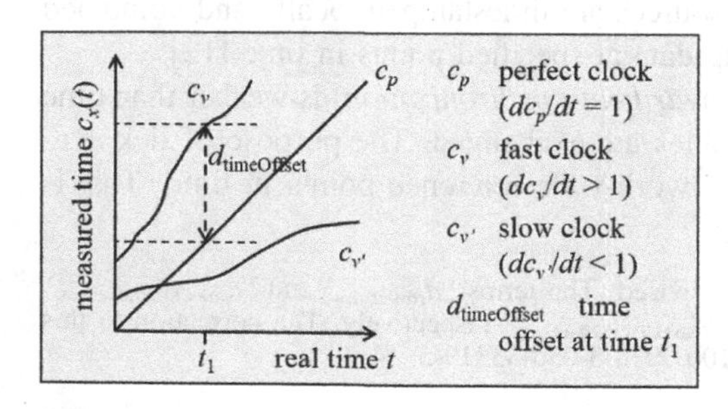

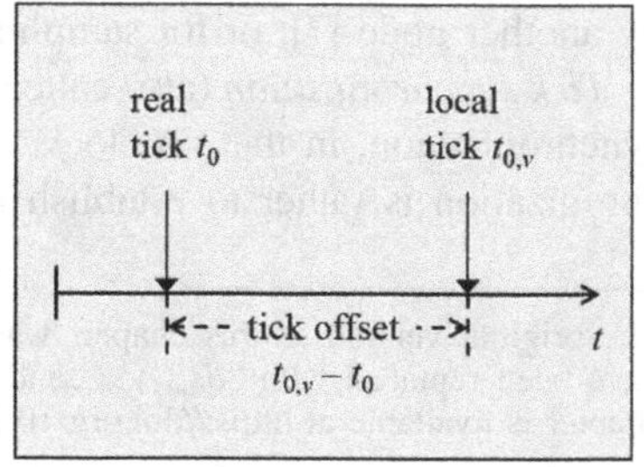

Fig. 2.1 Synchronization concepts (see [3], Fig. 1)

node v', i.e., $c_v(t) - c_{v'}(t)$. The *clock rate* of c_v at time t is the speed $c_v'(t)$ (first derivative) at which it progresses; for a perfect clock c_p, we have $c_p'(t) = 1$, for all $t \in Time$. *Clock skew* denotes the difference between the rate of a clock c_v and the perfect clock at time t, i.e., $c_v'(t) - 1$, or $c_v'(t) - c_{v'}'(t)$ relative to $c_{v'}$. Finally, *clock drift* is the rate change $c_v''(t)$ (second derivative) of clock c_v at time t, or $c_v''(t) - c_{v'}''(t)$ relative to some clock $c_{v'}$; for a perfect clock c_p, $c_p''(t) = 0$ for all $t \in Time$.

In addition to the terminology so far, we introduce concepts for tick synchronization. A *real tick* is a reference point in time, conceptually identified by its location t_0 in real time (see Fig. 2.1). A *local tick* is the point in time $t_{0,v}$ associated with a corresponding real tick t_0 by some node v. Since the only way to measure time is by physical clocks, node v may record $t_{0,v}$ as local time $c_v(t_{0,v})$, which is sufficient to refer to a global reference point in time locally. The *tick offset* of real tick t_0 for node v is then defined as $d_{\text{tickOffset}}(t_0) = t_{0,v} - t_0$, or as $t_{0,v} - t_{0,v'}$ relative to some node v'.

We can now define general requirements on synchronization protocols. For tick synchronization, the central requirement is on accuracy of the time basis, expressed by a tick offset $d_{\text{tickOffset}}(t)$ that is small and/or bounded for all $t \in Time$ during system operation, after an initial stabilization. It depends on the area of application what *small* means, and what upper bounds may be required. Analogous constraints on time offset $d_{\text{timeOffset}}(t)$ determine the required accuracy of time synchronization.

Another important requirement is on the duration until (re-) synchronization is achieved, expressed by a convergence delay d_{conv} that is small and/or bounded. As in case of tick and time offset, actual values for d_{conv} are application-specific. Furthermore, there may be constraints regarding the complexity of the synchronization protocol regarding computation, communication, storage, energy, and structure.[1] Finally, high robustness against topology changes due to node movements, node failures, and channel quality variations is desirable.

2.1.3 Areas of Operation

Tick and time synchronization are core functionalities of distributed real-time computer systems, enabling functionalities both on user level and on system level. *User level functionalities* requiring tick and/or time synchronization can be found, for instance, in control systems, where sensor sample system states and forward them to control nodes, which aggregate these data (data fusion) to determine steering values communicated to actuators. Typically, it is crucial that value sampling occurs at synchronized points in time, that steering values are applied in a coordinated fashion, and that values can be time stamped using a global time basis.

[1]Some protocols require, e.g., a tree or ring structure to be established before synchronization can start. This makes them highly vulnerable to topology changes.

In [7], the authors sketch a variety of user level functionalities in the areas of environment, home, and health that are based on tick and/or time synchronization. An example is forest fire monitoring, where sensors record, timestamp, and communicate temperature readings. These values are collected in sink nodes, combined, and evaluated to estimate locations, directions, and velocities of the fire. Regarding synchronization requirements, a reasonably small average time offset and convergence delay in the order of seconds would be sufficient. Furthermore, the protocol for time synchronization should be of low complexity, as energy may be scarce, and should exhibit some robustness, as some nodes may be destroyed by heat.

A crucial *system level functionality* of distributed real-time computer systems is global time slotting, which is needed for both computation and communication. Here, time is subdivided into intervals called time slots, which can then be assigned to tasks such that deadlines are met. Thus, global time slotting is a key to achieve predictable timing behavior.

In communication systems, time slots can be reserved exclusively for message transfers along a route in a way that maximal packet delays composed of waiting and transfer delays are assured. In communication technologies with shared medium access, this has the additional advantage that frame collisions[2] are avoided, enabling reliable message exchange. TDMA (time-division multiple access) is a well-known medium access method applying this scheme.

TDMA places high demands on synchronization accuracy and convergence delay. Consider the example in Fig. 2.2, where nodes v, v', and v'' are in mutual range. The figure also shows a sequence of time slots, with adjacent slots i and $i + 1$ assigned to nodes v and v' for transmission, respectively, and node v'' being the receiver. We assume that nodes are tick-synchronized, and that there is a tick offset $d_{\mathrm{tickOffset}} > 0$ µs. If nodes v and v' fully exploit their assigned slots for transmission, we have the situation that frames will collide for the period $d_{\mathrm{tickOffset}}$. This can only be avoided by either node v stopping earlier or node v' starting later. The problem here is that network nodes do not know the exact current tick offset. To avoid collisions, it is therefore required that an upper bound $d_{\mathrm{maxTickOffset}}$ that is known to all nodes can be determined and enforced,[3] and that nodes delay slot usage by this value[4] (see Fig. 2.2). This, however, may result in waste of bandwidth, as bandwidth that would be available due to a lower current tick offset $d_{\mathrm{tickOffset}} < d_{\mathrm{maxTickOffset}}$ is not used. Therefore, it is essential that $d_{\mathrm{maxTickOffset}}$ is both very small (e.g., in the order of microseconds) and bounded. In addition, convergence delay is required to be small (e.g., in the order of milliseconds) and bounded, and the synchronization protocol should be of low complexity and be robust against topology changes.

---

[2]Transmissions overlapping in time, frequency, and space are called *collisions*.

[3]The use of an average tick offset $d_{\mathrm{avgTickOffset}}$, which would be smaller than $d_{\mathrm{maxTickOffset}}$, is not sufficient, as it would not rule out collisions.

[4]If it is required that all nodes associate a received frame with the same slot, it is even necessary that sending ceases $d_{\mathrm{maxTickOffset}}$ before the slot ends.

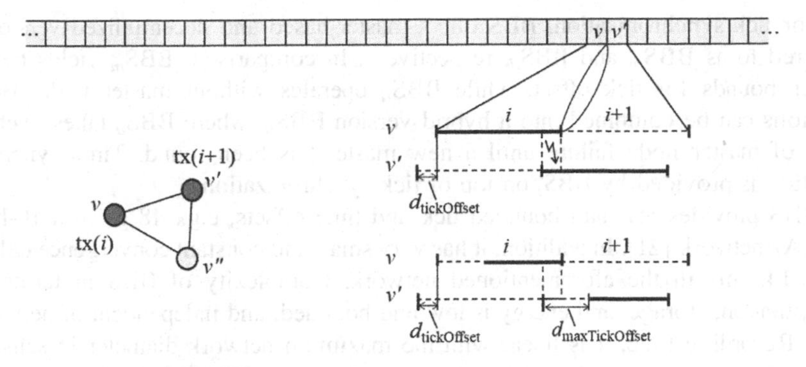

Fig. 2.2 Network-wide medium slotting: influence of tick offset

If nodes of a real-time system are battery-powered, energy is usually a scarce resource. To save energy, nodes can change between energy modes, following a joint time pattern, thereby extending network lifetime. Switching of hardware components (e.g., CPU, memory, and transceiver) to active mode when needed, and to sleep mode when not needed is called *duty cycling*. Here, it is essential that mode switching happens coordinated, which calls for accurate tick synchronization. To synchronize wake-up times, small and bounded tick offset, and convergence delay (e.g., in the order of milliseconds), low complexity and high robustness are required.

2.2 Black Burst Synchronization (BBS)

In this chapter, we present BBS (*Black Burst Synchronization*) [2–5], a deterministic protocol for internal tick and time synchronization in ad hoc networks with precise upper bounds for offsets and convergence delay. The protocol ProSync of the ProNet 4.0 protocol stack [1] is an implementation of BBS.

2.2.1 Overview of BBS

BBS (Black Burst Synchronization) [2–5] is a modular protocol for network-wide internal tick and time synchronization in wireless multi-hop ad hoc networks, located at Medium Access Control (MAC) layer (see Fig. 1.1). It classifies as a deterministic protocol, with upper bounds for tick and time offset, and for convergence delay, if the single-network property (see Sect. 1.3) holds. BBS defines the format of synchronization messages, rules for their exchange, and an algorithm to determine reference points in time locally and to adjust clocks. Messages are encoded with black bursts to render collisions non-destructive.

For tick synchronization, BBS offers master-based and decentralized versions, referred to as BBS_m and BBS_d, respectively. In comparison, BBS_m yields better upper bounds for tick offset, while BBS_d operates without master node. Both versions can be combined into a hybrid version BBS_h, where BBS_d takes over in case of master node failure until a new master has been found. Time synchronization is provided by BBS_t on top of tick synchronization.

BBS provides low and bounded tick and time offsets, e.g., 48 μs in a 10-hop MICAz network [21]. In addition, it has very small and constant convergence delay, e.g., 14.1 ms in the aforementioned network. Complexity of BBS in terms of computation, storage, and energy is low and bounded, and independent of network size. Regarding time, it is linear with the maximum network diameter in sensing hops. No overlay network structure is required to run BBS. The protocol is robust against topology changes caused, e.g., by node movements or node failures.

2.2.2 *Encoding of Bit Sequences with Black Bursts*

In wireless networks, frame collisions occur if several transmissions overlap in time, frequency, and space. As these frame collisions are usually destructive, MAC protocols adopt several measures to reduce the probability of their occurrence. Carrier Sense Multiple Access (CSMA) is one such measure, where nodes listen whether the medium is idle before starting transmission. Another measure is random backoff, i.e., nodes wait for a random time span before accessing a medium after it has become idle. For unicast transfers, RTS/CTS schemes are applied—a short Request-To-Send (RTS) frame is sent to a single-hop destination, which replies by a short Clear-To-Send (CTS) frame. This way, the hidden station problem is reduced to some degree. Nevertheless, even the combination of all measures cannot rule out destructive collisions.

To render frame collisions non-destructive, BBS uses the concept of *black bursts* [26] to encode bit sequences. A black burst is a period of transmission energy of defined length. Transmission of a black burst starts at a locally determined point in time, without prior medium arbitration. If two or more nodes transmit a black burst at almost the same time resulting in a collision, a node within sensing range of a subset of these nodes can still detect the period of energy on the medium, i.e., starting and ending time. Since this is the only information carried by a black burst, the collision is non-destructive.

Bit sequences can now be encoded with black bursts as follows: A logical 1 is encoded by transmitting a black burst of defined length $d_{BB} > 0$ μs; a logical 0 is encoded by not transmitting a black burst, i.e., $d_{BB} = 0$ μs. If several nodes in sensing range transmit a bit, a logical 1 will dominate a logical 0, which therefore is called recessive bit, while 1 is dominant. This feature can be used to encode a bitwise logical OR by sending bit sequences in an aligned way.

Figure 2.3 shows a scenario where nodes v and v' transmit bit sequences 101 and 100 of length 3 almost simultaneously, while a third node v'' in range of both v and

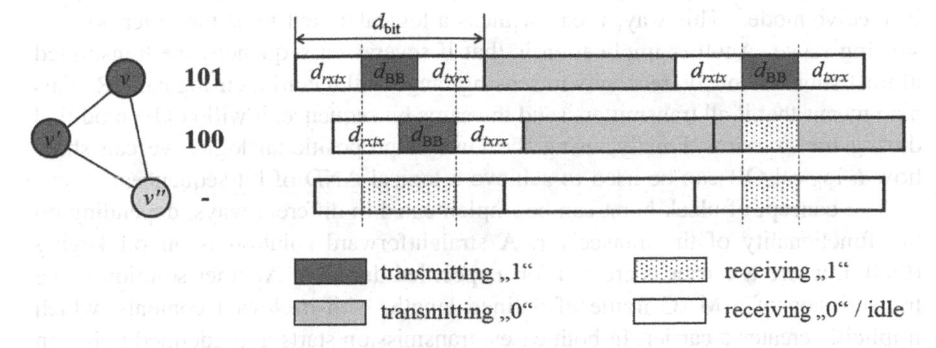

Fig. 2.3 Transmission of bit sequences encoded with black bursts (see [3], Fig. 2)

v' listens on the medium. Here, a bit time d_{bit}, i.e., the interval required to transmit a bit encoded as black burst, is composed of an interval d_{rxtx} to switch the transceiver from receive (*rx*) mode to transmit (*tx*) mode, the transmission interval of length $d_{\text{BB}} > 0$ μs (in case of a logical 1), and an interval d_{txrx} to switch the transceiver back to *rx* mode.

To transmit the bit sequences 101 and 100 (almost) simultaneously, 3 bit times are required. During the first bit time, nodes v and v' both send a logical 1 encoded as a dominant transmission. As there is a (small) tick offset, transmissions are not perfectly synchronized. Thus, the receiving node v'' detects a period of energy that is longer than d_{BB} and interprets this as the reception of a logical 1. During the second bit time, nodes v and v' both send a logical 0, encoded as "no transmission," which node v'' detects as idle medium and interprets as logical 0. During the third bit time, node v sends a logical 1, while v' sends a logical 0. Since 1 dominates 0, node v'' detects energy of length d_{BB} on the medium[5] and decides for a logical 1. Taken together, node v'' has correctly received the bit sequence 101 of node v, despite collision during the first bit time.

If transmissions of bit sequences overlap, it is essential that they are correctly timed (see [4]). Therefore, we require that they are sent almost simultaneously, with a displacement that is bounded in time, e.g., by the maximum tick offset $d_{\text{maxTickOffset}}$. Timing constraints on d_{bit} and d_{BB} have to ensure that overlapping black bursts can be recognized by receiving nodes, and that black burst receptions are finished within the current bit time of each receiving node.

Black burst encoding of bit sequences in the described way has a number of interesting implications that we exploit for BBS and another protocol called ACTP (Arbitration and Cooperative Transfer Protocol, see Chap. 6). One such implication is that a node "transmitting" a logical 0 is not transmitting physically, but can stay

[5]In an implementation, v'' may detect a shorter period of energy, which is due to the operation of the clear channel assessment (CCA) mechanism used for this purpose. Therefore, the duration d_{BB} has to be configured suitably.

in receive mode.[6] This way, it can witness a logical 1 sent by some other node in sensing range. Another implication is that if several bit sequences are transmitted almost simultaneously, receivers in sensing range will record their logical OR. This also means that if all transmitters send the same bit sequence, it will not be modified during transmission. Finally, using the laws of propositional logic, we can show how a logical OR can be used to achieve a logical AND of bit sequences.

The concept of black burst can be implemented in different ways, depending on the functionality of the transceiver. A straightforward solution is on–off keying (OOK), where a carrier is created for a specified duration. Another solution is the transmission of a MAC frame of defined length, with irrelevant contents, which implicitly creates a carrier. In both cases, transmission starts at predefined points in time, without prior clear channel assessment (CCA). To detect black bursts on the medium, the CCA mechanism can be used.

2.2.3 *Master-Based BBS* (BBS_m)

We now present *Master-Based BBS* (BBS_m for short), starting with a digest followed by a scenario, to foster an intuitive understanding. Then, we provide technical details, by semi-formalizing the algorithm of BBS_m, and by analyzing accuracy, performance, and complexity of the protocol.

2.2.3.1 Digest of BBS_m

Master-based BBS (BBS_m) is a self-contained protocol for tick synchronization in multi-hop wireless ad hoc networks. It assumes that a node acting as master has been determined.[7] For this outline, we further assume that the network is stable, i.e., all network nodes are operational and tick synchronized with a maximum tick offset.

To keep tick offset below an upper bound, the master node v_m periodically triggers resynchronization. A resynchronization phase consists of a constant number of rounds defined by the maximum network diameter in sensing hops.

At each local tick, i.e., at each reference point in time, v_m sends a master-tick frame (see Fig. 2.4). This frame is received by all nodes in sensing range of v_m, which associate local ticks with the start of reception. This completes synchronization round 1. At this point, node v_m and all nodes in sensing range of v_m are resynchronized.

[6] Different from wired communication, a node cannot listen while it is transmitting when using wireless communication with a single antenna.

[7] For the operation of BBS_m, any node can have the master role. If a master node fails, another node can take over.

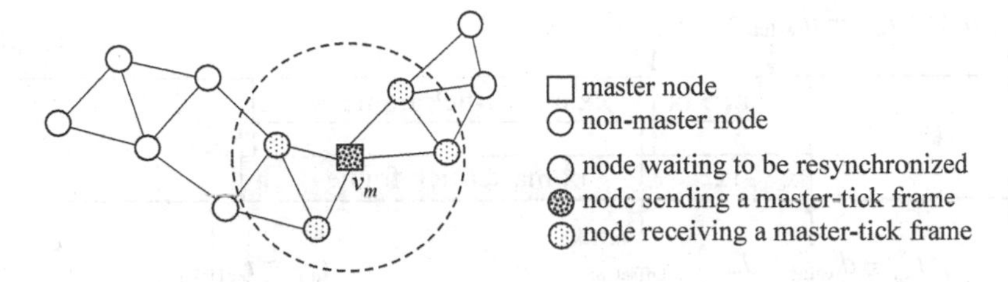

Fig. 2.4 BBS_m—synchronization round 1

Nodes receiving the master-tick frame in synchronization round 1 send a master-tick frame at the beginning of round 2, where rounds have a constant duration. Because sending nodes are already resynchronized, they send this frame almost simultaneously. Therefore, and because master-tick frames are encoded with black bursts, potential collisions are non-destructive.

Further synchronization rounds are needed until all nodes have received a master-tick frame. The number of rounds is given by the maximum network diameter in sensing hops, which is a configuration parameter. Local ticks are determined by backward calculation with high accuracy.

2.2.3.2 Scenario of BBS_m

We now present a complete scenario of BBS_m, for the same sensing topology as in Fig. 2.4, making the algorithm more precise. By convention, constants and variables denoting numbers, durations, and times have names *n*, *d*, and *t*, followed by a subscript, respectively. Let *V* be the set of network nodes, with $v_m \in V$ being the master node. Each node $v_i \in V$ records the current local tick in a variable $t_{lt,i}$. The constant n_{maxHops} is the maximum network diameter in sensing hops, a configuration parameter that must be chosen "suitably": if chosen too small, resynchronization leaves out nodes with a sensing distance greater than n_{maxHops} to the master node; if chosen too large, there is additional synchronization overhead. The parameter $d_{\mathrm{round\_}m}$ is the constant duration of a synchronization round, with a concrete value determined offline from the time needed to send, receive, and process a master-tick frame. Furthermore, d_{resInt} and $d_{\mathrm{maxTickOffset\_}m}$ denote the duration of the resynchronization interval and the maximum tick offset, respectively.

BBS_m performs tick (re-)synchronization in n_{maxHops} rounds, by broadcasting master-tick frames across the network. Round 1 is triggered by the master node v_m, which sends a master-tick frame carrying round number 1 as data at $t_{lt,m}$, i.e., at its next local tick, computed from the previous local tick by adding d_{resInt}, the duration of the resynchronization interval (see Fig. 2.5). Being encoded by a sequence of black bursts, the master-tick frame is sent without prior medium arbitration. For timely transmission, v_m starts switching to *tx* mode at $t_{lt,m} - d_{rxtx}$.

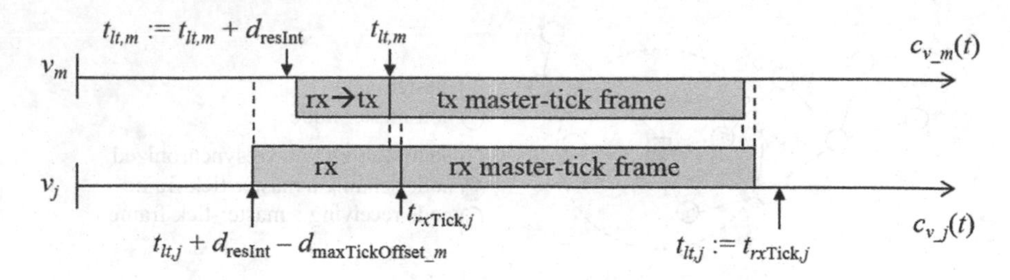

Fig. 2.5 BBS$_m$—timing in synchronization round 1 (see [3], Fig. 4)

All nodes $v_i \in V \backslash \{v_m\}$ start listening for a master-tick frame already at $t_{lt,i} + d_{\text{resInt}} - d_{\text{maxTickOffset\_m}}$ (see Fig. 2.5 for node v_j), because this is the earliest start of reception in case of maximal clock skew between v_m and v_i. Here, it is important to ensure that tick offset has a determinable upper bound; otherwise, a master-tick frame could be missed. Nodes v_j in sensing range of v_m perceive the start of tick frame reception at local time $t_{rxTick,j}$ (see Fig. 2.5). This time value is recorded as new local tick and assigned to $t_{lt,j}$ after reception of the master-tick frame is completed, which finishes synchronization round 1. At this point, v_m and all nodes v_j in sensing range of v_m are resynchronized.

Rounds 2 to n_{maxHops} are triggered by nodes $v_k \in V$ that have received a master-tick frame in the previous round n_{round}, $1 \leq n_{\text{round}} \leq n_{\text{maxHops}} - 1$, for the first time during the current resynchronization phase, i.e., by the nodes that have just been resynchronized. These nodes send a master-tick frame carrying $n_{\text{round}} + 1$ as data at $t_{rxTick,k} + d_{\text{round\_m}}$, i.e., with a constant delay after their first local tick reception. Obviously, master-tick frames of the same round may collide. However, sending them with a constant delay w.r.t. $t_{rxTick,k}$ implies that they are sent almost simultaneously by all nodes in mutual sensing range. Since master-tick frames of the same round carry the same round number, and because they are encoded by black bursts, collisions are non-destructive.

Nodes $v_k \in V$ receiving a master-tick frame in round n_{round} for the first time record the start of tick frame reception $t_{rxTick,k}$. By backward calculation, they determine their updated local tick as $t_{lt,k} := t_{rxTick,k} - (n_{\text{round}} - 1) \cdot d_{\text{round\_m}}$ (see Fig. 2.6). The term $(n_{\text{round}} - 1) \cdot d_{\text{round\_m}}$ expresses the ideal delay since $t_{lt,m}$, requiring perfect detection and forwarding of master-tick frames. In this case, backward calculation would yield $t_{lt,k} = t_{lt,m}$. However, due to variable delays d_{CCA} to recognize the start of tick frame reception, variable propagation delays d_{prop}, and clock skews, this ideal delay increases with each round, resulting in tick offsets $d_{\text{tickOffset\_m,k}} > 0$ µs (see Fig. 2.6).

Figure 2.7 illustrates the progress of resynchronization, with n_{maxHops} set to 7. In round 1, nodes $v_i \in V$ in sensing range of the master node v_m are resynchronized. These nodes then become sending nodes in round 2, propagating resynchronization one sensing hop further. In this round, two nodes perceive collisions of master-tick frames; however, as already explained, these collisions are non-destructive. In

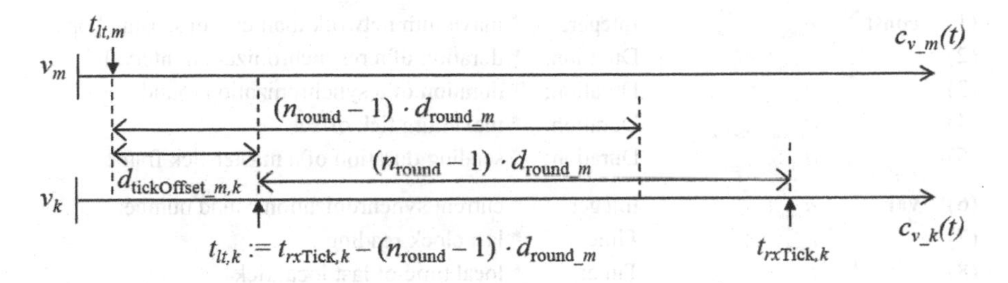

Fig. 2.6 BBS$_m$—timing in synchronization round n_{round} (see [3], Fig. 5)

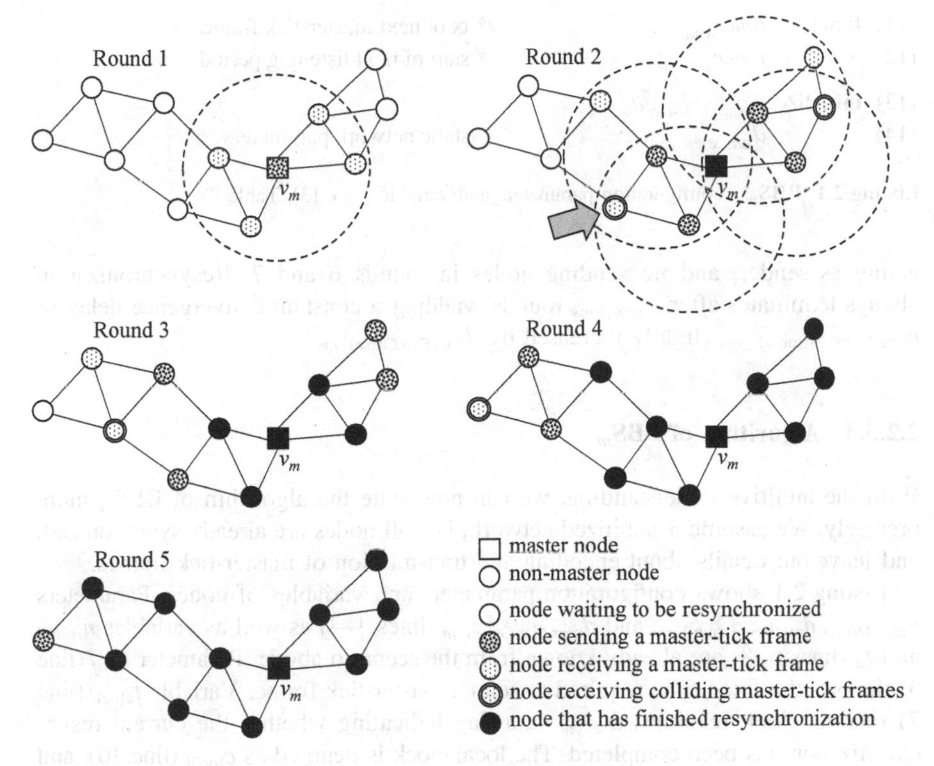

Fig. 2.7 BBS$_m$—synchronization rounds 1–5, with $n_{\text{maxHops}} = 7$

effect, the efficiency of BBS largely depends on the possibility to send synchronization frames almost simultaneously. In rounds 3 and 4, resynchronization is propagated to all remaining nodes. If sensing topology and placement of the master node were known to all nodes, resynchronization could stop after round 4. Since this is not realistic, the worst case—master node placed at the network border, network diameter equals n_{maxHops}—is to be considered, which means that the remaining rounds 5–7 have to be executed, too. Thus, in round 5, we have one node

```
(1)  const       n_maxHops:                  Integer;   /* maximum network diameter in sensing hops
(2)              d_resInt:                   Duration;  /* duration of a resynchronization interval
(3)              d_round_m:                  Duration;  /* duration of a synchronization round
(4)              d_maxTickOffset_m:          Duration;  /* maximum tick offset
(5)              d_mtf:                      Duration;  /* sending duration of a master-tick frame

(6)  var         n_round:                    Integer;   /* current synchronization round number
(7)              t_clock:                    Time;      /* last clock reading
(8)              t_lt:                       Time;      /* local time of last local tick
(9)              b_resync:                   Boolean;   /* flag: set after first tick frame reception

(10) clock       c_local;                               /* local clock

(11) timer       timer_txMasterTick;                    /* tx of next master-tick frame
(12)             timer_syncPhase;                       /* start of next listening period

(13) initialize  n_maxHops, d_resInt, d_round_m,
(14)             d_maxTickOffset_m, d_mtf;              /* static network parameters
```

Listing 2.1 BBS_m—configuration parameters and variables (see [3], Table 2)

acting as sender, and no sending nodes in rounds 6 and 7. Resynchronization always terminates after n_{maxHops} rounds, yielding a constant convergence delay of $n_{\text{maxHops}} \cdot d_{\text{round\_}m}$, slightly increased by $d_{\text{maxTickOffset\_}m}$.

2.2.3.3 Algorithm of BBS_m

With the intuitive understanding, we can now state the algorithm of BBS_m more precisely. We assume a stabilized network, i.e., all nodes are already synchronized, and leave out details about encoding and transmission of master-tick frames.

Listing 2.1 shows configuration parameters and variables of nodes. Parameters n_{maxHops}, d_{resInt}, $d_{\text{round\_}m}$, and $d_{\text{maxTickOffset\_}m}$ (lines 1–4) as well as variables n_{round} and t_{lt} (lines 6, 8) are already known from the scenario above. Parameter d_{mtf} (line 5) denotes the fixed time required to send a master-tick frame. Variable t_{clock} (line 7) records clock readings; b_{resync} is a flag indicating whether the current resynchronization has been completed. The local clock is denoted as c_{local} (line 10), and two timers are declared (lines 11, 12). Finally, all parameters are initialized with configuration-specific settings (lines 13, 14).

Listing 2.2 shows the code of the master node. When-clauses (lines 2, 7) define triggers of interrupt routines, i.e., events: When an event occurs, the associated code is executed. In line 2, this a timer event; in line 7, the interrupt is triggered by the network adapter when a master-tick frame is received. The timer $timer_{txMasterTick}$ triggers the next resynchronization phase; it is set during network stabilization (line 1) and after expiry (line 4). At the beginning of a resynchronization phase, the master node reads the current local time to update its local tick t_{lt} (line 3). It then

```
(1)  set timer_txMasterTick;                        /* start of next resynchronization phase
(2)  when timer_txMasterTick then{                  /* master-tick frame transmission triggered
(3)       t_clock := t_lt := c_local;               /* record current local time
(4)       set timer_txMasterTick := t_clock + d_resInt;  /* start of next resynchronization phase
(5)       broadcast masterTickFrame(1);             /* send master-tick frame
(6)  }
(7)  when receive masterTickFrame(n)                /* ignore received master-tick frames
(8)       then { }
```

Listing 2.2 BBS_m—behavior of the master node (see [3], Table 3)

broadcasts a master-tick frame, carrying round number 1 (line 5). Master-tick frames received subsequently are discarded (line 8), as the master node is already resynchronized.

Listing 2.3 shows the code of non-master nodes. The timer $timer_{syncPhase}$ triggers the listening period of the next resynchronization phase; it is set during network stabilization (line 1) and after resynchronization (lines 9, 10). When $timer_{syncPhase}$ expires, the flag b_{resync} is reset to express that resynchronization is still to be established (lines 2, 3). Further interrupts are triggered upon reception of a master-tick frame (line 4). If b_{resync} is false (line 5), the received frame is processed, otherwise, it is discarded. To establish resynchronization, the local start time of master-tick frame reception and the current round number are recorded (lines 6, 7). Then, the local tick is updated (line 8), and the timer $timer_{syncPhase}$ is set (lines 9, 10). If the maximum network diameter has not yet been reached (line 11), a master-tick frame transmission is scheduled at the beginning of the next round (line 12) and executed when the corresponding timer expires (lines 16, 17). Finally, the flag b_{resync} is set to indicate that resynchronization has been completed (line 13).

2.2.3.4 Format of Master-Tick Frames

Master-tick frames start with a dominant bit Master Present (MP), indicating the presence of a master and providing a start of frame indication. This is followed by the current round number n_{round}, encoded as $n_{round} - 1$ by a sequence of m bits (see Fig. 2.8). Here, the value of m depends on the value of $n_{maxHops}$, the maximum round number, and is calculated as follows:

$$m = \max(\lceil 1, \log_2 n_{maxHops} \rceil) \quad \text{(i)}$$

To protect master-tick frames against corruption, a checksum could be added. We will address redundancy in Sect. 2.2.8.

```
(1)  set timer_syncPhase;                          /* start of next resynchronization phase

(2)  when timer_syncPhase then                     /* timer expiry
(3)      b_resync := false;                        /* reset flag: not resynchronized

(4)  when receive masterTickFrame(n) then {        /* master-tick frame received
(5)      if (!b_resync) then {                     /* not resynchronized?
(6)          t_clock := c_local − d_mtf;           /* record local time of start of tick frame
(7)          n_round := n;                         /* record current round number
(8)          t_lt := t_clock − (n_round − 1) · d_round_m;  /* compute local tick
(9)          set timer_syncPhase :=                /* start of next resynchronization phase
(10)                 t_lt + d_resInt − d_maxTickOffset_m;
(11)         if (n_round < n_maxHops) then         /* more rounds?
(12)             set timer_txMasterTick := t_clock + d_round_m;  /* next master-tick frame transmission
(13)         b_resync := true;                     /* resynchronization completed
(14)     }
(15) }

(16) when timer_txMasterTick then                  /* master-tick frame transmission triggered
(17)     broadcast masterTickFrame(n_round + 1);   /* send master-tick frame
```

Listing 2.3 BBS_m—behavior of non-master nodes (see [3], Table 4)

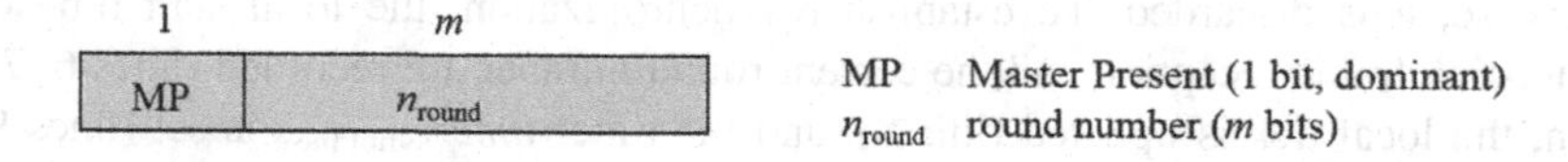

Fig. 2.8 Format of master-tick frames (see [3], Fig. 6)

2.2.3.5 Abstract Analysis of BBS_m

Accuracy, performance, and complexity of BBS_m during normal operation, i.e., assuming a stabilized network, can be analyzed based on configuration parameters, data structure, and algorithms as shown in Listings 2.1, 2.2, and 2.3. For a given hardware platform, concrete values can be determined (see Sect. 2.2.9).

Synchronization accuracy is a key performance indicator of synchronization protocols. As pointed out in [13] and [15], there are various sources reducing the accuracy of synchronization via message exchange, for instance, message creation delay, medium access delay, send and receive delay, and propagation delay. Delays consist of a fixed minimum delay and a variable portion, which is zero in case of constant delays. If the minimum delay can be determined, it can be taken into account and does not affect synchronization accuracy. However, variable portions contribute to inaccuracy.

To avoid message creation delays, master-tick frames are created in a timely manner. Because they are sent at predefined points in time without prior medium arbitration, medium access delay only consists of the duration to switch the transceiver from *rx* mode to *tx* mode, which is constant and therefore does not reduce synchronization accuracy. Send and receive delays are also constant, due to the fixed length of master-frames and constant bit times.

Nodes running BBS_m synchronize to the start of a master-tick frame, which is encoded by a sequence of black bursts. Here, the delays to detect the start of black bursts, i.e., energy on the medium, are variable and therefore contribute to synchronization inaccuracy. To detect energy, BBS uses the CCA (clear channel assessment) mechanism, which samples and integrates energy over a moving time window of fixed length. As soon as the amount of energy exceeds a given threshold, the medium is considered busy; otherwise, it is idle. Depending on received signal strength, a busy medium is detected earlier or later during the time window, leading to a variable CCA delay d_{CCA} in the order of microseconds. However, CCA delay d_{maxCCA} is bounded by the size of the time window, which is crucial for deriving worst-case upper bounds for tick offset.

Further factors contributing to synchronization inaccuracy are propagation delay and clock skew. In wireless ad hoc networks, the maximum sensing range depends on the communication technology, ranging, e.g., from 10 m to 1000 m. Given the speed of signals, this yields propagation delays $d_{maxProp}$ in the order of nanoseconds up to microseconds. We neglect the impact of clock skews, assuming that convergence delays are in the order of milliseconds.

The factors contributing to synchronization inaccuracy of BBS_m, i.e., d_{maxCCA} and $d_{maxProp}$, occur in every synchronization round. Therefore, the maximum base tick offset of BBS_m, $d_{maxBaseTickOffset\_m}$, that is, the offset after a resynchronization phase has terminated, is given by Eq. (2.1). Until the start of the next resynchronization phase, this offset deteriorates due to relative clock skews. Assuming a maximum clock skew $r_{maxClockSkew}$, the upper bound for this decrease is expressed as $r_{maxClockSkew} \cdot d_{resInt}$ for each pair of nodes, multiplied by 2 for all pairs of nodes, yielding the maximum tick offset $d_{maxTickOffset\_m}$ of Eq. (2.2). Equation (2.2) indicates that maximum tick offsets can be reduced by shortening the resynchronization interval d_{resInt}, however, at the cost of more overhead.

Statistical synchronization protocols determine average accuracies $\mathrm{d}_{avgTickOffset}$, e.g., during execution or by conducting real experiments, which typically are far below maximum offsets; here, maximum offsets could be measured, too, but different from BBS_m, they provide no evidence for worst-case offsets. In case of BBS_m, it is feasible to derive a worst-case maximum offset $d_{maxTickOffset\_m}$ offline, without measurements at runtime.

<u>Accuracy and performance</u>

Synchronization accuracy

$$d_{maxBaseTickOffset\_m} = n_{maxHops} \cdot (d_{maxCCA} + d_{maxProp}) \tag{2.1}$$

$$d_{maxTickOffset\_m} = d_{maxBaseTickOffset\_m} + 2 \cdot r_{maxClockSkew} \cdot d_{resInt} \tag{2.2}$$

Convergence delay

$$d_{\mathrm{conv}\_m} = n_{\mathrm{maxHops}} \cdot d_{\mathrm{round}\_m} + d_{\mathrm{maxTickOffset}\_m} \tag{2.3}$$

$$d_{\mathrm{round}\_m} = (1+m) \cdot d_{\mathrm{bit}\_m} + d_{\mathrm{proc}} \tag{2.4}$$

Relative overhead

$$o_{\mathrm{sync}\_m} = \frac{d_{\mathrm{conv}\_m}}{d_{\mathrm{resInt}}} \tag{2.5}$$

Another key performance indicator of synchronization protocols is *convergence delay*, i.e., the duration until resynchronization is achieved. To exploit an upper bound for tick offset, convergence delay must be bounded, too; otherwise, there would be uncertainty about the synchronization status of other nodes. Convergence delay d_{conv_m} of BBS$_m$ is bounded by the time needed to execute one resynchronization phase consisting of n_{maxHops} rounds, as shown in Eq. (2.3). Here, round duration d_{round_m} is constant, consisting of the time to transmit a master-tick frame of 1 + m bit length and a processing delay d_{proc} (see Eq. (2.4)). Both d_{bit_m} and d_{proc} have fixed durations, which is crucial to achieve an upper bound for d_{conv_m}. For a given hardware platform, the duration d_{proc} can be calculated based on the execution of the algorithms in Listings 2.2 and 2.3. We will address the calculation of d_{bit_m} in Sect. 2.2.9.

Given the fixed convergence delay d_{conv_m} and the duration of a resynchronization interval d_{resInt}, relative overhead of BBS$_m$ is given by Eq. (2.5).

Complexity

Communication	$c_{\mathrm{comm}\_m} = \mathrm{O}(\lvert V\rvert)$
Time	$c_{\mathrm{time}\_m} = \mathrm{O}(n_{\mathrm{maxHops}})$
Space	$c_{\mathrm{space}\_m} = \mathrm{O}(1)$
Structure	selection of some master node v_m

Apart from performance, the complexity of BBS$_m$ is of interest. From the algorithms in Listings 2.2 and 2.3, it is straightforward to argue that communication complexity c_{comm_m} is in the order of the number of network nodes, as during each resynchronization phase, each node sends exactly one master-tick frame. Given the previous analysis, it follows that time complexity c_{time_m} is in the order of n_{maxHops}, the maximum network diameter in sensing hops, because rounds have fixed durations. From Listing 2.1, it follows that space complexity c_{space_m} is constant and, in particular, independent of number of nodes and network diameter. Finally, apart from the selection of a master node v_m, there is no structural complexity.

2.2.4 Decentralized BBS (BBS$_d$)

Decentralized BBS (BBS$_d$) operates without master node. We present BBS$_d$ in several steps, starting with a digest followed by timing aspects and an analysis of its accuracy, performance, and complexity.

2.2.4.1 Digest of BBS$_d$

Decentralized BBS (BBS$_d$) is a self-contained protocol for tick synchronization in multi-hop wireless ad hoc networks. Unlike BBS$_m$, it operates without master node. For this outline, we assume that the network is stable, i.e., all network nodes are operational and tick-synchronized with a maximum tick offset.

The basic idea of BBS$_d$ is to resynchronize all nodes to the local tick of the node with the fastest clock. Resynchronization is performed periodically and started at the next local tick, i.e., a fixed duration after the current local tick. A resynchronization phase consists of a constant number of rounds defined by the maximum network diameter in sensing hops.

At their next local tick, all nodes send a decentralized tick frame (see Fig. 2.9). However, before doing so, they listen on the medium for a duration that is equal to the maximum tick offset. If a node detects a tick frame of another node, it corrects its local tick to the start of tick frame reception. This completes synchronization round 1.

To propagate the earliest local tick across the network, further synchronization rounds are needed. The number of rounds is determined by the maximum network diameter in sensing hops. The behavior in subsequent rounds is almost the same as in round 1. Local ticks are updated by backward calculation with high accuracy.

Operating BBS$_d$ produces frame collisions, which are non-destructive due to the encoding of decentralized tick frames with black bursts. Frame collisions may be observed by nodes with too slow clocks, i.e., nodes detecting tick frames of other nodes.

2.2.4.2 Timing Aspects of BBS$_d$

In this chapter, we provide some insights into the timing of BBS$_d$, thereby elaborating on the operation of the protocol. In synchronization round 1, nodes $v_i \in V$ start listening on the medium at $t'_{lt,i} - d_{\text{maxTickOffset\_}d}$, where $t'_{lt,i} = t_{lt,i} + d_{\text{resInt}}$ is the next local tick. If the medium remains idle until $t'_{lt,i} - d_{rxtx}$, nodes switch to transmit mode in order to start transmission of a decentralized tick frame at $t'_{lt,i}$. If the medium becomes busy, a node in sensing range has started sending its decentralized tick frame, and v_i records the start of tick frame reception $t_{rx\text{Tick},i}$ as its preliminary local tick $t*_{lt,i}$, which replaces $t'_{lt,i}$ in subsequent rounds.

The timing of BBS$_d$ in synchronization round 1 is illustrated in Fig. 2.10. At $t'_{lt,r} - d_{\text{maxTickOffset\_}d}$, node v_r starts listening on the medium, which remains idle.

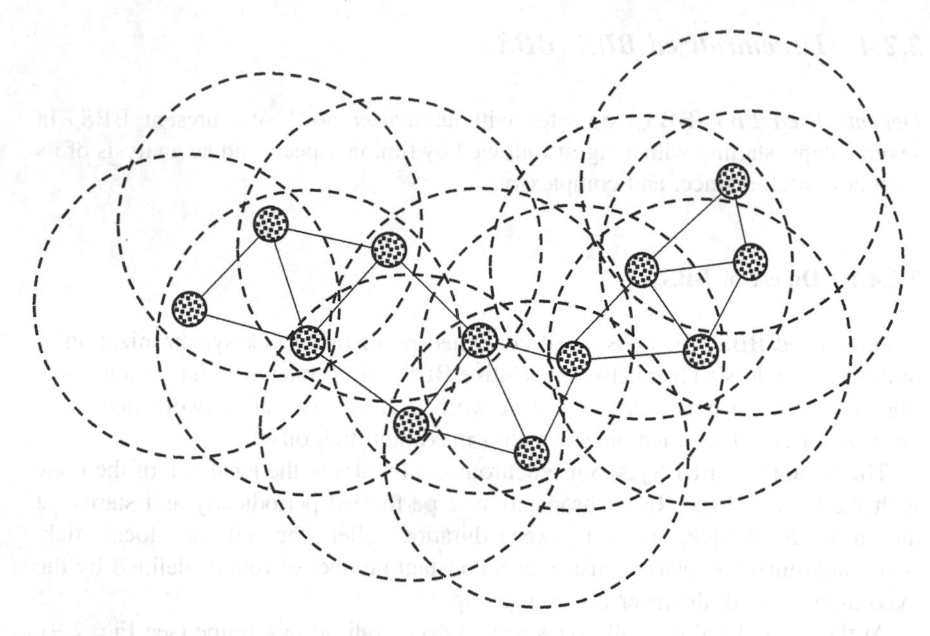

Fig. 2.9 BBS$_d$—synchronization round

Therefore, v_r starts switching to transmit mode at $t'_{lt,r} - d_{rxtx}$, for switching duration d_{rxtx}, and starts sending a tick frame at $t'_{lt,r}$. Nodes v_k and $v_{k'}$ in sensing range of v_r start listening with some tick offset, as their clocks are slower than the clock of v_r. As the medium remains idle, node v_k starts switching to *tx* mode at $t'_{lt,k} - d_{rxtx}$, and starts sending its tick frame at $t'_{lt,k}$. During mode switching and transmission, nodes cannot listen on the medium. Because the actual tick offset between v_r and v_k is smaller than d_{rxtx}, node v_k has already started mode switching and therefore cannot detect the tick frame sent by v_r in round 1 anymore.

The actual tick offset between v_r and $v_{k'}$ is greater than d_{rxtx}. Therefore, v_k can detect[8] the start of tick frame transmission of v_r, and records $t_{rxTick,k'}$ as its preliminary local tick $t*_{lt,k'}$. Then, it switches to transmit mode as planned and transmits its tick frame in round 1, which is needed to sustain the synchronization procedure. However, the start of subsequent rounds is determined based on $t*_{lt,k'}$.

The behavior of BBS$_d$ in subsequent rounds is rather similar: nodes $v_i \in V$ start listening on the medium $d_{\text{maxTickOffset\_}d}$ before a round starts and send a decentralized tick frame. If they detect the medium as busy before switching to *tx* mode, they record $t^*_{lt,i} := t_{rxTick,i} - (n_{\text{round}} - 1) \cdot d_{\text{round\_}d}$ as their preliminary local tick, which replaces $t'_{lt,i}$ in subsequent rounds. Thus, the local tick of the node with the

[8]To detect the medium as busy, CCA delay and propagation delay have to be taken into account, too, so the actual tick offset needed to detect a tick frame may be as large as $d_{rxtx} + d_{\text{maxCCA}} + d_{\text{maxProp}}$.

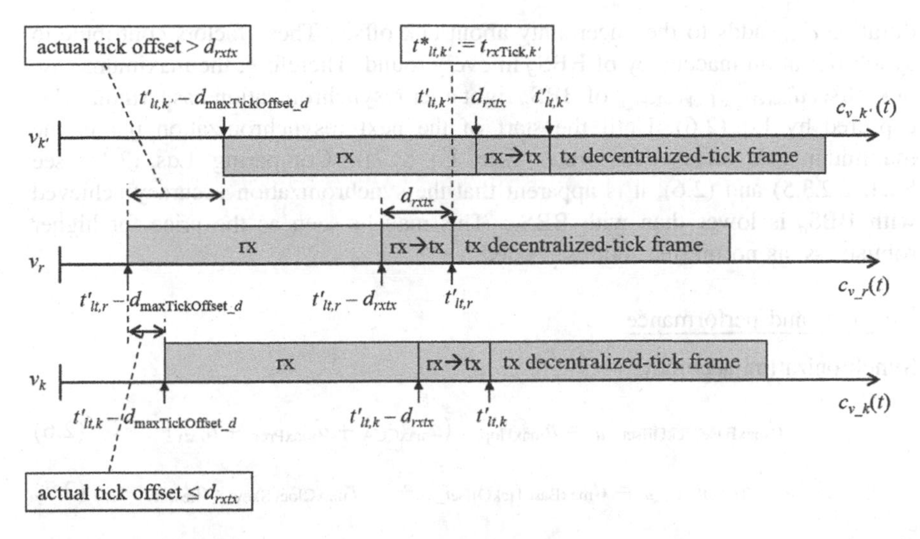

Fig. 2.10 BBS_d—timing in synchronization round 1 (see [3], Fig. 7)

fastest clock (modulo maximum tick offset) is propagated across the network. Resynchronization of BBS_d terminates after $n_{maxHops}$ rounds.

If all nodes in mutual sensing range are synchronized with a maximal tick offset of $d_{maxCCA} + d_{maxProp} + d_{rxtx}$, they will always start mode switching before detecting a decentralized tick frame. Thus, they do not get any feedback whether other nodes are still operational. Additional means are required to establish such feedback, e.g., combined use of BBS_d and BBS_m (hybrid BBS, see Sect. 2.2.5), or automatic topology detection based on message exchange.

Decentralized tick frames consist of a dominant bit Decentralized Sync Header (DSH), creating a reference point in time (Fig. 2.11). Different from BBS_m, no round number is sent. We will address redundancy in Sect. 2.2.8.

2.2.4.3 Abstract Analysis of BBS_d

Synchronization accuracy of BBS_d is bounded by the ability of nodes to detect the start of a decentralized tick frame, i.e., of a black burst, before switching to transmit mode, and by the switching delay. The upper bound to detect energy on the medium is the sum of maximum CCA delay d_{maxCCA} and maximum propagation delay $d_{maxProp}$. Since during mode switching, nodes cannot listen on the medium, and the

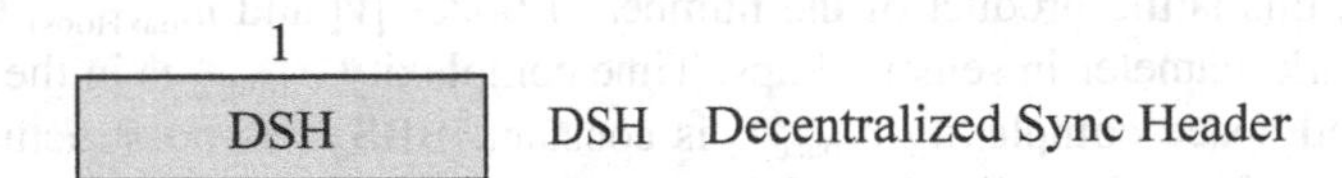

Fig. 2.11 Format of decentralized tick frames

duration d_{rxtx} adds to the uncertainty about tick offset. These factors contribute to synchronization inaccuracy of BBS$_d$ in every round. Therefore, the maximum base tick offset $d_{\text{maxBaseTickOffset\_}d}$ of BBS$_d$ right after resynchronization has terminated is captured by Eq. (2.6). Until the start of the next resynchronization phase, the maximum tick offset deteriorates (see Eq. (2.7)). Comparing Eqs. (2.1) (see Sect. 2.2.3.5) and (2.6), it is apparent that the synchronization accuracy achieved with BBS$_d$ is lower than with BBS$_m$. This may be seen as the price for higher robustness, as no master node is needed.

Accuracy and performance

Synchronization accuracy

$$d_{\text{maxBaseTickOffset\_}d} = n_{\text{maxHops}} \cdot (d_{\text{maxCCA}} + d_{\text{maxProp}} + d_{rxtx}) \tag{2.6}$$

$$d_{\text{maxTickOffset\_}d} = d_{\text{maxBaseTickOffset\_}d} + 2 \cdot r_{\text{maxClockSkew}} \cdot d_{\text{resInt}} \tag{2.7}$$

Convergence delay

$$d_{\text{conv\_}d} = n_{\text{maxHops}} \cdot d_{\text{round\_}d} \tag{2.8}$$

$$d_{\text{round\_}d} = d_{\text{maxTickOffset\_}d} + d_{\text{bit\_}d} + d_{\text{proc}} \tag{2.9}$$

Relative overhead

$$o_{\text{sync\_}d} = \frac{d_{\text{conv\_}d}}{d_{\text{resInt}}} \tag{2.10}$$

Complexity

Communication	$c_{\text{comm\_}d} = \mathrm{O}(\lvert V\rvert \cdot n_{\text{maxHops}})$
Time	$c_{\text{time\_d}} = \mathrm{O}(n_{\text{maxHops}})$
Space	$c_{\text{space\_}d} = \mathrm{O}(1)$
Structure	none

On an abstract level, the analysis of convergence delay and relative overhead of BBS$_d$ is quite similar to the analysis of BBS$_m$. Nevertheless, round and bit durations are to be adapted when selecting a specific communication technology and differ from the values of BBS$_m$, which is reflected in Eqs. (2.8), (2.9), and (2.10).

Communication complexity $c_{\text{comm\_}d}$ of BBS$_d$ is expressed by the number of tick frames sent in a single resynchronization phase. As each node sends a tick frame in each round, this is the product of the number of nodes $|V|$ and n_{maxHops}, the maximum network diameter in sensing hops. Time complexity $c_{\text{time\_}d}$ is in the order of n_{maxHops}, and space complexity $c_{\text{space\_}d}$ is constant. BBS$_d$ has no structural complexity. We omit the formalization of data structure and algorithm of BBS$_d$, as this does not provide additional insights.

2.2.5 Hybrid BBS (BBS_h)

Hybrid BBS (BBS_h) is composed of BBS_m and BBS_d. When a master node is present, all nodes are synchronized with $d_{maxTickOffset\_m}$. If the master node fails or if the network is partitioned due to topology changes, nodes are synchronized with $d_{maxTickOffset\_d}$, i.e., with less accuracy.

To compose BBS_m and BBS_d, synchronization rounds are merged, consisting of master part and decentralized part. During the master part, the rules of BBS_m (see Sect. 2.2.3) are applied, i.e., the master node v_m sends a master-tick frame in round 1, and nodes receiving a master-tick frame in round n_{round}, $1 \leq n_{round} < n_{maxHops}$, resynchronize their local clocks and send a master-tick frame at the beginning of round $n_{round} + 1$. During the decentralized part, the rules of BBS_d (see Sect. 2.2.4) are executed, i.e., all nodes send decentralized tick frames after listening, and resynchronize if they detect another tick frame. Synchronization round n_{round} of node $v_i \in V$ starts at $t_{lt,i} + (n_{round} - 1) \cdot d_{round\_h}$, where $t_{lt,i}$ is v_i's current perception of the local tick, possibly corrected during this resynchronization phase, and $d_{round\_h} = d_{round\_m} + d_{round\_d}$.

<u>Accuracy and performance</u>

Synchronization accuracy

$$d_{maxBaseTickOffset\_h} = \begin{cases} d_{maxBaseTickOffset\_m} & \text{if master present} \\ d_{maxBaseTickOffset\_d} & \text{otherwise} \end{cases} \tag{2.11}$$

$$d_{maxTickOffset\_h} = d_{maxBaseTickOffset\_h} + 2 \cdot r_{maxClockSkew} \cdot d_{resInt} \tag{2.12}$$

Convergence delay

$$d_{conv\_h} = n_{maxHops} \cdot d_{round\_h} \tag{2.13}$$

$$d_{round\_h} = (d_{bit\_m} + d_{proc}) + (d_{maxTickOffset\_d} + d_{bit\_d} + d_{proc}) \tag{2.14}$$

Relative overhead

$$o_{sync\_h} = \frac{d_{conv\_h}}{d_{resInt}} \tag{2.15}$$

If nodes receive a master-tick frame, this implies that a master node is present; consequently, the more accurate maximum master-tick offset $d_{maxTickOffset\_m}$ is established and applied. In addition, decentralized tick frames are exchanged. If the master node fails, this is detected by all other nodes during the next resynchronization phase, i.e., with small and bounded delay d_{resInt}. In this case, they immediately switch to decentralized tick synchronization, applying $d_{maxTickOffset\_d}$. As soon as another master node is activated, nodes resume master-based tick synchronization and reapply $d_{maxTickOffset\_m}$.

Without further measures, the overhead of BBS_h is the combined overhead of BBS_m and BBS_d, i.e., $o_{sync\_h} = (d_{conv\_m} + d_{conv\_d})/d_{resInt}$. However, it is possible to reduce this overhead by omitting the round number of master-tick frames (see Eq. (2.14)), as non-master nodes can also determine the current round number by counting decentralized tick frames.

During stable operation, it is even possible to derive the current round number from the reception time of master-tick frames. However, during stabilization (see Sect. 2.2.7), an explicit round number is needed. In addition, further redundancy to protect tick frames against the detection of false positives is desirable (see Sect. 2.2.8).

2.2.6 Time Sync BBS (BBS_t)

Time sync BBS (BBS_t) is based on a BBS module for tick synchronization, i.e., BBS_m, BBS_d, or BBS_h, which establishes network-wide reference points in time. To establish network-wide time synchronization, the local time value of a local tick is exchanged in a time frame. Upon reception of a time frame, nodes adjust their local clocks.

Figure 2.12 shows the operation of *master-based* BBS_t. Let v_m be the master node, and $t_{lt,m,n}$ be the nth local tick of v_m. The specific value of n is not important, as it is used here for presentation purposes only, to refer to subsequent local ticks. At $t_{lt,m,n}$, node v_m triggers a resynchronization phase, using BBS_m. Afterward, it sends a time frame containing the local clock reading $t_{lt,m,n}$ as reference time value, which is propagated by receiving nodes across the network. In addition, receiving nodes adjust their local clocks and correct the local time of the current local tick.

Figure 2.12 also shows the required actions of node v_i upon receipt of a time frame. The local clock c_{v_i} is set to the master time $t_{lt,m,n}$ plus the time that has

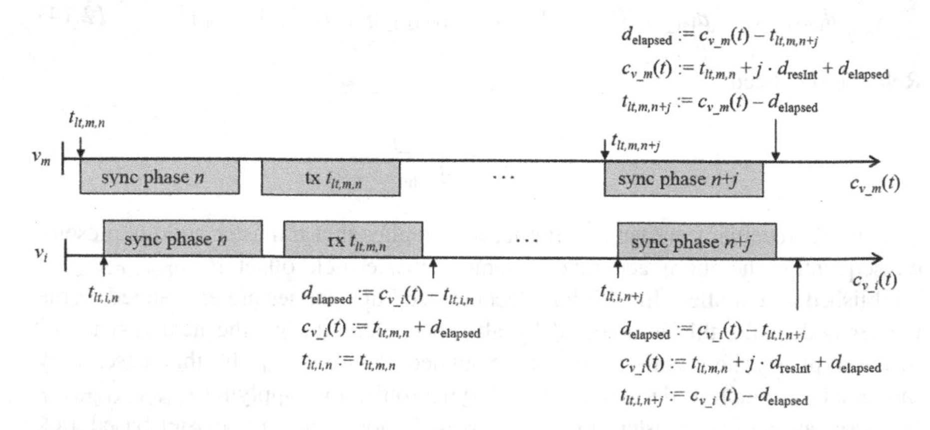

Fig. 2.12 BBS_t—master-based time synchronization

elapsed since the local tick $t_{lt,i,n}$. Furthermore, the current local tick $t_{lt,i,n}$ is aligned with the master tick. This provides the same accuracy for time synchronization as for tick synchronization, i.e., $d_{\text{maxTimeOffset\_}m} = d_{\text{maxTickOffset\_}m}$.

For time resynchronization, tick resynchronization and local computations are sufficient; no further time value exchange is required. Figure 2.12 shows the computations of master and non-master nodes, which are identical. Let j be the count of resynchronization phases since the time value exchange. After each resynchronization phase, nodes readjust their local clocks and correct their local ticks. To do so, nodes refer to the exchanged reference time value $t_{lt,m,n}$ and add the perfect elapsed time $j \cdot d_{\text{resInt}}$ until the current local tick $t_{lt,i,n+j}$ and the measured elapsed time since that tick. Then, the current local tick $t_{lt,i,n+j}$ is aligned with the master tick. This preserves the same accuracy for time synchronization as for tick synchronization.

To propagate a time frame network-wide, a regular frame could be broadcasted, using medium arbitration. This, however, has the drawback that collisions may occur, implying that no upper bound for convergence delay can be determined. Furthermore, time complexity is in the order of $|V|$. Therefore, to propagate time frames, we apply the same approach as for tick frames. Time frames are encoded using black bursts (see Sect. 2.2.2) and are composed of a dominant start-of-time-frame (SOTF) bit, followed by a time value. The number of bits n_{timeBits} required to encode this value depends on the representation of time.

At a specified point in time, e.g., right after completion of a tick resynchronization phase, the master node transmits a time frame. Receiving nodes adjust their local clocks and ticks and forward the time frame after a constant delay, i.e., almost simultaneously. Thus, collisions of time frames remain non-destructive. This is repeated until all network nodes are time synchronized, i.e., for n_{maxHops} rounds.

With this approach, time synchronization accuracy is the same as that of tick synchronization, i.e., $d_{\text{maxTimeOffset\_}m} = d_{\text{maxTickOffset\_}m}$. Convergence delay $d_{\text{conv\_}t}$ is bounded by $n_{\text{maxHops}} \cdot d_{\text{round\_}t}$, where $d_{\text{round\_}t} = (1 + n_{\text{timeBits}}) \cdot d_{\text{bit\_}t} + d_{\text{proc}}$. Since it is sufficient to exchange only one time frame, relative overhead on top of tick resynchronization is extremely low. However, it may be useful to repeat time frame exchange from time to time, to provide some redundancy. Complexity of master-based BBS$_t$ is similar to BBS$_m$.

Time synchronization can also be achieved with decentralized BBS$_t$. At a specified point in time, nodes $v_i \in V'$, $V' \subseteq V$, transmit time frames containing their local clock readings $t_{lt,i,n}$ as reference time values. Again, time frames are encoded with black bursts, so collisions are non-destructive. However, different from master-based BBS$_t$, time frames sent in the same round may have different values. Thus, as explained in Sect. 2.2.2, receivers record their logical OR. At this point, we slightly modify the algorithm for the exchange of bit sequences: nodes $v_i \in V'$ "sending" a logical 0, i.e., a recessive bit, and observing a logical 1 stop transmitting their own bit sequence and instead record and retransmit the rest of the bit sequence, thereby receiving and forwarding the actual clock value of some node. Nodes $v_j \in V \backslash V'$ in sensing range of a node $v_i \in V'$ record the entire tick frame.

This procedure is repeated in subsequent rounds until finally, the local clock value of the node with the fastest clock has been propagated, i.e., the fastest clock "wins."

2.2.7 Network Stabilization

So far, we have assumed a stabilized network, i.e., all network nodes are operational and tick-synchronized with a maximum tick offset. We now elaborate on achieving and maintaining stabilization despite dynamic sensing topology changes. Explanations address the operation of BBS_m; stabilization of BBS_d and BBS_h is handled analogously.

When a master node $v_m \in V$ is switched on, it listens on the medium for a defined number of resynchronization intervals. If the medium remains silent, v_m starts synchronization by sending master-tick frames; otherwise, v_m retires. When a non-master node $v_i \in V$ is powered up, it listens until detecting master-tick frames, and, after it has synchronized itself, starts transmitting master-tick frames itself, thereby joining the network. Here, it is important that the node can derive the current hop distance to the master node, which is contained as parameter n_{round} in the master-tick frame. Using the value of n_{round}, v_i can then compute its current local tick $t_{lt,i}$ by backward calculation (see Sect. 2.2.3.2).

If a master node $v_m \in V$ is no longer present, e.g., because it is switched off or is outside sensing range of all other nodes, none of the remaining nodes receives a master-tick frame in the next resynchronization phase. In this case, the network is not operational until a new master node is determined and active. In case of BBS_h, decentralized tick synchronization is still available and applied, with reduced synchronization accuracy. If a non-master node is no longer present, this will have no effect on tick synchronization unless the network is partitioned.

If sensing topology changes due to node movement occur, two situations are to be distinguished. If only the set of sensing links is modified, but the set of network nodes remains the same, tick synchronization is not affected as long as the maximum network diameter in sensing hops does not exceed n_{maxHops}. In particular, if the hop distance between a non-master node and the master node has changed, the new distance is given by the parameter n_{round} of the master-tick frame. The reason why link changes have no further impact is that BBS does not assume a particular network structure. However, if the topology change leads to network partitioning, only the partition hosting the master node remains stable. In partitions without master node, the situation is the same as in case of master node failure and treated accordingly.

If several networks synchronized with BBS get into mutual sensing range, there is a lack of tick synchronization among nodes belonging to different networks. This is detected by reception of tick frames at unexpected points in time, or possibly by collision of regular frames in reserved time slots. Nodes detecting another network send jamming frames, i.e., distinguishable energy patterns, which are forwarded by receiving nodes, and cease operation. It is then the task of a dedicated master node to reinitialize tick synchronization.

2.2.8 Redundancy

We now address redundancy measures to increase the robustness of BBS. In particular, we elaborate on the detection of bit errors of synchronization frames, false positives and negatives of black bursts, and timing errors.

A crucial property of BBS is that collisions of synchronization frames that are sent (almost) simultaneously are non-destructive. This is achieved by encoding bit sequences with black bursts (see Sect. 2.2.2), which therefore can be interpreted as a redundancy measure.

For the reliable detection of black bursts, their timing is crucial. First, the length d_{BB} of a black burst has to be larger than the maximum CCA delay d_{maxCCA} (constraint (2.16)); otherwise, a black burst may go undetected. Second, black bursts sent in the same bit time must overlap at the receiver, which places the stronger constraint (2.17) on d_{BB}.

$$d_{BB} > d_{\text{maxCCA}} \tag{2.16}$$

$$d_{BB} > d_{\text{maxCCA}} + d_{\text{maxBaseTickOffset\_}m} \tag{2.17}$$

Constraint (2.17) is actually required in the synchronization scenario shown in Fig. 2.13. Here, node v is in sensing range of v' and v'', which send their master-tick frames in the same round. At this point in time, the base tick offset accumulated at node v'' may differ from that of v' by at most $d_{\text{maxBaseTickOffset\_}m}$. In addition, the delay for the detection of the black burst of v'' has to be considered.

If the network is sufficiently dense, sensing topologies as in Fig. 2.13 do not occur. Rather, if a node v receives colliding master-tick frames, the sending nodes have been synchronized by the same node(s) in the round before. Thus, their tick offset will differ by at most d_{maxCCA}, meaning that constraint (2.17) can be reduced to (2.18), and finally to (2.16).

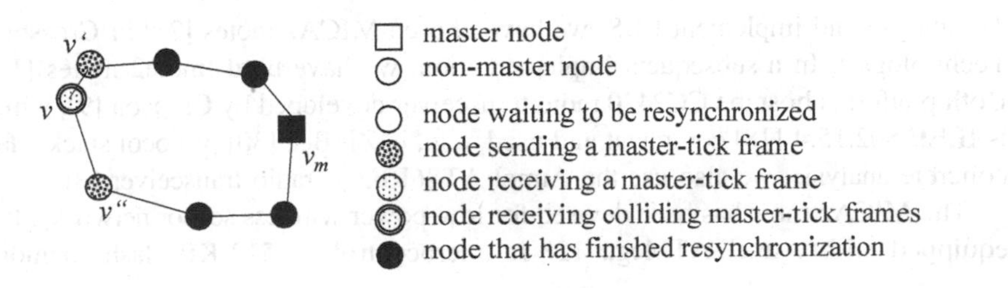

Fig. 2.13 BBS—synchronization scenario

$$d_{BB} > 2 \cdot d_{\mathrm{maxCCA}} \tag{2.18}$$

Additional timing constraints are required to ensure that black bursts are always assigned to the correct bit time, and that black bursts of different bit times do not overlap. Furthermore, the timing of the received energy pattern of synchronization frames can be checked for consistency. A detailed treatment of timing issues can be found in [4].

If a receiving node is located at the border of the sensing range to a sender, bit errors of synchronization frames can occur. In this case, false negatives may be detected, i.e., black bursts may go undetected, which is interpreted as the "reception" of a logical 0. Also, due to a noisy channel, false positives may be detected, which is interpreted as the reception of a logical 1.

To detect bit errors, several measures can be applied. First, checksums can be added to synchronization frames. For decentralized tick frames, it may be sufficient to repeat the DSH bit. Second, reception times can be checked. For instance, the start of reception of a master-tick frame carrying round number n_{round} is expected by node $v_i \in V$ at $t_{lt,i} + d_{\mathrm{resInt}} + (n_{\mathrm{round}} - 1) \cdot d_{\mathrm{round\_}m} \pm d_{\mathrm{maxTickOffset\_}m}$. If this constraint is not met, this indicates a bit error. In addition, the first bit of synchronization frames, which is always a logical 1 and therefore dominant, can be protected by an encoding with a black burst that has a different length $d'_{BB} > d_{BB}$. Again, timing constraints are required to ensure that this distinction always works.

2.2.9 *Analytical and Experimental Results*

To assess the accuracy, performance, energy efficiency, feasibility, and reliability of BBS, we now extend its abstract analysis (see Sects. 2.2.3–2.2.5) based on a real hardware platform, and report on the results of real experiments conducted with an implementation on that platform. We have decided against simulation experiments, because the effects to be studied are of physical nature and therefore difficult to capture completely and accurately in a simulation model.

2.2.9.1 Hardware Platform

To analyze and implement BBS, we have chosen MICAz motes [21] of Crossbow Technologies. In a subsequent implementation, we have used Imote2 motes [19]. Both platforms host the CC2420 radio transceiver developed by Chipcon [9], which is IEEE 802.15.4 [18] compliant and used, e.g., in ZigBee [30] protocol stacks. For concrete analysis, we also use the Atmel AT86RF230 radio transceiver [8].

The MICAz mote has been devised for low-power wireless sensor networks. It is equipped with Atmel ATMega 128 L microcontroller, 512 KB flash memory,

LEDs, a connector for various sensor boards, 2.4 GHz CC2420 radio supporting a transmission rate of 250 Kbps, and is powered by two AA batteries.

The CC2420 radio transceiver integrates PHY layer and parts of the MAC layer as specified by IEEE 802.15.4. PHY frames consist of preamble (3 bytes), start of frame (SOF) delimiter (1 byte), and MAC frame. When running with a data rate of 250 Kbps, PHY symbols encode four bits each, yielding a symbol duration $d_{\mathrm{sym}} = 16\ \mu\mathrm{s}$. Thus, sending one byte takes $d_{\mathrm{byte}} = 32\ \mu\mathrm{s}$.

For the implementation of BBS, we only use PHY layer functionality, as black bursts are sent without prior medium contention. To generate a black burst of defined length, we send regular PHY frames of minimal length, however, without interpreting their content. In our implementation, a PHY frame consists of preamble, SOF delimiter, and MAC length field with value 0, totaling 5 bytes. With $d_{\mathrm{byte}} = 32\ \mu\mathrm{s}$, the duration of a black burst is $d_{BB} = 160\ \mu\mathrm{s}$, which satisfies constraint (2.16).

To detect black bursts, we use the CCA (clear channel assessment) mechanism of the transceiver. The transceiver sets the CCA pin to 0 if the energy sampled over the last 8 symbols exceeds a defined threshold, or if it has detected a correct preamble and SOF delimiter. In case of high signal strength, this may occur already after one symbol. However, as we have to consider worst-case behavior, we obtain $d_{\mathrm{maxCCA}} = 8 \cdot d_{\mathrm{sym}} = 128\ \mu\mathrm{s}$ for the CC2420 transceiver ($d_{\mathrm{maxCCA}} = 1 \cdot d_{\mathrm{sym}} = 16\ \mu\mathrm{s}$ for the AT86RF230). To reduce further delays, the CCA pin of the transceiver is connected to a free external interrupt pin of the microcontroller, which avoids polling for black bursts. For clock skew, we have measured values $\mathrm{r}_{\mathrm{clockSkew}}$ of up to 10 ppm with the built-in MICAz processor quartz. In our concrete analysis, we assume a maximum clock skew $r_{\mathrm{maxClockSkew}}$ of 40 ppm.

2.2.9.2 Concrete Analytical Results

In this chapter, we extend the abstract analysis of BBS in Sects. 2.2.3–2.2.5. For this, we insert parameters from data sheets of the CC2420 and AT86RF230 transceivers, and calculate concrete values for accuracy and performance. In addition, we assess energy efficiency.

Synchronization accuracy

Table 2.1 shows the maximum tick offsets for BBS_m and BBS_d, for topologies with a maximum network diameter n_{maxHops} of 1, 4, and 10 hops, respectively. We have set the length of the resynchronization interval d_{resInt} to 1 s (1 and 4 hops) and 5 s (10 hops). Values d_{maxCCA} and d_{rxtx} are determined from the data sheets, and the $r_{\mathrm{maxClockSkew}}$ is set to 40 ppm. Inserting these values into Eqs. (2.1), (2.2), (2.6), and (2.7) yields the shown upper bounds for tick offset, which are also the upper bounds for time offset when operating BBS_t.

The maximum base tick offset achievable depends on symbol duration, the parameters of the CCA mechanism, and switching delays, and accumulates with

Table 2.1 Maximum tick (and time) offsets for BBS_m and BBS_d (see [3], Table 5)

	CC2420			AT86RF230		
$n_{maxHops}$	1	4	10	1	4	10
d_{resInt}	1 s	1 s	5 s	1 s	1 s	5 s
d_{maxCCA}	128 μs			16 μs		
d_{rxtx}	192 μs			17 μs		
$r_{maxClockSkew}$	40 ppm			40 ppm		
$d_{maxBaseTickOffset\_m}$	0.128 ms	0.512 ms	1.280 ms	0.016 ms	0.064 ms	0.160 ms
$d_{maxTickOffset\_m}$	0.208 ms	0.592 ms	1.680 ms	0.096 ms	0.144 ms	0.560 ms
$d_{maxBaseTickOffset\_d}$	0.320 ms	1.280 ms	3.200 ms	0.033 ms	0.132 ms	0.330 ms
$d_{maxTickOffset\_d}$	0.400 ms	1.360 ms	3.600 ms	0.113 ms	0.212 ms	0.730 ms

each hop. For the CC2420 transceiver, the resulting maximum CCA delay and switching delay are substantial, yielding maximum base tick offsets in the order of a millisecond. Here, the AT86RF230 performs about ten times better. By reducing CCA delay and switching delay, synchronization accuracy can be improved. On the other hand, accuracy decreases with increasing network size and/or resynchronization interval. In Sect. 2.2.10, we present ways to mitigate these effects.

We point out that the tick offsets in Table 2.1 are upper bounds, i.e., worst-case values. During operation, lower upper bounds may of course be observed, and certainly, average values would typically be far below measured upper bounds. The reason why measured worst cases are usually better is that d_{maxCCA} and $r_{maxClockSkew}$ are not stretched to their limits; however, they may also yield worse results, e.g., if $r_{maxClockSkew}$ is not chosen carefully.

To show this, we have conducted an experiment with five MICAz motes hosting the CC2420 transceiver in a line topology, running BBS_m. Parameters d_{resInt}, d_{maxCCA}, and $r_{maxClockSkew}$ are as in Table 2.1, and experiment duration was 4 h. Table 2.2 shows the results on measured upper bounds in comparison to calculated worst-case values. As expected, measured worst-case values are substantially smaller. However, to achieve deterministic timing behavior, the bounds from the offline analysis are to be used, as measured values cannot be exchanged among network nodes instantaneously and reliably.

Performance

We continue our concrete analysis of BBS by determining values for convergence delay and overhead, inserting data sheet values into Eqs. (2.3, 2.8, 2.13), (2.4, 2.9, 2.14), and (2.5, 2.10, 2.15). Table 2.3 summarizes the results. Parameters $n_{maxHops}$ and d_{resInt} are chosen as in Table 2.1; the number m of bits to encode n_{round} is determined using equation (i) (see Sect. 2.2.3.4). The bit duration for master bits is the sum of black burst duration d_{BB} and switching durations d_{rxtx} and d_{txrx}, and is independent of the network diameter (see Sect. 2.2.2). For decentralized bits, $d_{maxTickOffset\_d}$ is to be added; therefore, bit duration depends on network diameter, too.

Table 2.2 Calculated and measured maximum tick offsets for BBS_m (see [2], Table II)

	1 hop (ms)	2 hops (ms)	3 hops (ms)	4 hops (ms)
$d_{maxTickOffset\_m}$ (analysis)	0.208	0.336	0.464	0.592
$d_{maxTickOffset\_m}$ (measurement)	0.041	0.160	0.225	0.289

Table 2.3 Convergence delays and synchronization overheads for BBS_m, BBS_d, and BBS_h (see [3], Table 6)

	CC2420			AT86RF230		
$n_{maxHops}$ (m)	1 (1)	4 (2)	10 (4)	1 (1)	4 (2)	10 (4)
d_{resInt}	1 s	1 s	5 s	1 s	1 s	5 s
d_{BB}	160 μs			160 μs		
d_{rxtx}, d_{txrx}	192 μs, 192 μs			17 μs, 33 μs		
d_{bit_m}	0.54 ms			0.21 ms		
d_{bit_d}	0.94 ms	1.90 ms	4.14 ms	0.32 ms	0.42 ms	0.94 ms
d_{proc}	0.30 ms			0.30 ms		
d_{round_m}	1.39 ms	1.93 ms	3.02 ms	0.72 ms	0.93 ms	1.35 ms
d_{round_d}	1.56 ms	3.48 ms	7.64 ms	0.66 ms	0.85 ms	1.57 ms
d_{round_h}	2.49 ms	4.41 ms	8.89 ms	1.25 ms	1.44 ms	2.48 ms
d_{conv_m}	1.60 ms	8.32 ms	31.88 ms	0.82 ms	3.86 ms	14.06 ms
d_{conv_d}	1.56 ms	13.94 ms	76.44 ms	0.66 ms	3.42 ms	15.70 ms
d_{conv_h}	2.49 ms	17.63 ms	88.88 ms	1.25 ms	5.78 ms	24.80 ms
o_{sync_m}	0.16%	0.83%	0.64%	0.08%	0.39%	0.28%
o_{sync_d}	0.16%	1.39%	1.53%	0.07%	0.34%	0.31%
o_{sync_h}	0.25%	1.76%	1.78%	0.13%	0.58%	0.50%

With these settings, round durations, convergence delays, and synchronization overhead are determined (see Table 2.3). It turns out that BBS has excellent performance, with resynchronization phases in the order of milliseconds, and very low synchronization overhead. We point out that at the end of each resynchronization phase, maximum base tick offsets are reestablished, i.e., no statistics is to be collected. This makes BBS highly efficient compared to statistical synchronization approaches, which have to collect data points over a longer period of time before synchronization is established.

Energy efficiency

When using battery-powered nodes such as the MICAz, energy efficiency is an important concern. To extend the nodes' and thus the network's lifetime, nodes switch to idle or sleep mode when not needed, which is called *duty cycling* (see

Chap. 7). While this reduces energy consumption, it also downgrades the ability of the network to perform tasks.

To assess the energy efficiency of BBS, we determine the energy consumption of three activity profiles. The profile "idle" determines a base line, i.e., energy consumption when all hardware components of nodes are in idle mode. The profile "sync" adds energy consumption due to synchronization activities. Finally, the profile "duty cycle" considers energy consumption of nodes performing actual application tasks. Given a percentage of node activity for each of these profiles, the consumed energy is calculated as follows:

$$e_{\text{cons}}(o_{\text{act}}) = o_{\text{act}} \cdot (e_{\text{trAct}} + e_{\text{cpuAct}}) + (1 - o_{\text{act}}) \cdot (e_{\text{trIdle}} + e_{\text{cpuIdle}}) \tag{2.19}$$

Equation (2.19) distinguishes between energy consumption of transceiver and CPU in active and idle modes, respectively. Concrete values to be inserted can be found in data sheets of hardware platforms to be assessed. In the following, we will use values from the MICAz data sheet [21]. Thus, the transceiver draws a current of up to $e_{\text{trAct}} = 19.7$ mA in *tx* and *rx* mode, and of $e_{\text{trIdle}} = 20$ μA in idle mode. The CPU consumes $e_{\text{cpuAct}} = 8.65$ mA and $e_{\text{cpuIdle}} = 1$ mA in active and idle modes, respectively. External power is provided by two AA batteries with a capacity of 2000 mAh each.

We now compare the energy consumption of BBS_m and compare it to the beacon mechanism of ZigBee [30] providing single-hop synchronization. The resynchronization interval d_{resInt} is set to 5 s, beacon *tx* duration is calculated to be 1.024 ms, and the duty cycle is configured to be 4%. Table 2.4 shows the values for energy consumption and the expected network lifetime, for each activity profile.

The activity profile "idle" determines the base line, i.e., energy consumption of transceiver and CPU in idle mode. While the energy consumption of the CC2420 is almost negligible, the Atmel ATMega 128 L consumes 1 mA and therefore limits the lifetime of nodes to at most 81.7 days (assuming that the battery charge can be consumed in a linear way). With synchronization, energy consumption slightly increases, however, reducing the lifetime of nodes by less than one day. Here, ZigBee beacons are slightly more efficient, but can support single-hop synchronization only. With the full duty cycle of 4%, energy consumption doubles, reducing node lifetime to about 40 days. Without duty cycling, the nodes' lifetime would be less than 3 days.

Table 2.4 Energy consumption—comparison of BBS_m and ZigBee beacons

	BBS_m			ZigBee beacons		
Activity	Idle	Sync	Duty cycle	Idle	Sync	Duty cycle
o_{act}	0.00%	0.04%	4%	0.00%	0.02%	4%
e_{cons}	1.020 mA	1.031 mA	2.113 mA	1.020 mA	1.025 mA	2.113 mA
d_{life}	81.7 days	80.8 days	39.4 days	81.7 days	81.3 days	39.4 days

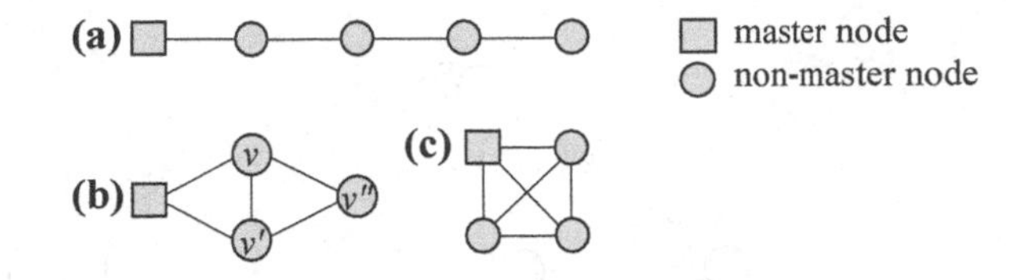

Fig. 2.14 Topologies for feasibility experiments (see [3], Fig. 9)

2.2.9.3 Experimental Results

In this chapter, we report on real-world experiments that were conducted to show feasibility and reliability of BBS_m and BBS_h [3, 6]. Further experiments had the purpose of calibrating the transceiver in order to increase reliability and accuracy [6]. Experiments were run in testbeds of MICAz [21] and Imote2 motes [19], both hosting the CC2420 radio transceiver [9].

Feasibility experiments

To show the feasibility of BBS_m and BBS_h, they were implemented and executed in representative small-scale topologies that cover scenarios occurring in larger networks [3]. Figure 2.14 shows three such topologies: The 4-hop line topology in (a) checks the multi-hop capability of BBS_m. Topology (b) produces master-tick frame collisions in round 2, when nodes v and v' resynchronized in round 1 send them (almost) simultaneously. Finally, topology (c) represents a single-hop network, where BBS_d is used as backup for BBS_m in case of master node failure.

Feasibility experiments were run on MICAz motes, over a period of 4 h. In all experiments, measured tick offsets were below the predicted maximum tick offsets, i.e., below $d_{\text{maxTickOffset\_}m}$ for experiments with a master node, and below $d_{\text{maxTickOffset\_}d}$ when no master node was present. When running BBS_h, master node failure was recognized by all other nodes during the next resynchronization phase, and restart of the master node led to the resumption of master-based synchronization, as expected.

Collision resistance and reliability of black bursts

For the operation of BBS, it is crucial that black bursts sent (almost) simultaneously are collision-resistant, i.e., that their collisions are non-destructive. In theory, destruction could be caused by signal cancellation. In practice, we assume that signal cancellation does not occur, and rather expect that overlapping black bursts cause additive interference at receivers. This assumption is supported by the chaotic nature of signal propagation, and by preventing nodes from sending identical signals.

When bit sequences are encoded with black bursts, two types of errors may occur, resulting in bit errors. A *false positive* is detected if a black burst is detected although nodes in sensing range are not transmitting. A *false negative* occurs if a black burst transmitted by some node in sensing range is not detected.

Collision resistance and reliability experiments (see [6]) were conducted in a controlled outdoor environment, with the single-network property (see Sect. 1.3) in force. The topology of the experiment in Fig. 2.15 shows 8 nodes (v_{s1},..., v_{s5}, v_c,

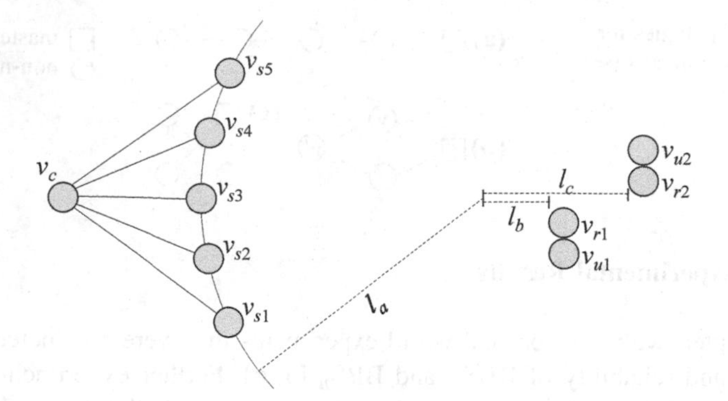

Fig. 2.15 Topologies for reliability and collision resistance experiments (see [6], Fig. 5.5)

Table 2.5 Variation of distances between senders and receivers (see [6], Table 5.1)

Experiment series	l_a	l_b	l_c	$l_a + l_b$	$l_a + l_c$
1	5 m	0 m	0 m	5 m	5 m
2	5 m	0 m	5 m	5 m	10 m
3	10 m	0 m	10 m	10 m	20 m
4	10 m	10 m	10 m	20 m	20 m
5	20 m	10 m	10 m	30 m	30 m

v_{r1}, v_{r2}) of type Imote2 and 2 software-defined radio nodes (v_{u1}, v_{u2}) of type USRP2 [14]. Up to 5 nodes v_{si} act as senders, coordinated by node v_c over wire to achieve overlapping black burst transmissions. Nodes v_{r1}, v_{r2}, v_{u1}, and v_{u2} act as receivers. The distance between senders and receivers is given by $l_a + l_b$ and $l_a + l_c$, respectively (see Fig. 2.15). In the experiments, l_a, l_b, and l_c are varied as shown in Table 2.5.

Experiments ran with five distance setups, with receivers placed in distances of 5, 10, 20, and 30 m of senders. For each distance setup, experiments with 1, 2, and 5 active senders were conducted. In each experiment, 60,000 dominant black bursts were sent in intervals of 10 ms, with recessive black bursts in-between.

Figure 2.16 shows the success rates of dominant black bursts, i.e., true positives, for receiver v_{r2}. Up to a distance of 20 m, success rates are 100%, independent of the number of senders. At a distance of 30 m and with 1 sender, the success rate drops to about 97.9%, indicating that the transceiver's sensitivity range has been reached. However, with more senders, the success rate increases and reaches 100% in case of 5 senders, which supports the expectation of additive interference and thus collision resistance. Success rates for recessive black bursts, i.e., true negatives, were 100% in all experiments, which supports the validity of the single-network property.

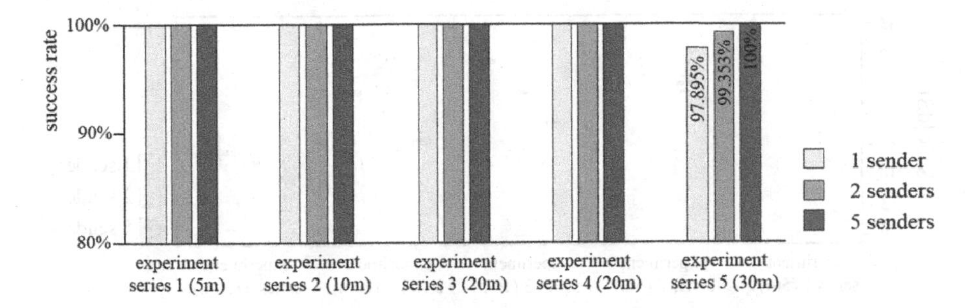

Fig. 2.16 Success rates of black burst receptions (true positives) of node v_{r2} (see [6], Fig. 5.6)

In addition to receivers v_{r1} and v_{r2} of type Imote2, software-defined radios v_{u1} and v_{u2} of type USRP2 recorded fine-grained energy levels. The measured signal-to-noise ratios (SNRs) during black burst transmissions are shown in Fig. 2.17. Here, black lines mark the median, boxes the first and third quartile, and whiskers the minimum and maximum values. Clearly, the SNR levels drop with increasing distance. Also, they never decrease with an increasing number of senders, which again supports the expectation of additive interference.

Further evidence for additive interference can be derived from durations of detected dominant black bursts. Measurements show that the duration of black bursts increases with the number of transmitters. One reason for this is that transmissions start with small offsets, as the synchronicity achieved through node v_c is not perfect. The main reason, however, is that black bursts with higher received signal strength indicators (RSSIs) are detected earlier by the CCA mechanism, leading to longer detected durations.

Transceiver calibration

For reliable detection of black bursts, calibration of the transceiver's CCA threshold is crucial. A high threshold reduces sensing range, and the probability to detect false negatives grows. With a low threshold, the probability for the detection of false positives rises, as noise may be misinterpreted as a black burst. Moreover, the probability to detect false negatives grows, too, because detected black bursts may be invalidated due to an unexpected length. Therefore, the CCA threshold should be configured well above noise level, but not higher than the RSS of correctly detected MAC frames.

In the experiments, sender v_s and receiver v_r were placed in a distance of 0.5 m. Black bursts were implemented as IEEE 802.15.4 compliant MAC frames, to enable detection as black burst (using the CCA mechanism) and reception as MAC frame (SFD detected and checksum correct). Output power was varied from −33 to −12 dBm (power levels 1–9). For each power level, 30,000 transmissions in intervals of 20 ms were performed. The CCA threshold of the transceiver was set to −97 dBm, which was just above the noise floor.

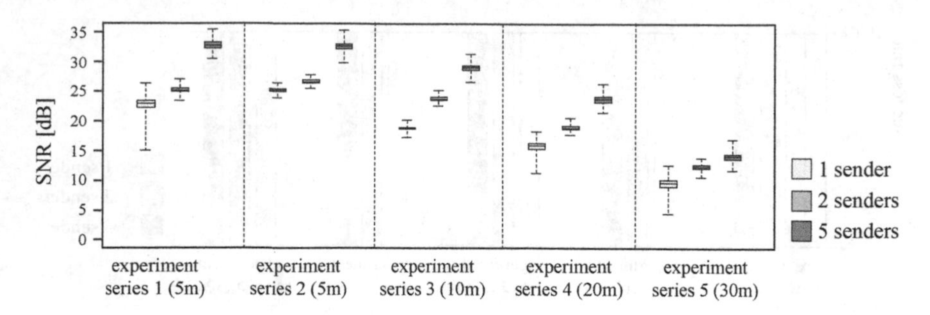

Fig. 2.17 Signal-to-noise ratio (SNR) during black burst transmissions recorded by v_{u1} (see [6], Fig. 5.7)

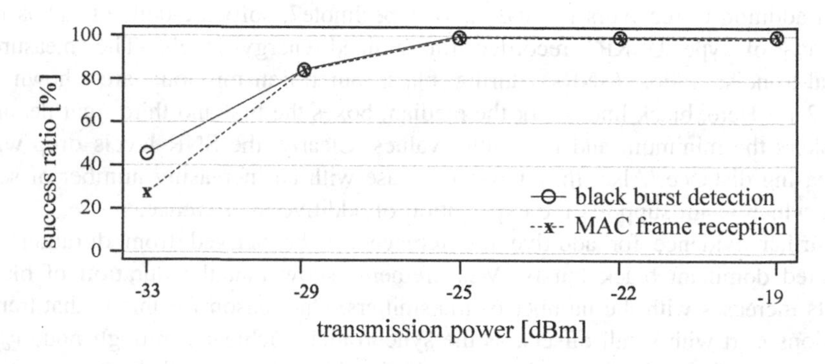

Fig. 2.18 Success ratios of black bursts and MAC frames (see [6], Fig. 5.4a)

Figure 2.18 shows the success ratios of black burst detections and MAC frame receptions, respectively. With an output power of −33 dBm, 44% black bursts are detected, whereas only 27% MAC frames are correctly received. This goes up to almost 100% with an output power of −25 dBm and higher. In the experiments, detection ratios of black bursts were never below the reception ratios of MAC frames and reached 100% for output powers of −22 dBm and higher. This confirms the usual assumption that sensing range is larger or equal to communication range.

Figure 2.19 shows the success ratio of MAC frames relative to valid black burst detections as function of received signal strength (RSS). For RSS of −90 dBm and higher, the ratio is roughly 1. Therefore, a CCA threshold of −90 dBm appears to be adequate to equalize sensing and communication range.

2.2.10 Optimizations

We now address three optimizations to reduce the usable upper bound for tick and time offset. The first optimization considers restricted node mobility. The second

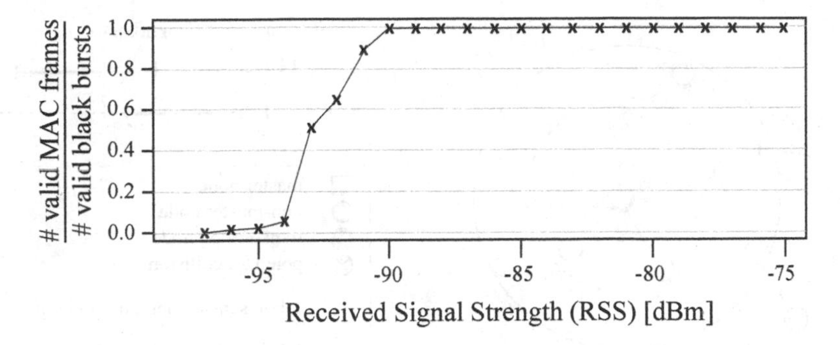

Fig. 2.19 Success ratio of MAC frames relative to valid black burst detections (see [6], Fig. 5.4b)

optimization applies current offsets instead of maximum offsets. The third optimization reduces maximum tick/time offset and convergence delay of the first synchronization round. Optimizations are complementary, i.e., they can be applied individually or together.

Optimization 1: Restricted node mobility

So far, we have considered network-wide maximum base tick and time offsets, determined, e.g., for BBS$_m$ (see Sect. 2.2.3.5) by

$$d_{\mathrm{maxBaseTickOffset}\_m} = n_{\mathrm{maxHops}} \cdot (d_{\mathrm{maxCCA}} + d_{\mathrm{maxProp}}) \tag{2.1}$$

This covers the extreme case that the network topology may change in an arbitrary way in-between two resynchronization phases. In practice, however, node mobility is limited, and network nodes may even be stationary. Assuming that there is no interference among nodes that are 2 or more sensing hops apart, this can be exploited to optimize the maximum base tick offset.

Figure 2.20 shows a scenario where nodes v_1 and v_3 are transmitting in adjacent time slots $i + 1$ and i, respectively. Before, they have been resynchronized by master node v_m in rounds 2 (v_1) and 4 (v_3), i.e., their maximum base tick offset is

$$d_{\mathrm{maxBaseTickOffset'}\_m} = 2 \cdot (d_{\mathrm{maxCCA}} + d_{\mathrm{maxProp}})$$

Since v_1 is resynchronized before v_3, it has the earlier local reference tick. In this scenario, it would be sufficient if node v_1 delays its transmission by

$$d_{\mathrm{maxTickOffset'}\_m} = d_{\mathrm{maxBaseTickOffset'}\_m} + 2 \cdot r_{\mathrm{maxClockSkew}} \cdot d_{\mathrm{resInt}}$$

In a single-hop network, it would even be sufficient to apply $d_{\mathrm{maxCCA}} + d_{\mathrm{maxProp}}$ as maximum base tick offset. Thus, the applicable maximum base tick offset in case of networks with stationary nodes is

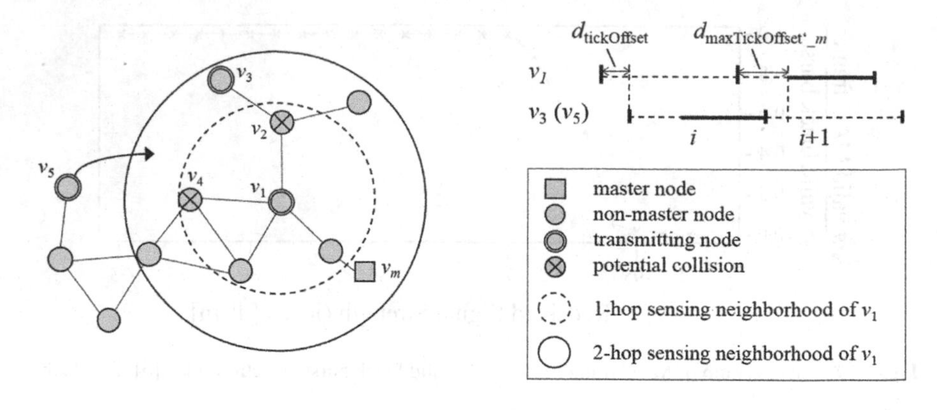

Fig. 2.20 Optimization 1—restricted node mobility

$$d_{\mathrm{maxBaseTickOffset'\_}m} = \min\{2, n_{\mathrm{maxHops}}\} \cdot (d_{\mathrm{maxCCA}} + d_{\mathrm{maxProp}}) \tag{2.20}$$

which substantially improves synchronization accuracy for networks with a diameter of more than two sensing hops.

Next, we consider the impact of limited node mobility. For instance, in a factory scenario with mobile robots, their maximum speed may be restricted for safety reasons. Let us assume a maximum node speed of 20 km/h (i.e., about 12 mph), a sensing range of 30 m, and a resynchronization interval of 5 s, then nodes can move at most one hop in-between resynchronizations. For example, in Fig. 2.20, node v_5 may move into one-hop distance of node v_4, which is in one-hop distance of node v_1. However, since v_5 has only been synchronized in round 6, the optimized maximum base tick offset derived for stationary nodes is not sufficient. Here, $n_{\mathrm{hopsPerResInt}}$, the upper bound for node mobility, is to be considered, yielding the optimized maximum base tick offset

$$d_{\mathrm{maxBaseTickOffset'\_}m} = \min(2 \cdot n_{\mathrm{hopsPerResInt}} + 2, n_{\mathrm{maxHops}}) \cdot (d_{\mathrm{maxCCA}} + d_{\mathrm{maxProp}}) \tag{2.21}$$

The factor 2 in Eq. (2.21) takes the scenario of nodes moving toward each other into account. In case of stationary nodes, i.e., $n_{\mathrm{hopsPerResInt}} = 0$, Eq. (2.21) yields the same result as Eq. (2.20).

We can now insert concrete values into Eq. (2.21). For a maximum sensing diameter $n_{\mathrm{maxHops}} = 10$ and the AT86RF230 transceiver [8], we have obtained $d_{\mathrm{maxBaseTickOffset\_}m} = 160$ µs without optimization. With optimization, this is reduced to 32 µs for stationary nodes, and to 64 µs for a maximum mobility of one hop per resynchronization interval.

Optimization 2: Current offsets

The second optimization is motivated by the observation that after each resynchronization, the maximal tick offset (in case of BBS_m) is only $d_{maxBaseTickOffset\_m}$, which then deteriorates gradually to $d_{maxTickOffset\_m}$ until the next resynchronization takes place, as captured by Eq. (2.2).

$$d_{maxTickOffset\_m} = d_{maxBaseTickOffset\_m} + 2 \cdot r_{maxClockSkew} \cdot d_{resInt} \tag{2.2}$$

Let $d_{elapsed}$ denote the time that has elapsed since the last resynchronization. Then, instead of $d_{maxTickOffset\_m}$, it is safe to apply $d_{maxCurrentTickOffset\_m}$:

$$d_{maxCurrentTickOffset\_m} = d_{maxBaseTickOffset\_m} + 2 \cdot r_{maxClockSkew} \cdot d_{elapsed} \tag{2.22}$$

Comparing Eqs. (2.2) and (2.22), the improvement is apparent: because the elapsed time since the last resynchronization is on average 50% of the resynchronization interval, the applicable maximum current tick offset is substantially smaller than the maximum tick offset.

Figure 2.21 illustrates this for a 4-hop network and values calculated for AT86RF230 transceivers. After an elapsed time of 24 ms, the maximum current tick offset is only 65.92 µs, compared to 864 µs without optimization. Combining optimizations 1 and 2 for a 4-hop network of stationary nodes, this value goes down to 33.92 µs.

If we assume that time is structured into slots, we can determine the waste caused by lack of synchronization accuracy. If the resynchronization interval d_{resInt} is structured into n_{slot} time slots, waste without optimization can be expressed as

$$o_{waste\_m} = \frac{n_{slot} \cdot d_{maxTickOffset\_m}}{d_{resInt}}.$$

With optimization, the waste is

$$o_{waste\_m,o2} = \frac{\sum_{i=1}^{n_{slot}} \left(d_{maxBaseTickOffset_m} + 2 \cdot r_{maxClockSkew} \cdot \frac{d_{resInt}}{n_{slot}} \cdot i \right)}{d_{resInt}}$$
$$= \frac{n_{slot} \cdot d_{maxBaseTickOffset_m} + 2 \cdot r_{maxClockSkew} \cdot \frac{d_{resInt}}{n_{slot}} \cdot \frac{n_{slot}(n_{slot}+1)}{2}}{d_{resInt}}$$

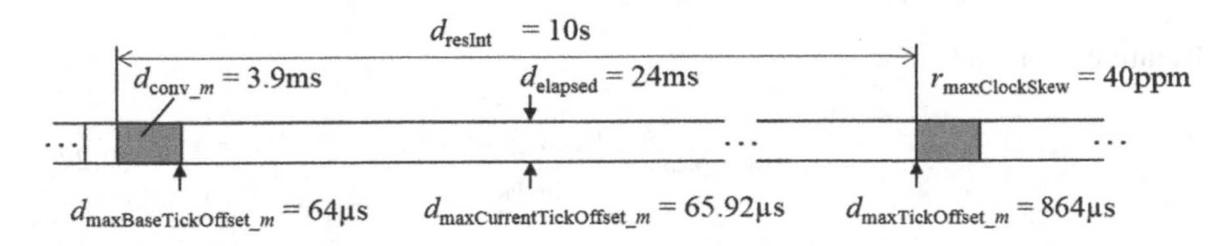

Fig. 2.21 Optimization 2—current offsets

For the example in Fig. 2.21 and n_{slot} = 2000, i.e., a slot length of 5 ms, we get $o_{\text{waste\_}m}$ = 17.28% (Eq. (2.2)), while $o_{\text{waste\_}m,o2}$ is only 9.28% (Eq. (2.22)). For n_{slot} = 5000, we even have $o_{\text{waste\_}m}$ = 43.2%, compared to $o_{\text{waste\_}m,o2}$ = 23.2%. Thus, optimization 2 increases usable bandwidth substantially.

Optimization 3: First synchronization round

When running BBS$_m$, only the master node v_m is sending a master-tick frame in synchronization round 1. This implies that in round 1, no master-tick frame collision can occur. For this reason, nodes in communication range of v_m can detect valid MAC frames used to implement black bursts. This observation enables optimization of synchronization accuracy and convergence delay.

Instead of transmitting a sequence of black bursts to encode master-tick frames, it is sufficient to send one regular MAC frame carrying synchronization round number 1. Thereby, the duration of the first synchronization round is reduced to $d_{\text{round1\_}m,o3} = d_{\text{bit\_}m} + d_{\text{proc}}$, which shortens convergence delay and reduces relative overhead (see Eqs. (2.25) to (2.27)).

To monitor the state of the medium, the CCA mechanism of the CC2420 transceiver offers two modes. With energy detection mode, it reports a busy medium when the energy level measured over a period of 8 symbol durations, i.e., 128 μs, exceeds a threshold. With carrier sense mode, it signals the medium as busy when detecting a valid IEEE 802.15.4 carrier, i.e., a signal with valid modulation and spreading characteristics. Both modes can be operated individually and combined.

Accuracy and performance

Synchronization accuracy

$$d_{\text{maxBaseTickOffset\_}m,o3} = d_{\text{maxSFD}} + d_{\text{maxProp}} + (n_{\text{maxHops}} - 1) \cdot (d_{\text{maxCCA}} + d_{\text{maxProp}}) \tag{2.23}$$

$$d_{\text{maxTickOffset\_}m,o3} = d_{\text{maxBaseTickOffset\_}m,o3} + 2 \cdot r_{\text{maxClockSkew}} \cdot d_{\text{resInt}} \tag{2.24}$$

Convergence delay

$$d_{\text{conv\_}m,o3} = d_{\text{round1\_}m,o3} + (n_{\text{maxHops}} - 1) \cdot d_{\text{round\_}m} + d_{\text{maxTickOffset\_}m,o3} \tag{2.25}$$

$$d_{\text{round\_}m} = (1+m) \cdot d_{\text{bit\_}m} + d_{\text{proc}} \tag{2.4}$$

$$d_{\text{round1\_}m,o3} = d_{\text{bit\_}m} + d_{\text{proc}} \tag{2.26}$$

Relative overhead

$$o_{sync\_m,o3} = \frac{\mathrm{d}_{\text{conv\_}m,o3}}{\mathrm{d}_{\text{resInt}}} \tag{2.27}$$

Compared to pure energy detection, carrier sensing is far more accurate and, according to the CC2420 data sheet, can locate the start of a valid MAC frame with a tolerance of 3 μs only (instead of 128 μs for energy detection). The CC2420 transceiver signals medium detection with the carrier sense mode at a dedicated SFD output pin. This improves synchronization accuracy of the first round to $d_{\mathrm{maxSFD}} + d_{\mathrm{maxProp}}$ (see Eqs. (2.23) and (2.24)).

2.3 Related Work

To place Black Burst Synchronization (BBS) into context, we now survey and assess related work. Due to the large number of publications on synchronization, this survey is not comprehensive, but is rather intended to identify important differences of BBS compared to other approaches. Broad surveys are provided, for instance, in [27, 28] and [25].

2.3.1 Remote Clock Reading Method

In [10], Cristian has published a straightforward approach to time synchronization, to which we refer here as *RCRM* (*remote clock reading method*). The algorithm works as follows (see Fig. 2.22):

1. At local time $c_{v\_i}(t_1)$, client v_i sends a time request message req_t to a time server v_m, requesting a current timestamp.
2. After receiving req_t, v_m reads its local clock to determine the current timestamp $t_{ts} = c_{v\_m}(t_2)$ and returns a time response message $resp_t$ containing t_{ts}.
3. Client v_i receives message $resp_t$ at local time $c_{v\_i}(t_3)$ and (re-)synchronizes with v_m by setting its local clock to $t_{ts} + (c_{v\_i}(t_3) - c_{v\_i}(t_1))/2$.
4. To resynchronize, steps 1–3 are repeated periodically.

Synchronization accuracy of RCRM highly depends on round-trip time, reaction delay of the time server, and asymmetry of transfer delays. Ideally, the timestamp should be determined at $t_{2'} = t_1 + (t_3 - t_1)/2$ (see Fig. 2.22). In fact, it is determined

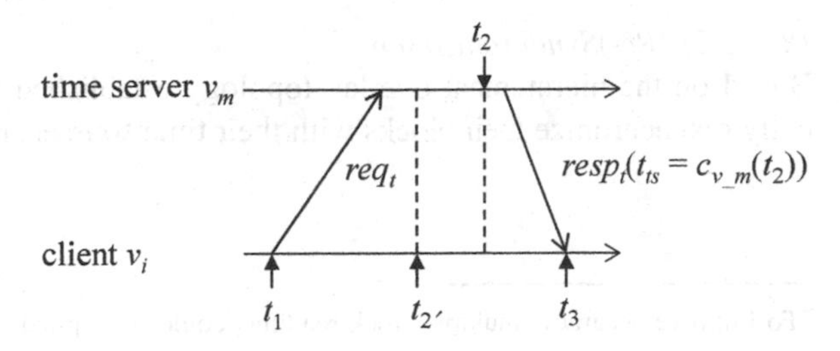

Fig. 2.22 Operation of RCRM

somewhere between t_1 and t_3, yielding a potential time offset $d_{\mathrm{potTimeOffset}} = \pm(t_3 - t_1)/2$. Since $d_{\mathrm{potTimeOffset}}$ depends on the delay $d_{\mathrm{req\_t,resp\_t}}$ between sending of req_t and reception of resp_t, which is unbounded, a maximum time offset cannot be determined. While a small delay $d_{\mathrm{req\_t,resp\_t}}$ reduces $d_{\mathrm{potTimeOffset}}$, it will still be large compared to $d_{\mathrm{maxTimeOffset\_m}}$ as achieved by BBS_t, which depends on delays for the detection of channel activity and signal propagation only.

If single clock readings are used for resynchronization,[9] convergence delay depends on the time needed to resynchronize the entire network, i.e., the period between two resynchronizations, provided communication between clients and time server is sufficiently reliable; nevertheless, message loss may occur, e.g., due to collisions. In multi-hop networks, clients need routing information, which adds further structural complexity to the approach.

2.3.2 *Timing-Sync Protocol for Sensor Networks*

In [15], Ganeriwal et al. have published a time synchronization protocol that improves RCRM significantly, referred to as *TPSN* (*Timing-sync Protocol for Sensor Networks*). The algorithm works in two phases:

Phase 1: Level discovery
In phase 1, a hierarchical overlay topology is established, where each node is associated with a level i and comes to know the identity of its predecessor in the hierarchy, which becomes its time source in phase 2. Phase 1 is started by a dedicated time reference node v_m, associated with level 0.

1. To start phase 1, v_m broadcasts a Level Discovery Packet containing its identity, e.g., its node address, and its level 0.
2. A node $v_i \neq v_m$ receiving a Level Discovery Packet for the first time assigns itself the next level and records the identity of the sender as its predecessor in the hierarchy. Then, v_i broadcasts a Level Discovery Packet containing its identifier and its level.
3. If a node $v_j \neq v_m$ has not received a Level Discovery Packet within a defined time span, it broadcasts a Level Request Packet. When receiving a Level Request Packet, a node broadcasts a Level Discovery Packet. Among received Level Discovery Packets, node v_j selects one with the lowest level, assigns itself the next higher level and records the sender as its predecessor.

Phase 2: (Re-)Synchronization
Based on the hierarchical overlay topology established in phase 1, nodes periodically resynchronize their clocks with their time sources, i.e., with their predecessors

---

[9] To improve accuracy, multiple clock readings could be applied.

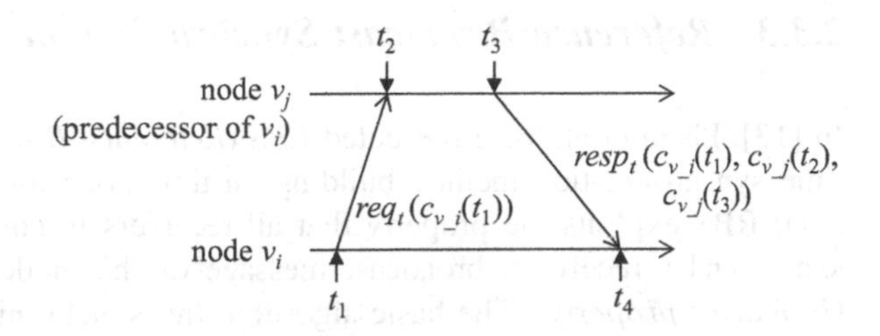

Fig. 2.23 Operation of phase 2 of TPSN

in the hierarchy. As node v_m has no predecessor, it acts as time reference node. As soon as a node has received a Level Discovery Packet, it can start and repeat phase 2, which works as follows (see Fig. 2.23):

1. At local time $c_{v\_i}(t_1)$, node v_i sends a message[10] req_t containing $c_{v\_i}(t_1)$ and v_i's level to its predecessor in the hierarchy, called v_j.
2. When receiving a message req_t from a node of the next level, node v_j records its local reception time $c_{v\_j}(t_2)$, which can be expressed as (i) $c_{v\_j}(t_2) = c_{v\_i}(t_1) + d_{\mathrm{timeOffset}\_i,j} + d_{\mathrm{trans}}$. Here, $d_{\mathrm{timeOffset}\_i,j}$ denotes the current time offset between nodes v_i and v_j; d_{trans} is the transfer delay composed of medium access delay, sending delay, propagation delay, and processing delay.
3. At $c_{v\_j}(t_3)$, v_j returns a response resp_t containing $c_{v\_i}(t_1)$, $c_{v\_j}(t_2)$, and $c_{v\_j}(t_3)$.
4. When receiving resp_t at $c_{v\_i}(t_4)$, which can be expressed as (ii) $c_{v\_i}(t_4) = c_{v\_j}(t_3) - d_{\mathrm{timeOffset}\_i,j} + d_{\mathrm{trans}}$, v_i computes[11] its time offset (iii) $d_{\mathrm{timeOffset}\_i,j} = ((c_{v\_j}(t_2) - c_{v\_i}(t_1)) - (c_{v\_i}(t_4) - c_{v\_j}(t_3))/2$ and corrects its clock, thereby (re-)synchronizing with its time source v_j.
5. To resynchronize, v_i repeats steps 1–4 periodically.

TSPN assumes the same transfer delay d_{trans} for both message exchanges (see Equations (i) and (ii)). This is crucial for determining $d_{\mathrm{clockOffset}\_i,j}$. Without this assumption, d_{trans} cannot be eliminated in equation (iii). In practice, the achievable synchronization accuracy is reduced by the difference between transfer delays, which cannot be determined during operation. Nevertheless, the authors report high accuracy in their experiments, with a measured average base time offset of 17 μs per hop in a 5-hop IEEE 802.15.4 network.

To synchronize the entire network, the clock value of the reference node has to be propagated across all n_{level} levels. Covering one level per resynchronization interval of duration d_{resInt}, average convergence delay can be computed as $d_{\mathrm{conv\_TPSN}} = n_{\mathrm{level}} \cdot d_{\mathrm{resInt}}$, provided communication between clients and time server is sufficiently reliable; nevertheless, message loss may occur, e.g., due to collisions, so there actually is no upper bound for $d_{\mathrm{conv\_TPSN}}$. Structural complexity is substantial, as a hierarchical overlay topology is to be computed and maintained.

[10] In [15], this message is called *pulse*, $resp_t$ is called *ack*.

[11] By resolving equations (i) and (ii) to $d_{\mathrm{timeOffset}\_i,j}$, adding, and dividing by 2.

2.3.3 Reference Broadcast Synchronization

In [13], Elson et al. have presented *RBS* (*Reference Broadcast Synchronization*), a time synchronization method building on time zone formation and clock conversion. RBS exploits the property that all receivers in communication range of the same sender receive a broadcast message of this node at about the same time (*broadcast property*). The basic algorithm (re-)synchronizes nodes as follows:

1. A *sender* periodically broadcasts a reference beacon to a set of receivers in communication range.
2. Each *receiver* records the reception time by reading its local clock.
3. All receivers exchange the observed reception times of reference beacons.
4. Each receiver computes the clock offsets to all other receivers.

To improve accuracy between resynchronizations, the clock skew between pairs of receivers is determined, based on clock offsets over a longer period.

RBS only synchronizes nodes receiving references beacons of the same sender. This excludes the sender, for which the broadcast property does not hold. Although it could be argued that the sender should know the sending time, the uncertainty about the actual start of transmission in case of shared medium access is still quite high compared to the differences of reception times. Therefore, to synchronize the sender, too, another node in its communication range has to take the sender role.

Network-wide time synchronization requires sender coverage, i.e., nodes are selected as senders such that all nodes act as receivers of at least one sender. Furthermore, the network has to be sufficiently dense. Each sender establishes a *time zone* consisting of a set of synchronized nodes. In Fig. 2.24, nodes v_1, v_7, and v_9 act as senders, resulting in three time zones z_1, z_7, and z_9. In time zone z_1, nodes v_2 to v_7 are synchronized. As sender, v_1 remains unsynchronized, but also is a receiver in time zone z_7 and therefore synchronized with v_5, v_6, v_8, v_9, and v_{10}. In summary, all nodes are synchronized in at least one time zone. Without time zone z_7, i.e., with only two senders v_1 and v_9, both v_1 and v_9 would remain unsynchronized.

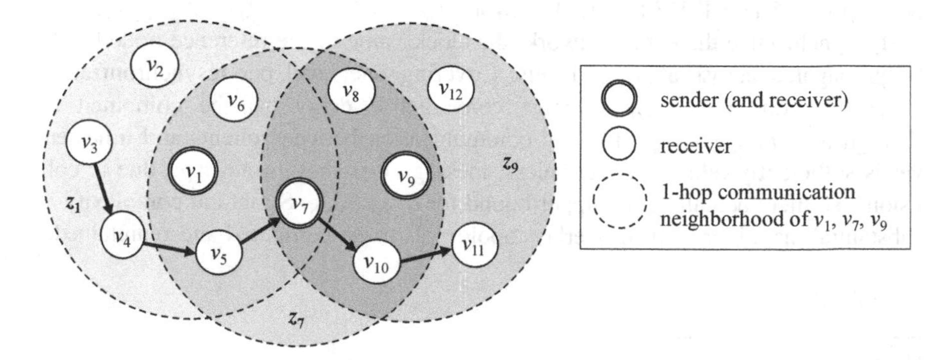

Fig. 2.24 Operation of RBS

RBS can be applied, for instance, to convert timestamps when routing packets. In Fig. 2.24, node v_3 records and timestamps events, and then forwards these observations to node v_{11} via route $<v_3, v_4, v_5, v_7, v_{10}, v_{11}>$. Here, each node v_{i+1} along this route converts the received timestamp to its own clock by adding the clock offset between v_{i+1} and the previous hop v_i. For this to work, it is necessary that receiver v_{i+1} and predecessor v_i belong to a common time zone, which is the case for the route shown in Fig. 2.24. However, route $<v_3, v_1, v_5, v_7, v_{10}, v_{11}>$ would not be feasible, as v_1 belonging to time zone z_7 only cannot convert the timestamp of v_3 belonging to z_1 only. Thus, route discovery has to consider time zones, which mixes network functionalities that should be independent.

By exploiting the broadcast property, RBS can achieve very high average synchronization accuracy if message receptions are timestamped early. In [13], the authors report a measured average base clock offset of 6 µs per hop. Moreover, clock skew is considered to improve accuracy between resynchronizations. Drawbacks are high communication overhead up to $O(n^2)$, if all nodes act as senders, lack of robustness against loss of reference beacons and observation frames due to collisions, and therefore high convergence delay without upper bound. The need to establish sender coverage requires a sufficiently dense network and leads to high structural complexity. Finally, topology changes may impair sender coverage, resulting in unsynchronized nodes.

2.3.4 Syncob

In [20], Krohn et al. have published *Syncob*, a decentralized tick synchronization approach located on PHY layer that has similarities with BBS_d. Nodes running Syncob exchange decentralized tick frames called synchronization packets, which are encoded as a sequence of black bursts called sync-symbols.[12]

Figure 2.25 (adapted from Fig. 9 of [20]) shows a simplified flow chart describing the operation of Syncob. When not synchronized, nodes try to detect synchronization packets (called sync-symbols in the original figure) and accept an existing synchronization on reception. Otherwise, they start transmitting synchronization packets, but remain not synchronized until detecting incoming packets.

When receiving a synchronization packet, nodes are synchronized. As before, they try to detect synchronizations packets to resynchronize with other nodes and transmit synchronization packets themselves. The difference to the unsynchronized case is that these packets are now sent at aligned points in time, similar to the exchange of decentralized tick frames in a given round of BBS_d. If no tick frames are received for some time, nodes return to the unsynchronized state.

[12]In Sect. III of [20], it is stated that Syncob supports time synchronization, by exchanging reference points in time via beacon frames. Later, only sync-symbols are exchanged, which is sufficient for tick synchronization. Time synchronization and, in particular, the resetting of local clocks based on the reception of time values are not further addressed.

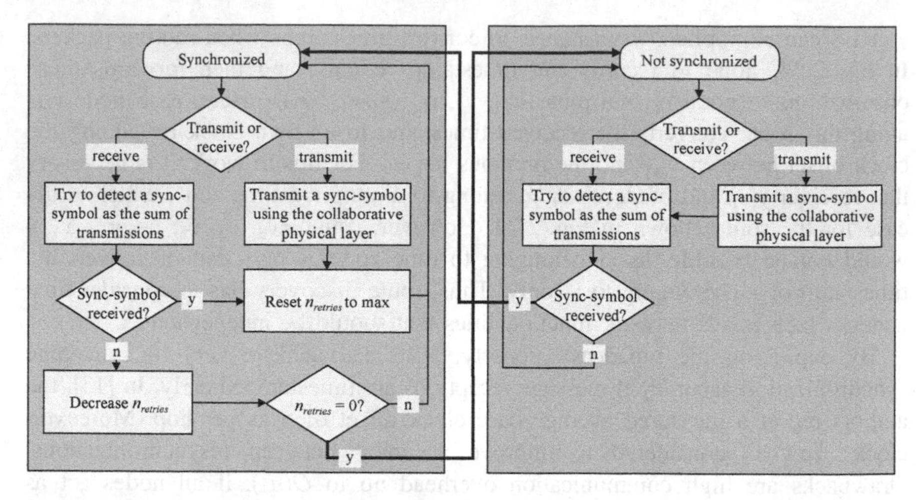

Fig. 2.25 Operation of Syncob (see [20], Fig. 9 [The flow chart is adapted from Fig. 9 in [20], with some corrections])

Figure 2.25 needs some interpretation. For instance, the decision whether to transmit or receive is taken non-deterministically, with the intention to transmit sufficiently often in order to maintain synchronization among nodes. However, if the decision to transmit is taken, the variable n_{retries} is reset to the maximum, implying that synchronization is never lost even if no synchronization packets are received. This obviously needs some refinement.

Operation of Syncob in multi-hop networks is straightforward and briefly addressed in [20]. The authors point out some limitations considered as future work.

Krohn et al. state that they have implemented Syncob on the pPart particle sensor platform [23]. The built-in radio transceiver supports on–off keying, where a carrier is created for a specified duration. With a carrier detection delay of 0.2 μs only, tick offsets between nodes in single-hop distance are extremely small. It remains, however, unclear how reliable this detection works in practice.

An important concern of Syncob is the non-deterministic decision about when to transmit sync-symbols. If single-hop neighbors transmit or receive simultaneously, they do not resynchronize. This may even lead to situations where nodes become unsynchronized, although synchronization would be feasible. Thus, deterministic upper bounds for tick offset and convergence delays cannot be derived.

2.3.5 BitMAC

In [24], Ringwald and Römer describe a customized MAC protocol for wireless sensor networks called *BitMAC*, with an approach for tick synchronization that has

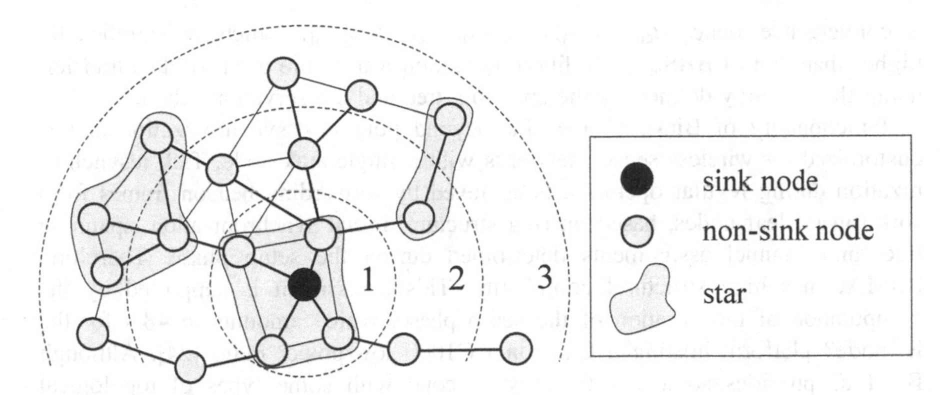

Fig. 2.26 Setup phase of BitMAC

similarities with BBS_m. Here, the sink node of the sensor network takes the role of the master node.

Before commencing regular operations, BitMAC executes a complex setup phase (see Fig. 2.26). First, a ring structure and initial synchronization are established. Ring formation is started by the sink node v_m, which broadcasts a beacon frame containing the current ring level n_{level}, which is 1. When a non-sink node v_i receives a beacon frame for the first time, it assigns itself ring level $n_{\text{level}} + 1$, and determines a local tick by backward calculation. After a fixed delay, v_i broadcasts a beacon frame with its ring level. This part of the setup phase terminates after n_{maxLevel} beacon rounds.

Figure 2.26 shows a topology with $n_{\text{maxLevel}} = 3$ resulting ring levels. Obviously, there are nodes receiving beacon frames from more than one non-sink node in the same beacon round. To prevent these collisions from being destructive, beacon frames are encoded as sequence of black bursts (on–off keying) and sent (almost) simultaneously.

Based on the ring structure, nodes exchange data needed to determine a spanning tree with the sink node as root (see Fig. 2.26) and to run a coloring algorithm, where colors represent communication channels. The spanning tree is then decomposed into a set of connected stars, consisting of a parent node of a given level i, $1 \leq i < n_{\text{maxLevel}}$, and child nodes of the next level $i + 1$ in single-hop distance of the parent node. Next, by running the coloring algorithm, channels are assigned to each star such that communication within a star is not disturbed by nodes belonging to other stars. Since parent nodes on levels i, $1 < i < n_{\text{maxLevel}}$, are also child nodes on level i–1, the protocol has to ensure that during regular operation, channels are switched in an alternating way.

After this setup phase, BitMAC commences regular operations. Based on the initial synchronization, time is decomposed into operation rounds. The parent of a star starts a round by transmitting a beacon frame, which resynchronizes its children. This resynchronization is carried from sink node to all leaf nodes over a duration of n_{maxLevel} operation rounds with constant duration d_{opRound}, thus yielding a convergence delay $d_{\mathrm{conv\_BitMAC}} = n_{\mathrm{maxLevel}} \cdot d_{\mathrm{opRound}}$, which is significantly higher than that of BBS_m. In addition, operation rounds are used for data transfer, using the topology defined by the spanning tree and a reservation scheme.

Functionality of BitMAC goes far beyond pure tick synchronization and is customized for wireless sensor networks with a single sink node. Tick resynchronization during regular operation is achieved by forwarding beacon frames from sink toward leaf nodes, based on ring structure, initial synchronization, spanning tree, and channel assignments determined during the setup phase. Therefore, BitMAC has high structural complexity. This assessment is supported by the computation of the duration of the setup phase, which amounts to 48 s for the BTnode3 platform hosting a Chipcon CC1000 low-power radio [24]. Although BitMAC provides some functionality to cope with some types of topological changes, the setup phase may have to be repeated from time to time.

A major concern is that BitMAC is highly prone to transmission errors. Bit errors in beacon frames or in frames containing data to determine a spanning tree, to run the coloring algorithm, and to assign channels will lead to malfunctioning. Especially if nodes are located at the border of their communication range, the RSSI output of the receiving node may not reflect the transmitted bit sequence reliably.

2.3.6 Comparison

In Table 2.6, we compare the synchronization approaches surveyed and discussed in Sect. 2.3 to BBS. We assess each protocol qualitatively w.r.t. a number of criteria, based on the information available from the published material.

- The first criterion is accuracy, with sample values for precision per hop and an indication whether these values are upper bounds. While Syncob, which is located on PHY layer, provides excellent precision, only BitMAC[13] and BBS provide upper bounds.
- Regarding time complexity, BitMAC and BBS perform best: The time for resynchronization depends on the maximum network diameter d only, while in all other approaches, it depends on the number of nodes n. Structural complexity is substantial except for Syncob and BBS, which can operate without overlay topologies.
- BBS delivers a very low convergence delay, as synchronization rounds are executed consecutively. For Syncob, this delay depends on the non-deterministic decision when to transmit sync-symbols, and thus is hard to assess. BitMAC resynchronizes one level per operation round. Only BitMAC and BBS provide an upper bound.

[13]The upper bounds reported for BitMAC are not the result of an analysis, but have been measured during experiments; thus, they may be higher.

Table 2.6 Comparison of synchronization approaches (see [3], Table 1)

Protocol	Accuracy (precision/hop; deterministic)	Time / structural complexity	Convergence delay (deterministic)	Robustness against topology changes	Experiments (PHY layer; network diameter)
RCRM Sect. 2.3.1	very low (ms; no)	O(*n*) / time server, routing	high (no)	low	–
TPSN Sect. 2.3.2	high (17 μs; no)	O(*n*) / node hierarchy	high (no)	low	IEEE 802.15.4 5 hops
RBS Sect. 2.3.3	very high (6 μs; no)	O(n^2) / sender coverage	high (no)	low	IEEE 802.11 4 hops
Syncob Sect. 2.3.4	very high (0.2 μs; no)	O(*n*) / -	low (no)	high	specialized 1 hop
BitMAC Sect. 2.3.5	high (20 μs; yes)	O(*d*) / ring structure, stars	low (yes)	low	specialized 2 hops
BBS Sect. 2.2	high (16 μs; yes)	O(*d*) / -	very low (yes)	high	IEEE 802.15.4 4 hops

- Robustness against topology changes largely depends on the structural complexity. Most protocols establish overlay topologies before starting resynchronization; therefore, their structural complexity is high. Here, Syncob and BBS perform best.
- With the exception of RCRM, experiments with real hardware are reported. Some experiments use hardware conforming to existing standards, e.g., IEEE 802.11 or IEEE 802.15.4, on top of PHY or MAC layer. Syncob and BitMAC use specialized hardware supporting on–off keying, which may explain the high precision. All experiments can be classified as small-scale.

In summary, only BitMAC and BBS provide upper bounds for precision, which are derived by experiments for BitMAC, and analytically for BBS. Both protocols have a deterministic convergence delay and low time complexity. However, BitMAC has high structural complexity, and therefore is not robust against topology changes.

2.4 Conclusions

In this chapter, we have explained the concepts and requirements of tick and time synchronization, have presented our synchronization protocol Black Burst Synchronization (BBS) for multi-hop ad hoc networks, and have surveyed and assessed selected related work. We have argued that time synchronization is essential for the operation of distributed real-time computer systems, and in particular for data fusion or synchronized value sampling. For network-wide medium slotting and duty cycling, tick synchronization is already sufficient. This makes tick and time synchronization a core functionality.

After reaching network stabilization, BBS provides the following properties: tick and time synchronization with low upper bounds for tick offset, time offset, and convergence delay, low complexity of computation, storage, time, and structure, and robustness against topology changes. We have implemented BBS on commercial hardware platforms (MICAz and Imote2), using the CC2420 transceiver and its IEEE 802.15.4 PHY layer functionality, and have analytically and experimentally assessed its accuracy, performance, and reliability. By optimizing the jitter of the clear channel assessment mechanism and by reducing switching times, the achievable accuracy and performance of BBS could even be improved.

BBS exploits three key insights. First, nodes can synchronize time slots by agreeing on sufficiently accurate network-wide reference ticks, without synchronizing their local clocks. This reduces the amount of synchronization data to be exchanged among nodes. Second, time synchronization can be achieved on top of tick synchronization, without much additional exchange of synchronization data. Third, by encoding synchronization messages with black bursts, potential collisions are non-destructive. This reduces convergence delay substantially, as nodes can transmit messages simultaneously, and enables deterministic operation.

Some protocols reported in the literature achieve better accuracy than BBS. However, they often are more complex in terms of computation, time, and/or storage, which is a drawback in the context of resource-limited nodes. Furthermore, they often lack upper bounds for clock offset and convergence delay, require the establishment of an overlay topology, and/or depend on a modulation scheme that is only supported by specialized transceivers.

Literature

Chair for Networked Systems

1. Gotzhein R (2014) ProNet 4.0—a wireless real-time communication system for industry 4.0. White Paper, Networked Systems Group, Department of Computer Science, University of Kaiserslautern, 2014, http://vs.informatik.uni-kl.de/publications/2014/Go14/whitePaperEN-ProNet4.0.pdf. Last Accessed 27 Aug 2019
2. Gotzhein R, Kuhn T (2008) Decentralized tick synchronization for multi-hop medium slotting in wireless Ad Hoc networks using black bursts. In: Proceedings of the 5th annual IEEE communications society conference on sensor, mesh, and ad hoc communications and networks (SECON 2008), San Francisco, USA, June 16–20, 2008, pp 422–431
3. Gotzhein R, Kuhn T (2011) Black burst synchronization (BBS)—a protocol for deterministic tick and time synchronization in wireless networks. Comput Netw 55(13):3015–3031 Elsevier

4. Engel M, Christmann D, Gotzhein R (2014) Implementation and experimental validation of timing constraints of BBS. In: Krishnamachari B, Murphy AL, Trigoni N (eds) 11th European conference on wireless sensor networks (EWSN 2014), Oxford, United Kingdom, Feb 17–19, 2014. Springer LNCS 8354, pp 84–99
5. Gotzhein R, Kuhn T (2011) Method, computer program product and system for the tick synchronization of nodes in a wireless multi-hop network. European Patent Office, Az EP 2195949, Aug 31, 2011 (date of granting)
6. Christmann D (2015) On the development of a wireless binary countdown protocol and the applicability of SDL to such time-critical systems. Ph.D. Thesis, Computer Science Department, University of Kaiserslautern

Further References

7. Akyildiz IF, Su W, Sankarasubramaniam Y, Cayirci E (2002) Wireless sensor networks: a survey. Comput Netw 38(4):393–422 Elsevier
8. Atmel Corporation (2009) AVR low power transceiver AT86RF230, 5131E-MCU Wireless-02/09, http://ww1.microchip.com/downloads/en/DeviceDoc/doc5131.pdf. Last Accessed 27 Aug 2019
9. Chipcon AS (2019) CC2420 Zigbee-ready RF transceiver, http://www-inst.eecs.berkeley.edu/~cs150/Documents/CC2420.pdf. Last Accessed 27 Aug 2019
10. Cristian F (1989) Probabilistic clock synchronization. Distrib Comput 3:146–158 Springer
11. https://en.wikipedia.org/wiki/DCF77. Last Accessed 27 Aug 2019
12. Dorf RC, Bishop RH (2016) Modern control systems, 13th edn. Pearson Education
13. Elson J, Girod L, Estrin D (2002) Fine-grained network time synchronization using reference broadcasts. In: Proceedings of the fifth symposium on operating systems design and implementation (OSDI 2002), Boston, MA, USA, Dec 2002
14. Ettus Research, USRP (2013) Universal software radio peripheral 2, https://www.ettus.com/product/category/USRP-Networked-Series/. Last Accessed 27 Aug 2019
15. Ganeriwal S, Kumar R, Srivastava MB (2003) Timing-sync protocol for sensor networks. In: SenSys'03. Los Angeles, CA, USA
16. https://en.wikipedia.org/wiki/Global_Positioning_System. Last Accessed 27 Aug 2019
17. van Hoesel LFW, Havinga PJM (2008) Collision-free time slot reuse in multi-hop wireless sensor networks. In: Conference on intelligent sensors, sensor networks and information processing, Melbourne, Australia, Dec 5–8, 2008, pp 101–107
18. IEEE Standards Association: IEEE Std 802.15.4–2011: IEEE Standard for Local and Metropolitan Area Networks, Part 15.4: Low-Rate Wireless Personal Area Networks (LR-WPANs), Sep 2011, 314p
19. MEMSIC Inc (2019) Imote 2 datasheet. https://vs.cs.uni-kl.de/downloads/Imote2NET_ED_Datasheet.pdf. Last Accessed 27 Aug 2019
20. Krohn A, Beigl M, Decker C, Riedel T (2007) Syncob: collaborative time synchronization in wireless sensor networks. In: Fourth international conference on networked sensing systems, June 6–8, 2007
21. Crossbow Technology Inc (2019) MICAz wireless measurement system, document part Nr 6020-0060-04 Rev A http://www.openautomation.net/uploadsproductos/micaz_datasheet.pdf. Last Accessed 27 Aug 2019
22. Mills DL (1991) Internet time synchronization: the network time protocol. IEEE Trans Commun COM-39(10):1482–1493
23. Particle Web Site (2019) University of Karlsruhe, http://particle.teco.edu/. Last Accessed 27 Aug 2019
24. Ringwald M Römer R (2005) BitMAC: a deterministic, collision-free, and robust Mac protocol for sensor networks. In: Proceedings of the second European workshop on wireless sensor networks, Feb 2005

25. Sarvghadi MA, Wan T.-C. (2016) Message passing based time synchronization in wireless sensor networks: a survey. Int J Distrib Sens Netw 12(5)
26. Sobrinho JL, Krishnakumar AS (1999) Quality of service in ad hoc carrier sense multiple access networks. IEEE J Sel Areas Commun 17(8):1353–1368
27. Sundararaman B, Buy U, Kshemkalyani AD (2005) Clock synchronization for wireless sensor networks: a survey. Ad Hoc Netw 3(3):281–323
28. Swain AR, Hansdah RC (2015) A model for the classification and survey of clock synchronization protocols in WSNs. Ad Hoc Netw Elsevier 27:219–241
29. Ye W, Heidemann JS, Estrin D (2002) An energy-efficient mac protocol for wireless sensor networks. In: 21st IEEE conference on computer communications (INFOCOM 2002), New York, USA, June 23–27, 2002, pp 1567–1576
30. ZigBee™ Alliance (2019) Zigbee specification, Version 1.0, June 2005, URL: www.zigbee.org, 378 p. Last Accessed 27 Aug 2019

Chapter 3
Global Time Slotting

Global time slotting is a key functionality for distributed real-time computer systems and in particular for deterministic medium access. In this chapter, we explain the foundations of global time slotting, present our flexible solution *ProSlot* of the ProNet 4.0 protocol stack [2, 3], survey and assess related work, and draw conclusions.

3.1 Foundations

In this chapter, we provide the context of global time slotting, explain concepts and requirements, and address areas of operation.

3.1.1 Context

Time slotting is essential for the operation of distributed real-time computer systems and applied in many other areas as well. The idea of time slotting is to structure time into a sequence of intervals called *(time) slots*. Slots can then be used as scheduling units, for instance, in production, maintenance, logistics, venue management, and, of course, in computing. Operating system schedulers assign computation slots called time slices to control CPU usage of tasks. Communication systems build on time slots to control medium access.

To enable deterministic operation of communication systems, a network-wide perception of time structuring, i.e., *global time slotting*, is required. The basis for this is established by tick synchronization (see Chap. 2), which provides global reference points in time of specified accuracy. Time slots can be associated with specific purposes. In communication systems, time slots are typically used to schedule medium access, for instance, exclusive access, shared access, or shared

R. Gotzhein, *Real-time Communication Protocols for Multi-hop Ad-hoc Networks*, Computer Communications and Networks,
https://doi.org/10.1007/978-3-030-33319-5_3

access restricted to a defined subset of competing nodes or messages. Time slots may also be used to schedule resynchronization and as idle intervals to save energy.

3.1.2 Concepts and Requirements

The objective of global time slotting is to structure time into synchronized intervals. This requires network-wide tick synchronization. Time slotting is a common technique to enable the deterministic operation of wired and wireless communication systems.

- In the automotive domain, FlexRay [5] has become an important technology for x-by-wire systems such as steering and braking. Here, time is structured into a hierarchy of time slots. On top level, time is subdivided into groups of 64 subsequent communication cycles of equal length. Each communication cycle is decomposed into 4 segments, some of which are further substructured into static slots, mini slots, and/or dynamic slots.
- IEEE 802.15.4 [6], a wireless technology standard used by ZigBee [9], structures time into a sequence of super frames, which are decomposed into active and inactive period. Active periods are then subdivided into 16 super frame slots of equal length, which are composed into distinct intervals for contention and contention-free medium access. In addition, backoff slots and guaranteed time slots are used.

While the notions used to refer to types of time slots are different, the basic concepts of structuring time are common to these and other technologies. We can identify the following parameters of time slotting:

- *Slot hierarchy*: Structuring of time is usually done in a hierarchical way, starting with long time intervals, which may then be decomposed into smaller slots step-by-step.
- *Slot types*: Slots may be assigned different purposes, on different levels of the slot hierarchy as well as on a given level, which is made explicit by introducing slot types.
- *Number of slots*: Slots may be substructured into a number of smaller slots, which may be fixed or bounded.
- *Length of slots*: Slots have a length in time, which may be fixed, bounded, or variable.

Global time slotting is the basis for providing timely and reliable medium access and for synchronizing the operation of nodes. The parameters of time slotting, i.e., slot hierarchy, slot types, number and length of slots, are derived from the intended use on application and system level.

- In control systems, sensor values are sampled periodically, with periods determined by timing requirements to achieve control objectives such as system stability. In each period, measured values are sampled and communicated to the controller, which determines steering values sent to actuators.

- Many control system theories assume that measured values are transferred reliably at specified points in time. This can be achieved by reserving suitable time slots exclusively.
- With time slotting, it is possible to support maximum reaction delays. Here, the interval between reserved time slots must be chosen such that the lengths of this interval plus event handling delays are below the specified reaction delay. A drawback in the case of sporadic events with short maximum reaction delays is that this consumes many time slots, which are wasted if no event occurs.
- If several control systems are operated over the same communication system, different sampling periods may exist, which may render time slotting more complicated.
- On system level, tick and time resynchronization is to be scheduled periodically, with periods derived from required synchronization accuracy.
- Energy saving is another system level functionality that requires global time slotting. Here, hardware devices such as CPU or transceiver are switched to sleep or idle mode for defined intervals.

In addition to the considerations so far, flexibility of global time slotting is important. For instance, fixing the number and/or length of time slots reduces flexibility. Furthermore, a strict pattern of active and inactive periods, as found in existing communication technologies, impedes the provision of maximum reaction delays. In general, a flexible global time slotting supports a variable number of slots of different slot types, lengths, and placements.

In summary, we can state that the global time slotting of the communication medium is an important functionality that requires careful planning.

3.1.3 Areas of Operation

Global time slotting supports the operation of distributed systems both on user level and on system level. In networked production and control systems, time slots are reserved to provide predictable timing behavior and reliable message exchange. Time slotting is also needed for periodic resynchronization and for duty cycling.

Figure 3.1 shows an abstract communication schedule, based on global time slotting. Here, 4 macroslots Ms_i of different sizes n_{Ms_i} are distinguished, each consisting of mini slots ms_k of fixed length. Mini slots are exclusively assigned to message types[1]; e.g., mini slot ms_3 of macroslot Ms_1 is assigned to message type

[1]In the example, we assume that addressing is based on message types: A message type identifier is uniquely associated with a sending node, and one or more receiving nodes.

mini slot / macro slot	ms_1	ms_2	ms_3	ms_4	ms_5	ms_6	ms_7	ms_8	ms_9	ms_{10}	ms_{11}	ms_{12}	…	ms_{36}	…	ms_{72}
Ms_1 (sampling)			id_7													
Ms_2 (sampling)							id_{23}									
Ms_3 (resync)	id_r	id_r									id_{35}		…			
Ms_4 (sampling)										id_{12}			…	id_{52}	…	

Fig. 3.1 Time slotting and transmission schedule

id_7. Sizes of macroslots are derived from sampling periods and maximum reaction delays on application level (Ms_1, Ms_2, Ms_4), from the resynchronization period (Ms_3), and from the required length of a mini slot d_{ms} for sending a message and receiving an acknowledgment. For instance, macroslots Ms_1 and Ms_4 consist of $n_{Ms\_1} = 6$ and $n_{Ms\_4} = 72$ mini slots, yielding a length of $d_{Ms\_1} = 6 \cdot d_{ms}$ and $d_{Ms\_4} = 72 \cdot d_{ms}$, respectively.

Sampling periods and therefore macroslots are repeated periodically, and overlay with each other. This means that the reservation of ms_3 of Ms_1 is repeated every 6 mini slots and overlays with slots ms_3 and ms_9 of Ms_2, slots ms_3, ms_9, ms_{15}, ms_{21}, ms_{27}, and ms_{33} of Ms_3, etc. Thus, these mini slots cannot be assigned to messages in their macroslots.

In Fig. 3.1, sampling periods are chosen such that smaller periods fit into larger periods for a multiple of whole numbers times. In other words, macroslots Ms_i, $1 < i \leq 4$, can be arranged as harmonic chain: $d_{Ms\_i} = k_i \cdot d_{Ms\_i\text{-}1}$, where $k_i \in \mathbb{N}$. This has the advantage that reservations can be made with the smallest possible waste of mini slots. Consider, for instance, macroslots $Ms_{1'}$ and $Ms_{2'}$ of length $d_{Ms\_1'} = 4 \cdot d_{ms}$ and $d_{Ms\_2'} = 6 \cdot d_{ms}$, with ms_1 of $Ms_{1'}$ reserved for some message id_k. This reservation not only consumes ms_1 and ms_5 of $Ms_{2'}$, but also ms_3. The reason is that the macroslots cannot be arranged as harmonic chain, resulting in substantial waste.

The transmission schedule in Fig. 3.1 can also be used to determine idle periods. Only nodes involved in a scheduled communication have to be active; all other nodes can switch to sleep mode, in order to save energy.

A drawback of the time slotting in Fig. 3.1 is that mini slots have a fixed size. If frames of different sizes are to be sent, it is possible to reserve $n > 1$ consecutive mini slots for the transmission of larger frames. However, this will produce waste, as the nth mini slot of each reservation will not be fully used. To minimize waste, macroslots could be structured into very short micro slots, which are then composed into mini slots of flexible size adapted to frame lengths. This, however, would destroy the structure of macroslots, as mini slots of different super slots would no longer be aligned. We will revisit this problem in Sect. 3.2.

3.2 Global Time Slotting in ProNet 4.0

Global time slotting is a basic functionality of the protocol stack ProNet 4.0 [1] and performed by the protocol component ProSlot [2, 3]. Based on the analysis of general timing requirements on application and system level, we have decided to structure time physically and then compose physical time slots into virtual time regions. We show that these design decisions provide a high degree of flexibility.

3.2.1 Physical Time Slotting

Physical time slotting divides time into a sequence of super slots consisting of micro slots (see Fig. 3.2). The rationale behind this time structuring is the following:

- The duration of super slots depends on sampling periods of applications and resynchronization period. To keep bandwidth waste small, sampling and resynchronization periods should be chosen such that they can be arranged as a harmonic chain (see Sect. 3.1.3), with the longest period defining the (fixed) duration of super slots.
- Each super slot is decomposed into consecutively numbered micro slots of fixed length. Micro slots are later composed into virtual time regions of different type and variable length. As a time region consists of an integer number of micro slots, we can expect that to achieve a defined length, there is an average waste of 50% of a micro slot per time region. To keep this waste small, micro slots should be very short, e.g., in the order of a few microseconds.

In summary, physical time slotting establishes a fixed, stable, and strictly periodic time structure.

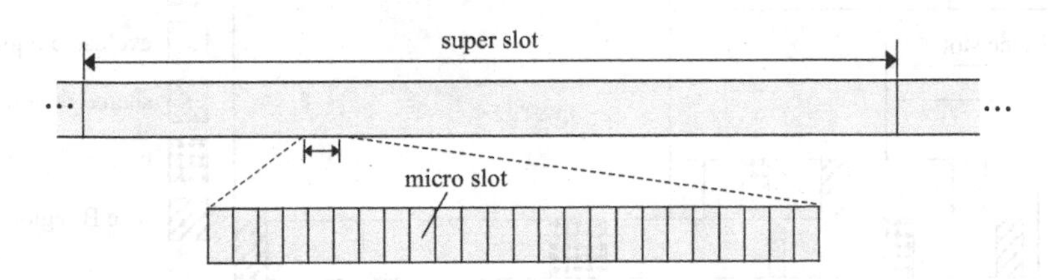

Fig. 3.2 Physical time slotting

3.2.2 Virtual Time Slotting

Virtual time slotting builds on physical time slotting and combines consecutive micro slots into *virtual time regions* of different types. For tick resynchronization, *sync regions* are formed and placed at the beginning of each resynchronization interval (see Fig. 3.3). The size of a sync region depends on the amount of resynchronization to be performed. In case of BBS (Black Burst Synchronization; see Chap. 2), it is determined by the constant convergence delay d_{conv}. The size d_{resInt} of a resynchronization interval—also called *resync slot*—depends on the required tick and clock accuracy.

Figure 3.4 illustrates the versatility of virtual time slotting. Starting points are sampling slots 1 and 2 of the application, and a resync slot on system level. In the example, these slots can be arranged as harmonic chain, to reduce bandwidth waste (see Sect. 3.1.3). Each sampling and resync slot has its own virtual time slotting, with the following slot types:

- *Sync regions* are formed for (re-)synchronization purposes and contain a sync slot, as already explained.
- *Exclusive regions* host contention-free traffic and are substructured into exclusive slots that can have different length.

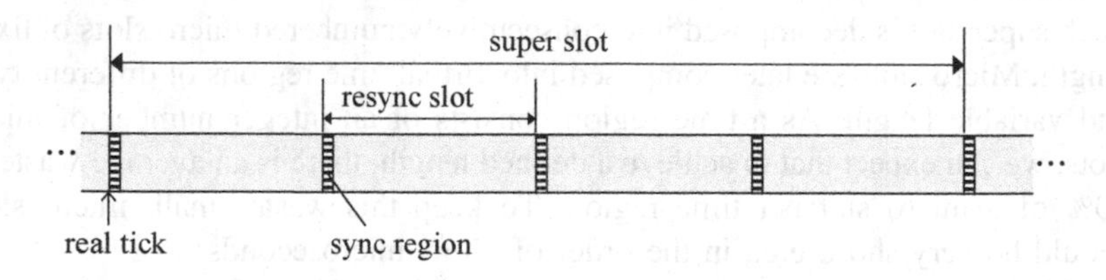

Fig. 3.3 Virtual time slotting—sync regions

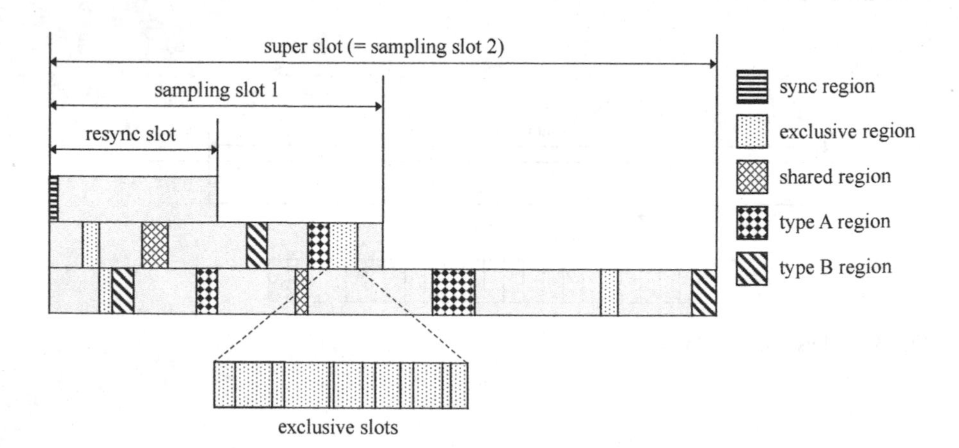

Fig. 3.4 Global virtual time slotting

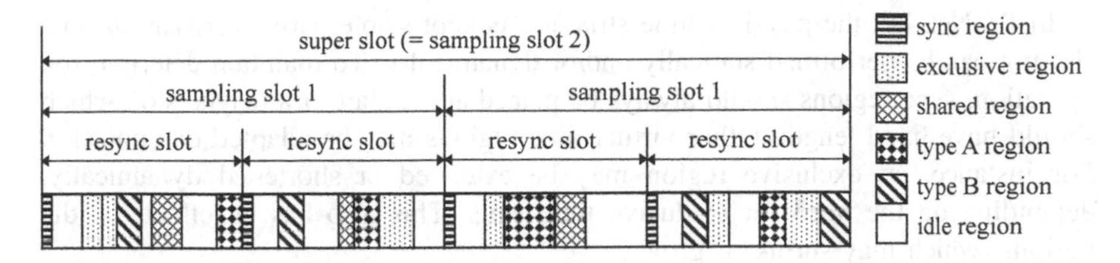

Fig. 3.5 Global virtual time slotting (cont'd)

- *Shared regions* are formed for general contention-based traffic, applying, e.g., random backoff schemes and frame priorities.
- Type A and type B regions are placeholders for other kinds of traffic, e.g., restricted contention (see Sect. 5.2.3) or multi-hop arbitration (see Chap. 6).

It is important that virtual time regions of sampling slots and resync slots do not overlap. Figure 3.5 continues the example in Fig. 3.4. Here, resync slot and sampling slot 1 are repeated, to fill the super slot. Then, all virtual time regions are projected into a single diagram. This shows that the virtual time slotting in Fig. 3.4 is feasible. Furthermore, it shows that there are periods in time that are not yet consumed. We compose the corresponding micro slots into idle regions, which can be used for energy saving.

3.2.3 Flexibility

The combination of physical and virtual time slotting provides a high degree of flexibility, while reducing bandwidth waste due to unusable slot portions to a minimum. Physical time slotting establishes a fixed, stable, and strictly periodic time structure, consisting of super slots and (very short) micro slots. Virtual time regions of different types and lengths can then be formed and placed into super slots, by combining consecutive micro slots and defining the starting point as the first of these micro slots. Depending on their usage, virtual time regions can also be substructured into virtual slots of different length, as shown in Fig. 3.4 for an exclusive region. In Chap. 2, sync regions (called synchronization phases) have been substructured into sync slots (called synchronization rounds). Again, this substructuring exploits the physical time slotting, by combining consecutive micro slots and defining starting points of virtual slots.

Flexibility is further enhanced by the possibility to place virtual slots at arbitrary points in time of a super slot, without considering idle periods in the first place. This is illustrated in Fig. 3.5, where idle regions are derived after all other virtual time regions have been placed according to application and system level requirements. This minimizes waiting delays, because the striving for long idle periods does not dominate the need for short sampling periods and reaction delays.

In ProNet 4.0, the physical time structure is kept stable. However, virtual time slotting can be performed statically and/or dynamically. To maintain deterministic operation, sync regions should always be placed at the start of a resync slot, which should have fixed length. Other virtual time regions may be adapted dynamically. For instance, an exclusive region may be extended or shortened dynamically, depending on the need for exclusive time slots. This also has an effect on idle regions, which may shrink or grow.

3.3 Related Work

To place the global time slotting performed by ProSlot into context, we now survey and discuss related work. In particular, we address how the time structure of other wireless communication technologies can be mapped to the time slotting of ProSlot, which further proves its flexibility.

3.3.1 IEEE 802.15.4

IEEE 802.15.4 [6] is a wireless personal area network (PAN) technology standard for short distances supporting time slotting, which is also used by ZigBee [9]. Time slotting is possible in so-called beaconed mode, where a master station referred to as PAN coordinator periodically transmits beacon messages to nodes in communication range, indicating the start of a super frame. Thus, time slotting is restricted to single-hop networks.

Figure 3.6 shows the hierarchical structuring of time defined by IEEE 802.15.4. On top level, time is decomposed into super frames, consisting of active and inactive period. The active period is further subdivided into 16 super frame slots of equal length, which are composed into dynamic intervals called contention access period (CAP) and contention free period (CFP). At the beginning of the active period, a beacon message is sent, signaling the start and structure of the current super frame to nodes in communication range. In the CFP, up to 7 consecutive super frame slots are composed into Guaranteed Time Slots (GTSs) for exclusive medium access.

Super frame slots have a minimum length of aBaseSuperframeDuration defined by a number $n_{\text{sym}} = 60$ of symbol durations d_{sym}. For instance, when transmitting on the ISM band, transmission rate and encoding yield a value $d_{\text{sym}} = 16$ µs. By multiplying this minimum length with 2^{SO} (SO is the super frame order), super frame slots and thereby the active period can be extended, yielding the super frame duration SD[2].

[2]Actually, SD is the duration of the active period, but referred to as super frame duration.

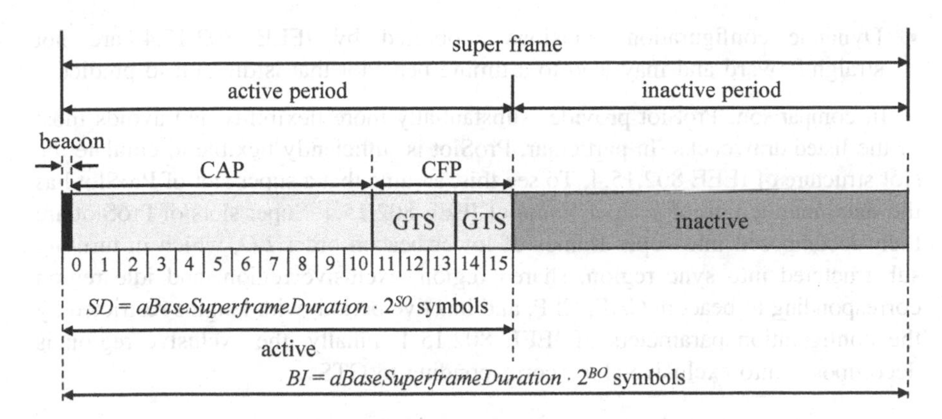

Fig. 3.6 Time slotting in IEEE 802.15.4—beaconed mode (see [6], Fig. 8)

The length of the inactive period is defined as length of the beacon interval *BI* minus *SD*. *BI* is given as the minimum super frame slot duration multiplied by 2^{BO} (*BO* is the beacon order, with $BO \geq SO$). This implies a length of $2^{BO-SO} - 1$ times the duration of the active period, which is a very coarse super frame structure yielding long waiting delays. On the other extreme, if $BO = SO$, nodes are always active.

To cope with varying communication requirements, some of the above configuration parameters can be adapted dynamically. In particular, beacon order *BO*, super frame order *SO*, and the number of super frame slots assigned to CAP and CFP can be modified by the PAN coordinator for each super frame and communicated in the beacon message. Similarly, reservations of GTSs can be changed during execution.

Though time slotting in IEEE 802.15.4 supports energy-saving, exclusive reservations, medium contention, and dynamic adaption of super frame configuration and reservations, it has a number of drawbacks.

- With IEEE 802.15.4, beaconing and thus time slotting are only feasible in single-hop networks.
- A maximum of only 7 super frame slots can be assigned to the CFP, meaning that at any point in time, only up to 7 exclusive reservations can be supported.
- Super frame slots are of equal length and rather large. Thus, if a GTS consisting of one or more such slots is formed, on average 50% of a super frame slot are wasted.
- Dimensioning of super frames, active periods, and inactive periods is rather inflexible and coarse, due to the use of sizes to the power of 2.
- The size of an inactive period always is $2^{BO-SO} - 1$ times the duration of the active period. In case of a duty cycle (fraction of active period and beacon interval) of 3.125%, achievable by $SO - BO = 5$, for instance, the maximum reaction delay is increased to 32 times the length of the active period.

- Dynamic configuration decisions supported by IEEE 802.15.4 are not straightforward and may lead to a timing behavior that is difficult to predict.

In comparison, ProSlot provides substantially more flexibility and avoids most of the listed drawbacks. In particular, ProSlot is sufficiently flexible to emulate the slot structure of IEEE 802.15.4. To see this, assume that a super slot of ProSlot has the maximum length of a super frame of IEEE 802.15.4. Super slots of ProSlot are then decomposed into super frames of lower beacon order *BO*, which in turn are substructured into sync region, shared region, exclusive region, and idle region corresponding to beacon, CAP, CFP, and inactive period, with lengths restricted by the configuration parameters of IEEE 802.15.4. Finally, the exclusive region is decomposed into exclusive slots, corresponding to GTSs.

3.3.2 WirelessHART

WirelessHART [4, 8] is a technology devised for wireless networking of intelligent field devices, especially for process control in factories.[3] It extends the wired Highway Addressable Remote Transducer (HART) protocol developed by the company Rosemount and has been standardized by the HART Communication Foundation (HCF).

Communication in WirelessHART is based on global time slotting, which is controlled by a master node v_{m} taking the role of a network manager. Time slotting is performed in two steps.

- First, time is structured physically into slots of fixed duration $d_{\mathrm{slot}} = 10$ ms. Slots are numbered consecutively, by assigning an Absolute Slot Number (ASN) with a value range of 0–$2^{40} - 1$. As there is no reset of ASN, there is no further super slot structure.
- Next, physical time slots are composed into virtual time regions called super frames sf_i of fixed length d_{sf_i}, derived from scan periods i of the application. Thus, the number n_{sf_i} of physical time slots of a super frame sf_i is $d_{sf\_i}/d_{\mathrm{slot}}$.

Figure 3.7 shows an example of time slotting and slot scheduling in WirelessHART, adapted from [4]. On application level, 4 scan periods of 1, 4, 8, and 16 s are distinguished, yielding super frames 0–3 with these durations. We observe that super frames can be arranged as harmonic chain (see Sect. 3.1.3), which reduces bandwidth waste. Given a slot duration of 10 ms, we obtain 100, 400, 800, and 1600 physical slots.

The figure also shows part of a slot schedule, where physical time slots ts_{i} are exclusively assigned for transmissions $v_{\mathrm{i}} \rightarrow v_{\mathrm{j}}$ between pairs of nodes. In

[3]Another communication technology in this application field is ISA 100.11a [7]. Since time slotting of ISA 100.11a has strong similarities with WirelessHART, we omit a detailed presentation.

super frame 0 (scan period = 1 sec)

Ch. Offset	ts_0	ts_1	ts_2	ts_3	ts_4	ts_5	ts_6	…	ts_{99}
0	$v_2 \rightarrow v_1$	$v_2 \rightarrow v_5$		$v_5 \rightarrow v_m$		$v_1 \rightarrow v_m$			
1	$v_3 \rightarrow v_1$	$v_3 \rightarrow v_4$			$v_4 \rightarrow v_m$	$v_1 \rightarrow v_m$			

super frame 1 (scan period = 4 sec)

Ch. Offset	ts_0	ts_1	ts_2	ts_3	ts_4	ts_5	ts_6	ts_7	…	ts_{399}
1			$v_4 \rightarrow v_m$							

super frame 2 (scan period = 8 sec)

Ch. Offset	ts_0	ts_1	ts_2	ts_3	ts_4	ts_5	ts_6	ts_7	ts_8	…	ts_{799}
1				$v_5 \rightarrow v_m$							

super frame 3 (scan period = 16 sec)

Ch. Offset	ts_0	ts_1	ts_2	ts_3	ts_4	ts_5	ts_6	ts_7	ts_8	ts_9	…	ts_{1599}
2	$v_6 \rightarrow v_1$	$v_6 \rightarrow v_5$	$v_1 \rightarrow v_m$					$v_5 \rightarrow v_m$		$v_1 \rightarrow v_m$		

Fig. 3.7 Time slotting and schedule in WirelessHART (adapted from [4], Fig. 7.8)

Sect. 3.2.2, we have pointed out that it is important that after projecting all virtual time regions—super frames of WirelessHART—into a single diagram, slot assignments do not overlap. In the example, this constraint is not satisfied. For instance, ts_0 is assigned three times. To support multiple slot usage, the capability of WirelessHART to use several active channels concurrently is exploited. Given an active channel table, an index into this table is computed from ASN, channel offset (see Fig. 3.7, first column) and table size. For different channel offsets and a sufficiently large number of active channels, this computation yields different channels.

Global time slotting in WirelessHART supports multi-hop networks. Furthermore, time slotting is performed synchronously on multiple channels, which increases the communication bandwidth. Another plus factor is a certain degree of flexibility achieved by the combination of physical and virtual time structuring, and the possibility to adapt slot schedules at runtime. On the other hand, there are a number of drawbacks limiting flexibility:

- Physical and virtual time structures are static. In particular, physical time slots have a fixed length d_{slot} = 10 ms, which results in bandwidth waste if messages fitting into a slot have different length. Furthermore, messages not fitting into a single slot have to be split for transmission.
- Only exclusive slots are supported; other slot types, e.g., contention slots, are not permitted.[4]

[4]Synchronization is achieved as a byproduct of regular message exchange; therefore, explicit sync slots are not required.

In comparison, ProSlot provides substantially more flexibility and avoids the listed drawbacks. In particular, ProSlot is sufficiently flexible to emulate the slot structure of WirelessHART.[5] For this, physical time slotting can be adapted by using micro slots of 10 ms each. Furthermore, sampling slots of ProSlot (see Fig. 3.4) can be directly derived from super frames of WirelessHART. Finally, sampling slots are configured as exclusive regions, consisting of exclusive slots of one micro slot each.

3.4 Conclusions

In this chapter, we have explained the concepts and requirements of global time slotting, have presented our time slotting protocol ProSlot of the protocol stack ProNet 4.0 [1], and have surveyed and compared existing wireless technologies. The idea of time slotting is to structure time into well-defined intervals called slots, which are used by communication systems to control medium access. For instance, by assigning time slots exclusively, deterministic medium access can be granted.

We have argued that our approach of combining fixed, strictly periodic physical time slotting and variable, weakly periodic virtual time slotting provides a high degree of flexibility. This minimizes waiting delays and enables customized idle periods. We have shown how by configuring ProSlot suitably, the less general time slotting schemes of IEEE 802.15.4 (beaconed mode) and of WirelessHART can be emulated.

Literature

Chair for Networked Systems

1. Gotzhein R (2014) ProNet 4.0—a wireless real-time communication system for Industry 4.0. White Paper, Networked Systems Group, Department of Computer Science, University of Kaiserslautern. http://vs.informatik.uni-kl.de/publications/2014/Go14/whitePaperEN-ProNet4.0.pdf. Last accessed: 27 Aug 2019
2. Christmann D, Braun T, Engel M, Gotzhein R (2016) BiPS—a real-time-capable protocol framework for wireless sensor networks. In: Proceedings of the 6th international conference on pervasive and embedded computing (PEC 2016), Lisbon, Portugal, 25–27 Jul 2016, pp 17–27
3. Engel M, Kramer C, Braun T, Christmann D, Gotzhein R (2017) BiPS—a real-time-capable protocol framework for wireless networked control systems and its application. In Obaidat M,

[5] A difference here is that in WirelessHART, no explicit sync regions are needed, as resynchronization is based on reception of regular messages. On the other hand, this solution mixes functionalities.

Cabello E (Eds) E-business and telecommunications (ICETE 2017), Communications in computer and information science, vol 990. Springer, pp 313–336

Further References

4. Chen D, Nixon M, Mok A (2010) WirelessHART™—real-time mesh network for industrial automation. Springer
5. FlexRay Consortium (2005) FlexRay communications system protocol specification version 2.1, Revision A
6. IEEE Standards Association (2011) IEEE Std 802.15.4–2011: IEEE standard for local and metropolitan area networks, part 15.4: low-rate wireless personal area networks (LR-WPANs), Sep 2011, 314p
7. International Electrotechnical Commission (2012) Industrial communication networks—wireless communication network and communication profiles—ISA 100.11a (IEC 62734 ed 1.0), Geneva, Switzerland, March 2012
8. International Electrotechnical Commission (IEC) (2010) Industrial communication networks—wireless communication network and communication profiles—WirelessHART (IEC 62591 ed 1.0), Geneva, Switzerland, April 2010
9. ZigBee™ Alliance (2019) ZigBee specification, version 1.0, June 2005, 378p. www.zigbee.org. Last accessed: 27 Aug 2019

Chapter 4
Automatic Topology Detection

Information about network topology is essential for the operation of wireless networks in general, and of networked control systems in particular. In this chapter, we explain the foundations of topology detection, present our protocol[1] *Automatic Topology Discovery Protocol* [2] (*ATDP*) for topology detection in TDMA-based wireless multi-hop networks, survey and assess related work, and draw conclusions.

4.1 Foundations

In this chapter, we provide the context of topology detection, explain concepts and requirements, and address areas of operation.

4.1.1 Context

In wireless networks, information about network topology is of utmost importance. Knowledge about communication, interference, and sensing topologies enables more efficient and reliable use of the wireless medium. Manually measuring and configuring topology information is very cumbersome, even in stationary environments. Rather, it is crucial to detect topology automatically.

Topology information, or, more general, network status information, is needed for various networking functionalities, e.g., for routing, clustering, and network management. The objective of routing is to discover and operate routes between sets of source and destination nodes. Clustering divides a network into groups of

[1]The protocol ProTop of the ProNet 4.0 protocol stack is an implementation of ATDP.

The original version of this chapter was revised: Figures 4.4 and 4.5 have been replaced. The correction to this chapter is available at https://doi.org/10.1007/978-3-030-33319-5_14

R. Gotzhein, *Real-time Communication Protocols for Multi-hop Ad-hoc Networks*, Computer Communications and Networks,
https://doi.org/10.1007/978-3-030-33319-5_4

nodes supporting, e.g., network scaling, hierarchical routing, and energy management. Network management addresses performance monitoring, quality of service provision, and analysis of network failures.

Typically, routing and clustering protocols comprise the collection of network status information, which makes them self-contained. Different protocols may need different views on the network, e.g., link states, aggregated path states, or n-hop neighbor sets. To collect and update network status information, management messages are exchanged.

For routing, the network status provides information about the communication topology, i.e., about links for successful message exchange. In wireless networks, there is the problem that nodes outside communication range of a receiving node may still interfere and thereby render a message exchange unsuccessful. To address this problem, the network status should comprise information about the interference topology, i.e., about nodes in interference range, too. A decision which nodes may cause interference can be made, for instance, based on previous knowledge about their transmission intervals, which is feasible in TDMA-based wireless multi-hop networks with transparent slot assignment. In a similar way, sensing links, i.e., links where transmission activities can be perceived, may be detected, which is useful for the transfer of messages encoded by black bursts (e.g., see Chap. 6) or to relax the hidden station problem (see Sect. 4.1.2).

4.1.2 Concepts and Requirements

The objective of topology detection is to collect and disseminate information about the nodes and links of a network. This information can then be used by other functionalities, e.g., routing and clustering.

In the literature, different types of communication models are used:

- *Physical models*, also called radio propagation models, such as free-space path loss or two-ray ground reflection.
- *Stochastic models* based, for instance, on geometric properties, such as the random disk model.
- *Graph-based models*, with nodes and edges, where the presence and absence of edges model the existence of links with reliability 1 and 0, respectively.

As we focus on deterministic protocols, we use graph-based models to describe network topologies. Of course, as in all types of models, this is an abstraction from real-world phenomena. Formally, a graph-based model $G = (V, E)$ is given by a set V of nodes and a set $E \subseteq V \times V$ of edges.

Among graph-based models, we distinguish the following types of topologies:

- A *communication topology* is modeled as a graph $G_C = (V, E_C)$, where links $e = (v_i, v_j)$ express that v_j is in communication range of v_i, i.e., v_j can successfully receive messages from v_i.

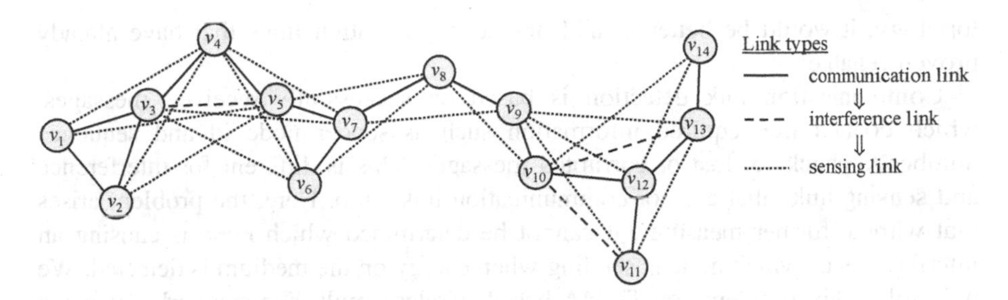

Fig. 4.1 Graph-based symmetrical topology (see [13], Fig. 1)

- An *interference topology* is modeled as a graph $G_I = (V, E_I)$, where links $e = (v_i, v_j)$ express that v_j is in interference range of v_i, i.e., v_i can interfere with a reception at v_j.
- A *sensing topology* is modeled as a graph $G_S = (V, E_S)$, where links $e = (v_i, v_j)$ express that v_j is in sensing range of v_i, i.e., v_j can determine whether v_i is currently transmitting.

We assume that a node v_j in communication range of v_i is also in interference and sensing range of v_i, and that a node v_j in interference range of v_i is also in sensing range of v_i. The graphs G_C, G_I, and G_S can be merged into a combined network topology $G = (V, L, E)$, where $L = \{c, i, s\}$ is a set of labels distinguishing between different types of links. We require that the complete communication subgraph[2] $G_C = (V,\{c\}, E_C)$ is connected, i.e., for all pairs of distinct nodes, there is a path of communication links.

Figure 4.1 shows a symmetrical graph-based topology. Solid lines represent communication links, which are interference and sensing links, too. To simplify presentation, only communication links are shown in these cases. Additional interference links, which are also sensing links, are depicted as dashed lines. Finally, sensing links not yet captured as communication or interference links are shown as dotted lines.

The topology in Fig. 4.1 has the additional feature that for each communication link (v_s, v_r, c), nodes v_i in interference range of v_r are also in sensing range of v_s, i.e., (v_i, v_r, i) implies (v_i, v_s, s). It follows that there are no hidden stations in this topology; so, carrier sensing before transmission is sufficient. To check for this feature, knowledge about communication, interference, and sensing topologies is required.

Automatic topology detection and dissemination can be achieved by systematic exchange of messages. In principle, a communication link (v_i, v_j, c) can be added when v_j successfully receives a message from v_i and removed when there has been no such reception for some time (concept of soft state). However, for a stable

[2]Obtained by reducing G to the set of all nodes and all communication links.

topology, it would be better to add only communication links that have already proven reliable.

Communication link detection is based on successfully received messages, which contain the required information such as sender node id and sequence number to check for lost or corrupted messages. This is different for interference and sensing links that are not communication links, too. Here, the problem arises that without further measures, it cannot be determined which node is causing an interference, or which node is sending when energy on the medium is detected. We will solve this problem for TDMA-based wireless multi-hop networks, using a transparent assignment of exclusive time slots during system startup.

4.1.3 Areas of Operation

Topology detection supports the operation and efficient usage of wireless multi-hop networks. Routing protocols as well as clustering protocols require knowledge about communication topologies. Space-division multiple access (SDMA), where nodes with transmissions not impairing each other may use the medium simultaneously, is based on additional knowledge about interference topologies. Communication via black burst encoded bit sequences (see Chaps. 2 and 6) exploits sensing topologies.

The objective of routing is to discover and operate routes between sets of source and destination nodes, which requires information about the communication topology. Depending on the type of routing, this information may be global or aggregated. Global information consists of the entire communication topology known to one or more nodes and may comprise additional link properties such as delay and throughput. Aggregated information reduces global information to abstract parameters, such as global topology to hop count between pairs of nodes or link delays to path delays.

The objective of network clustering is to divide a network into groups of nodes, to support, e.g., network scaling and energy management. Network scaling is a technique for the management of large networks, where, generally speaking, dedicated nodes act on behalf of others. On application level, this is needed for the operation of distributed and/or replicated service registries. On system level, clustering is the basis for hierarchical routing, which reduces routing traffic and the size of routing tables. Energy management is a technique for extending network lifetime, e.g., by rotating tasks between nodes of a cluster for even energy consumption. Depending on clustering objectives, different information such as network density of n-hop neighborhood may be deduced from knowledge about communication topology.

In TDMA-based wireless multi-hop networks, SDMA is a well-known technique to increase throughput. If nodes are sufficiently far apart, they may use the same time slot for exchanging messages. Figure 4.2 shows the one-hop communication and interference neighborhoods of node v_5. Obviously, transmissions of v_5 would

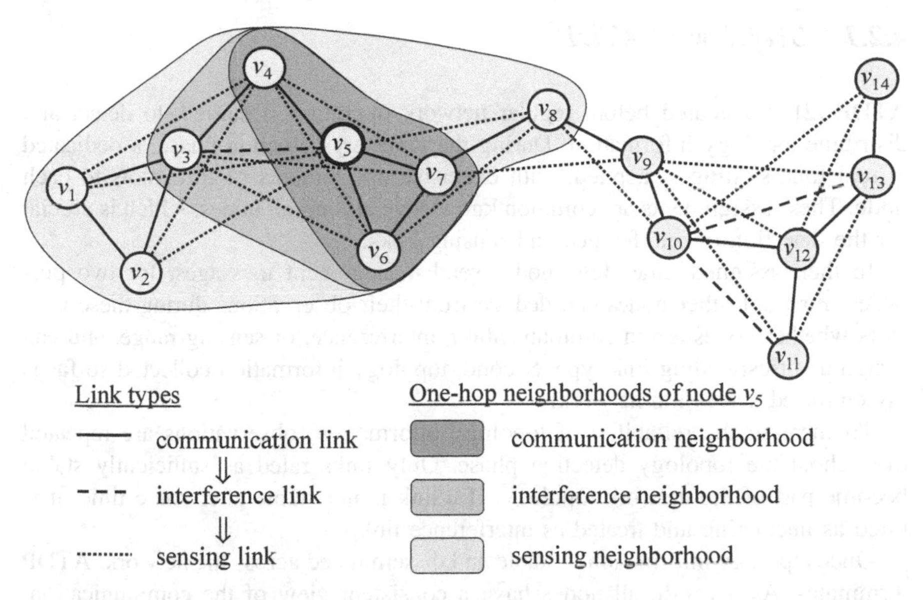

Fig. 4.2 One-hop neighborhoods of node v_5 (see [13], Fig. 1)

interfere with receptions of v_3, v_4, v_6, and v_7 from nodes other than v_5. However, nodes outside interference neighborhood of v_5 can communicate simultaneously without destructive collisions,[3] thereby increasing usable bandwidth.

Knowledge about the sensing topology can be exploited passively, to relax the hidden station problem, and actively, to transmit messages encoded by black bursts. Figure 4.2 shows the one-hop sensing neighborhood of node v_5. Here, nodes v_1 to v_4 and v_6 to v_8 can detect whether v_5 is using the medium and refrain from using it simultaneously, thereby avoiding possibly destructive collisions. In addition, node v_5 can reach these nodes with a single transmission of a message encoded by black bursts, a feature applied, for instance, by Black Burst Synchronization [(BBS), see Chap. 2] and Arbitration and Cooperative Transfer Protocol [(ACTP), see Chap. 6].

4.2 Automatic Topology Discovery Protocol (ATDP)

In this chapter, we present *Automatic Topology Discovery Protocol* (*ATDP*) [2], a deterministic protocol for the detection and dissemination of communication, interference, and sensing topologies in TDMA-based wireless multi-hop networks. The protocol ProTop of the ProNet 4.0 protocol stack [1] is an implementation of ATDP.

[3]Transmissions overlapping in time, frequency, and space are called *collisions*.

4.2.1 Overview of ATDP

ATDP [2] is executed before regular network operation is started, to detect and distribute topology information. During the topology detection phase, a dedicated virtual time slotting is applied, with exclusive assignments of time slots to each node. These assignments are common knowledge among all nodes, which is crucial for the detection of interference and sensing links.

In their assigned time slots, nodes send management messages, for two purposes. First, all other nodes can deduce from their observations during these time slots whether nodes are in communication, interference, or sensing range, and can assign a corresponding link type. Second, topology information collected so far is disseminated across the network.

To increase the reliability of topology information, observations are repeated throughout the topology detection phase. Only links rated as sufficiently stable become part of the network topology. If a link is not stable after some time, it is rated as fluctuating and treated as interference link.

Once topology information is stable and disseminated across the network, ATDP terminates. As a result, all nodes have a consistent view of the communication, interference, and sensing topologies. Here, we assume that ATDP is operated in a stable environment, i.e., there are no significant topology changes after the topology detection phase. If the environment is not stable, ATDP could be continued during regular network operation, to detect and disseminate topology changes.

4.2.2 Virtual Time Slotting of the Topology Detection Phase

In the following, we assume that nodes $v_i \in V$ have unique numerical node ids $i \in \{0, \ldots, n-1\}$, which may be different from the node address. It is not required that all values of $\{0, \ldots, n-1\}$ are used as node ids; however, a highest node id $n-1$ equal to or not significantly larger than $|V|-1$ yields a shorter topology detection phase. Based on node ids, the exclusive assignment of topology detection slots is determined.

Figure 4.3 shows a virtual time slotting (see also Chap. 3) of the topology detection phase. To establish this time slotting, tick synchronization (cf. Chap. 2) is required. Therefore, the scheme depicted in Fig. 4.3 contains resynchronization slots, with a sync region at the beginning of each resync slot. Apart from sync regions, time intervals are free and can be used for topology detection.

In the example in Fig. 4.3, a dedicated management slot for topology detection is introduced, which has the same size as a super slot. The size of this management slot is chosen as a whole-numbered multiple of a resync slot, to reduce waste (see Sect. 3.1.3). Topology detection regions are then placed in such a way that there is no overlap with sync regions and are preceded by a termination region.

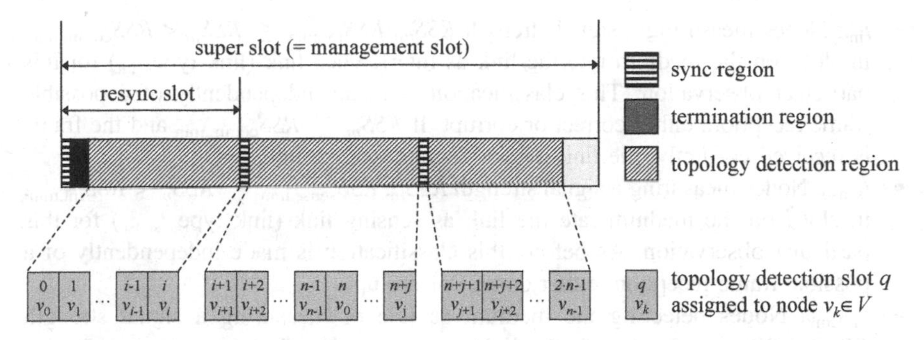

Fig. 4.3 Virtual time slotting of the topology detection phase

Topology detection regions of a management slot are decomposed into topology detection slots, with consecutive slot numbers 0 to $m_{\text{fact}} \cdot n - 1$, where $n - 1$ is the highest numerical node id, and $m_{\text{fact}} > 0$. The size of these slots is chosen such that they fit management messages. Slots are then exclusively assigned to nodes, such that slot i, $0 \leq i < m_{\text{fact}} \cdot n$, is assigned to node v_k, with $k = i \bmod n$. Thus, in a management slot, each node gets an equal number of m_{fact} slot assignments.

As virtual time slotting, highest node id and slot assignment rule are known to all nodes, each node can determine which node is expected to send a message in a given topology detection slot, and where this slot is located in time. In Fig. 4.3, the topology detection regions of a management slot are decomposed into $2 \cdot n - 1$ slots, assigned to nodes $v_k \in V$ as shown.

Before the first topology detection region, a termination region is placed. In this region, a special protocol for the network-wide collision-protected exchange of bit sequences is used to terminate ATDP once the topology information is stable and disseminated across the network.

4.2.3 *Link Types*

The objective of ATDP is to determine stable links between pairs v_j, $v_k \in V$ of nodes and to disseminate this information across the network. Stable links are derived from a series of individual observations in topology detection slots. In its assigned slots i, $0 \leq i < m_{\text{fact}} \cdot n$, each node v_j, with $j = i \bmod n$, locally broadcasts MEASURE messages, which contain its node id and topology data. All other nodes v_k, $k \neq i \bmod n$, listen on the medium and associate a link type with each observation of link $e_{j,k} = (v_j, v_k)$. For individual observations, we distinguish the following link types $t \in T_{\text{link}}$:

- t_{comm}: Nodes v_k, $k \neq i \bmod n$, receiving a frame in slot i correctly and with received signal strength $RSS_m \geq RSS_{\text{comm,min}}$ classify the link $e_{j,k}$, with $j = i \bmod n$, as communication link (link type t_{comm}) for this particular observation.

- t_{int}: Nodes measuring a signal strength RSS_m, $RSS_{int,min} \leq RSS_m < RSS_{comm,min}$, in slot i on the medium rate the link as interference link (link type t_{int}) for this particular observation. This classification is made independently of a possible frame reception, either correct or corrupt. If $RSS_m \geq RSS_{comm,min}$ and the frame is received corruptly, the link is rated as interference link, too.
- t_{sense}: Nodes measuring a signal strength RSS_m, $RSS_{sense,min} \leq RSS_m < RSS_{int,min}$, in slot i on the medium rate the link as sensing link (link type t_{sense}) for this particular observation. As before, this classification is made independently of a possible frame reception, either correct or corrupt.
- t_{noLink}: Nodes detecting the medium as idle or measuring a signal strength $RSS_m < RSS_{sense,min}$ classify the link as non-existing (link type t_{noLink}). In this case, node v_j is out of range, not switched on, or node id v_j is not used, i.e., $v_j \notin V$.
- t_{fluct}: This link type can be assigned only after a series of observations, to links that have not stabilized after some time.

We point out again that since virtual time slotting, highest node id, and slot assignment rule are known to all nodes, each node can determine which node is expected to send a message in a given topology detection slot, and where this slot is located in time.

Thresholds $RSS_{comm,min}$, $RSS_{int,min}$, and $RSS_{sense,min}$ are hardware-specific, taking the sensitivity of the transceiver into account. We will address this aspect when presenting results of real experiments.

4.2.4 Link-State Graph

To render topology information reliable, observations are repeated until all link states have become stable. The process of link-state stabilization is modeled, for all nodes $v_k \in V$ and all node ids 0, …, $n - 1$, by the *link-state graph* in Fig. 4.4. During the topology detection phase, each link $e = (v_j, v_k)$ is in one of the states *unstable*, *stabilizing*, *stable*, or *fluctuating*.

Initially, a link is in state *unstable*. The start transition of the link-state graph resets counters c_{event} and $c_{unstable}$, which count the number of observations and the number of transitions from other states back to *unstable*, respectively. A link e stays in state *unstable* for a number n_{ignore} of observations *event*(e, t), independent of the link type t associated with these observations. This mechanism is intended to filter out frequent oscillations of the link state while nodes are powered up.

As soon as $c_{event} = n_{ignore}$ holds, the link state changes to *stabilizing*, with the link type associated with the current observation determining link type t_e, and c_{event} being reset. A link remains in this state as long as further observations *event*(e, t) yield the same link type, i.e., $t_e = t$, for a maximum number $n_{stabilizing}$ of events. When $c_{event} = n_{stabilizing}$ is reached, the link state changes to *stable*.

After reaching state *stable*, a link remains in that state as long as the same link type is observed. However, if a link type $t_e \neq t$ is detected, the link state changes,

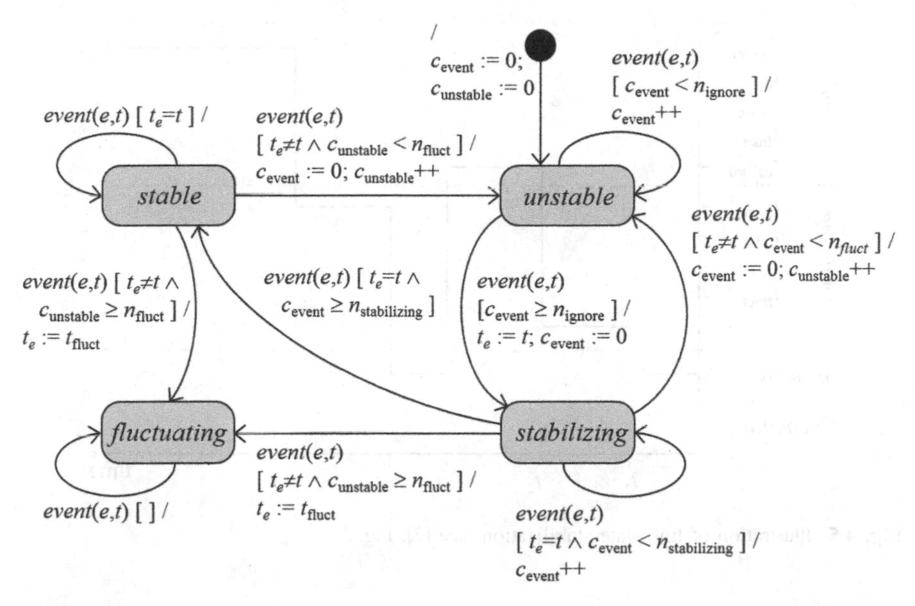

Link types

$T_{\text{link}} = \{ t_{\text{comm}}, t_{\text{int}}, t_{\text{sense}}, t_{\text{noLink}}, t_{\text{fluct}} \}$

Link events

event(e,t)	event for link *e* of type $t \in T_{\text{link}}$

Link parameters

$n_{\text{stabilizing}}$	number of consecutive identical link events in state *stabilizing* until a link is considered *stable*
n_{ignore}	number of link events to ignore after transition to state *unstable*
n_{fluct}	number of transitions to state *unstable* before a link is considered *fluctuating*

Link variables

t_e	current type of link *e*
c_{event}	event counter
c_{unstable}	counter of transitions to state *unstable*

Fig. 4.4 Link-state graph (see [2], Fig. 2)

depending on the link history. If the number of changes to state *unstable* is below an upper bound n_{fluct}, the link state changes to *unstable*, with the counter c_{unstable} being incremented; otherwise, the link state becomes *fluctuating*, with the link type set to t_{fluct}.

The state *fluctuating* is a terminal link state, as there are no transitions back to other link states. When reaching this state, the link type of $e = (v_j, v_k)$ is set to t_{fluct}, to express that link *e* is not reliable. For this reason, it is used neither as communication link nor as sensing link. Instead, it is treated as interference link, i.e., transmissions by node v_j may result in interference at node v_k. This corresponds to the worst case, as the existence of an interference link may prevent other

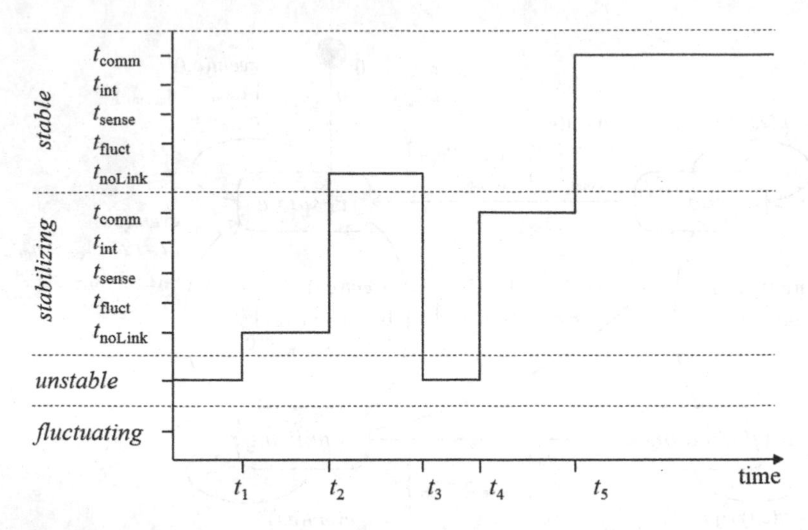

Fig. 4.5 Illustration of link-state stabilization (see [2], Fig. 3)

communication and sensing links from being used in case of exclusive slot assignments, reducing the potential of SDMA. In this sense, this conceptual decision is conservative.

In state *stabilizing*, there are further transitions to states *fluctuating* and *unstable*. If in state *stabilizing*, the observed link type t is different from the current link type t_e, the link state changes back to *unstable*, if the number of changes to state *unstable* is below the upper bound n_{fluct}, and to *fluctuating* otherwise.

The process of link stabilization according to the link-state graph in Fig. 4.4 is illustrated in Fig. 4.5. After the initial interval in state *unstable*, the link state changes to *stabilizing* at t_1, with link type t_{noLink}, and then to state *stable* at t_2. This corresponds to a scenario where the sender node of the link is either out of range of the receiver node or not active. At t_3, the sender node is detected, resulting in a transition back to state *unstable*. This is followed by state changes to *stabilizing* at t_4 and *stable* at t_5, with the link type now set to t_{comm}.

We note that link-state stabilization as specified by the link-state graph in Fig. 4.4 is rather strict: only if the same link type is observed for $n_{stabilizing}$ consecutive observations, stabilization occurs. In other words, if there are single observations from time to time yielding different link types, a link ends up in state *fluctuating*. Furthermore, link types are treated in a disjoint way during link detection. For instance, if a link is in state *stabilizing* with link type t_{sense}, the state changes to *unstable* or *fluctuating* if link type t_{int} or t_{comm} is observed. Altogether, this may lead to situations where all links are *fluctuating* or non-existing, leaving communication, interference, and sensing topologies empty. In our experiments, such situations did not occur; however, they may give rise to a refinement of the link-state graph. For instance, stabilization with link type t_{sense} may be permitted even if some observations yield link type t_{int} or t_{comm}, without changing the link state back to *unstable*.[4] We leave this topic for further study.

4.2.5 Dissemination of Topology Information

In its assigned topology detection slots, each node locally broadcasts MEASURE messages, which contain its node id and topology data. For every link $e = (v_j, v_k)$, the following information is provided:

- Node ids of sender v_j and receiver v_k: Links may be asymmetrical; therefore, ratings for link (v_j, v_k) may differ from link (v_k, v_j).
- Link type t_e: $t_e \in T_{\text{link}} = \{t_{\text{comm}}, t_{\text{int}}, t_{\text{sense}}, t_{\text{noLink}}, t_{\text{fluct}}\}$
- Link-state sequence number seq_e: This number is defined by node v_k detecting link (v_j, v_k) and increased every time the link state changes. Thus, seq_e characterizes the up-to-dateness of link-state information contained in a MEASURE message and helps other nodes to detect and ignore outdated link-state information.
- Signal strength $RSS_{\text{avg},e}$: If $t_e = t_{\text{comm}}$, the average signal strength measured on this link is propagated. Other protocols such as routing algorithms may use this information to give preference to stronger links.

MEASURE messages also contain a checksum, to detect corrupt messages. For the detection of interference and sensing links, it may be necessary to add padding bytes in order to obtain a fixed message length.

Given the maximal node id of $n - 1$, topology information about n^2 link states has to be disseminated repeatedly during the topology detection phase. It can be expected that even for a relatively small network, this information does not fit into a single MEASURE message. Therefore, we split up this information and focus on the dissemination of differences to previous link states, i.e., most recent changes are sent first. As MEASURE messages are sent in exclusively assigned slots, we assume that no frame collisions occur, if the single-network property (see Sect. 1.3) holds.

4.2.6 Termination

Once a stable topology, where all links are in states *stable* or *fluctuating*, has been determined and disseminated, ATDP terminates. The decision about termination is

[4]Note that for the *operation* of links, we still assume that $G_C \subseteq G_I \subseteq G_S$, i.e., a node v_j in communication range of v_i is also in interference and sensing range of v_i, and that a node v_j in interference range of v_i is also in sensing range of v_i.

based on local termination criteria and global agreement of all nodes. A node is ready to terminate if its locally detected links are in states *stable* or *fluctuating* for at least $n_{\text{maxHops}}/m_{\text{fact}}$ topology detection slots, and during this interval, no changes from other nodes have been received. Here, n_{maxHops} denotes the maximum network diameter in communication hops, and m_{fact} is the number of slot assignments per node in a management slot (see Sect. 4.2.2). This way, it is ensured that all potential updates have been propagated across the network.

To reach global agreement on termination, nodes indirectly signal readiness to terminate by exchanging NOTERM messages[5] in termination regions, which are placed before the first topology detection region of each management slot (see Sect. 4.2.2). NOTERM messages are black burst encoded frames (see Sect. 2.2.2) consisting of a dominant bit.[6] If sent, a NOTERM message indicates that the local termination criteria of a node are not yet satisfied.

All nodes not agreeing to terminate send NOTERM messages (almost) simultaneously in round 1 of each termination region. Due to the special transmission procedure, collisions of NOTERM messages are non-destructive. Nodes agreeing to terminate remain silent and listen on the medium in round 1, and thus are able to receive NOTERM messages from nodes in range. In subsequent rounds i of the same termination region, $1 < i \leq n_{\text{maxHops}}$, nodes that have sent or received a NOTERM message in the previous round $i - 1$ send a NOTERM message. This ensures that at the end of the termination region, all nodes are aware whether there still is at least one node that does not yet agree on termination. If no NOTERM message is received during the termination region, ATDP terminates.

Regular termination of ATDP occurs, if a complete communication topology, i.e., a topology with communication paths between all pairs of nodes is detected. This detection may be delayed if nodes are switched on one-by-one. For a given configuration, an upper bound for the duration of regular termination can be determined. Regular termination cannot be guaranteed without complete communication topology, as topology data cannot be exchanged reliably among all nodes; therefore, local termination criteria cannot be satisfied. In this case, ATDP uses a timeout for termination. Further nodes could then be placed before restarting ATDP until connectivity is finally achieved.

4.2.7 Experimental Results

To assess the feasibility and reliability of ATDP, we have implemented the protocol on a real hardware platform and have conducted real experiments. We have decided against simulation experiments, because the effects to be studied are of physical

[5]To exchange NOTERM messages, ATDP uses Arbitrating and Cooperative Transfer Protocol (ACTP), which is presented in detail in Chap. 6.

[6]In an implementation, this is to be complemented by redundancy measures.

nature and therefore difficult to capture completely and accurately in a simulation model.

Hardware Platform

For the implementation of ATDP, we have used Imote2 motes [9], hosting the CC2420 radio transceiver [6]. Furthermore, we have used available implementations of BBS (see Chap. 2) for tick synchronization, and of ACTP (see Chap. 6) for the exchange of NOTERM messages.

For evaluation purposes, our implementation of ATDP provides debug information to the UART interface, which is recorded by a PC connected to the Imote2 board. This debug information includes an output of the current local topology data at the beginning of each management slot. Additionally, link state and type changes are reported when they occur. Thus, the topology detection process can be monitored during execution.

We have set thresholds for signal strength as follows: $RSS_{comm,min}$ = –82 dBm, $RSS_{int,min}$ = –85 dBm, and $RSS_{sense,min}$ = –89 dBm. Given the transceiver's minimal sensitivity of about –90 dBm, these thresholds are expected to produce valid link type assessments. To discover interference and sensing links, we have used the clear channel assessment) (CCA) mechanism of CC2420, configured with a CCA threshold of $RSS_{sense,min}$. The CCA mechanism triggers a hardware interrupt when a change of the medium state is detected. During interrupt processing, ATDP reads the observed signal strength from the received signal strength indicator (RSSI) register of the transceiver and determines the link type of a given observation.

Feasibility

To assess the feasibility of ATDP, we have conducted a series of real experiments, with different node placements inside our university building. To satisfy the single-network property (see Sect. 1.3), we have used an IEEE 802.15.4 channel that showed no overlaps with occupied IEEE 802.11 channels. In all experiments, ATDP has terminated within a few minutes, with meaningful detected topologies. Figure 4.6 shows one experimental scenario, with five nodes placed in four rooms, and the topologies detected by ATDP. All nodes are connected by paths of communication links, yielding a complete communication topology. We note that the links between nodes v_7 and v_{16} are asymmetrical: (v_7, v_{16}) is an interference link, whereas (v_{16}, v_7) is a communication link.

Reliability

To assess the reliability of link detection, we have validated the thresholds of the link rating mechanism. For this, we have conducted extensive experiments on the detected topology shown in Fig. 4.6.

First, we have checked the reliability of detected communication links. For each communication link, 48,100 regular MAC frames with a payload of 120 bytes each were sent during a period of 5 min. On average, only 0.02% of these frames were lost or corrupted, with a maximum combined loss and corruption rate of 0.13%. This shows that the choice of $RSS_{comm,min}$ = –82 dBm has been suitable.

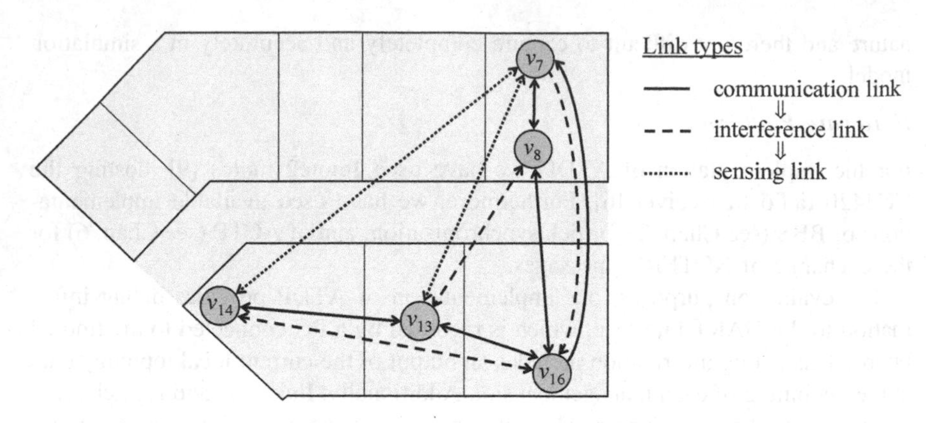

Fig. 4.6 Floor plan, placements of nodes, and topologies detected by ATDP (see [2], Fig. 4)

Next, we have validated interference links. For this purpose, we have devised a test system running on three nodes: one sender, one receiver, and one interferer. Sender and interferer are synchronized and concurrently send regular MAC frames of 120 bytes of payload. The receiver records the numbers of frames correctly received from the sender and the interferer, respectively, and the numbers of corrupted and lost frames. The results of these experiments are shown in Fig. 4.7.

Looking at the communication between nodes v_8 and v_{16}, we observe that the impact of interfering node v_{14} is extremely high. Here, only 3% of all frames sent by v_8 were correctly received; instead, 17% of frames sent by the interferer were received correctly, 17% of all frames were received corruptly, and 63% were received neither correctly nor corruptly and therefore counted as lost. This shows that (v_{14}, v_{16}) is indeed an interference link.

Figure 4.7 also shows that in three cases, the interferer did not have significant impact on the reliability of the communication link. For instance, 99.99% of the

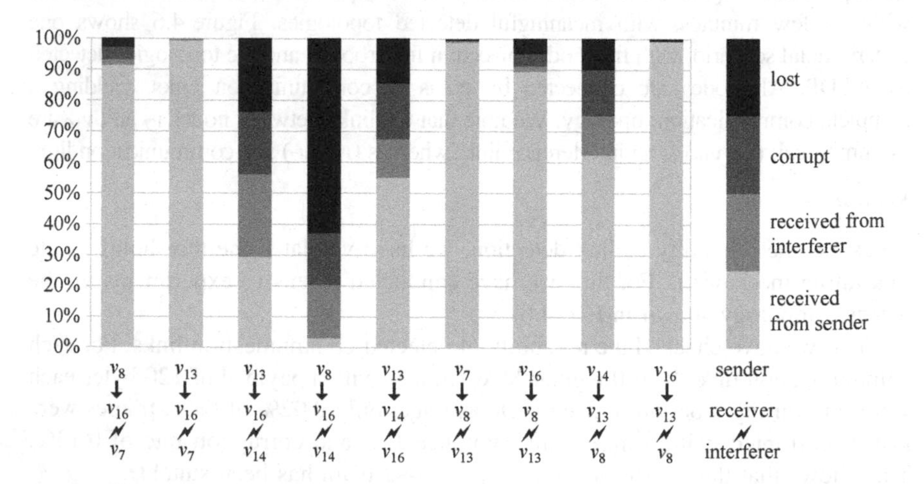

Fig. 4.7 Ratio of frames received correctly, corruptly, and lost in the presence of interference by nodes in interference range (see [2], Fig. 5)

Table 4.1 Frames received correctly, corruptly, and lost in the presence of interference by nodes in sensing range (see [2], Table 1)

Experiment sender → receiver ↯ interferer	Received correctly	Received corruptly	Lost
$v_{13} \rightarrow v_{14}$ ↯ v_7	48,100		
$v_{16} \rightarrow v_{13}$ ↯ v_7	48,100		
$v_{14} \rightarrow v_{13}$ ↯ v_7	48,092	6	2
$v_8 \rightarrow v_7$ ↯ v_{13}	48,100		
$v_{16} \rightarrow v_7$ ↯ v_{13}	48,063	25	12
$v_8 \rightarrow v_7$ ↯ v_{14}	48,100		
$v_{16} \rightarrow v_7$ ↯ v_{14}	48,099		1

frames sent by node v_{16} were correctly received by v_{13}, despite concurrent transmissions by v_8. However, the same interference link had significant impact on the reliability of communication from node v_{14} to node v_{13}, with only 63% of the frames correctly received from the intended sender. In summary, we observe that although interferers did not always have significant impact, there is at least one experiment where the interferer disturbed a communication link. Therefore, all detected interference links should be considered when applying SDMA.

Finally, we have validated the impact of sensing links on the reliability of communication links. The experimental setup was similar to the validation of interference links. However, this time, interfering nodes were in sensing range of the receiver. Table 4.1 shows the results of these experiments. In most cases, all frames of the sending node were correctly received. Only in three cases, one or more frames were received corruptly or lost, with a maximum combined loss and corruption ratio of 0.0008%. This again shows that these links have been correctly classified.

4.3 Related Work

In this chapter, we survey work that addresses topology detection in wireless ad hoc networks. Our survey is not intended to be comprehensive, but to identify a number of protocols performing communication and interference topology detection and to point out differences to ATDP. We are not aware of other protocols addressing the detection of sensing topologies.

4.3.1 Routing Protocols

Topology information is needed for route discovery in multi-hop networks. Routing protocols usually incorporate functionality for topology detection. Depending on the type of protocol, the degree of detail varies. In link-state protocols, each node collects data about its local links and exchanges them with all other nodes, yielding a complete global topology, as in ATDP. Different from ATDP, this global topology is usually restricted to the set of communication links. In distance vector protocols, nodes only collect aggregated topology information, e.g., the shortest hop distance to a destination node and the next hop on the shortest path. As this is substantially less information compared to ATDP, we will omit these protocols in our survey. A comprehensive survey about routing protocols for ad hoc networks published until about 2010 can be found in [5].

Global State Routing (*GSR*) [7] is a well-known link-state protocol for wireless ad hoc networks. Each node $v_i \in V$ maintains the following network status information:

- A_i: neighbor set containing ids of all neighbors v_j of v_i.
- TT_i: topology table, containing, for each node $v_j \in V$, link-state information LS_j reported by v_j and sequence number SEQ_j generated by v_j to mark the version of LS_j.
- $NEXT_i$: next hop table, indicating, for each $v_j \in V$, the next hop v_k on the shortest path to v_j.
- D_i: distance table, indicating, for each $v_j \in V$, the length of the shortest path.

Nodes $v_i \in V$ learn about their neighbors when receiving messages containing the sender id. These neighbors are then added to the neighbor set A_i, determining the set of local links. GSR assumes that links are symmetrical, i.e., if a node v_i receives from v_j, then v_j will also receive from v_i.

Periodically, each node v_i copies the neighbors recorded in A_i to its local link state LS_i and increments the sequence number SEQ_i. It then broadcasts a network status message containing its id v_i and its topology table TT_i to its neighbors. When receiving a network status message from a neighbor, v_i updates its topology table TT_i, considering only those link states LS_j that have a newer sequence number. Thus, local link-state information of all nodes is propagated across the network, yielding a complete global topology.

Periodically, based on TT_i, each node v_i computes shortest paths to all other nodes v_j, using a slightly modified version of Dijkstra's algorithm and a weight function initially assigning 1 to all node pairs (v_i, v_j), with $v_j \in A_i$, and $\propto$ to all other node pairs. As the complete topology is available, this computation can be performed locally, yielding tables $NEXT_i$ and D_i.

Compared to ATDP, topology detection of GSR has a number of drawbacks. First, link detection is based on single observations only. Node ids are added to the neighbor set A_i already after a single successful message reception. This implies that even weak links are recorded, if the sending node has been successful at least once.

Furthermore, links are assumed to be symmetrical, which cannot be taken for granted in wireless networks.

GSR explicitly addresses wireless networks with mobile nodes. This means that the neighborhood of a node may change, implying the need to remove nodes from the neighbor set A_i. In the algorithm, there is a function addressing this; however, it remains unclear how non-neighbors are detected. A solution could be to assign and update timestamps with entries in A_i and to delete outdated entries.

GSR computes shortest path routes. With this metric, links covering long distances to the destination node are preferred, as this yields shorter paths. However, in wireless networks, such links tend to be weak. Therefore, preferred usage of these links in route establishment may result in unreliable routes and possibly increased traffic due to frequent retransmissions.

Finally, different from ATDP, GSR is restricted to determining the communication topology. As the originators of interference and energy cannot be identified, interference and sensing topologies remain undetected.

Other link-state routing protocols for wireless ad hoc networks, such as *Optimized Link State Routing* (*OLSR*) [10], work in similar ways as GSR. Differences are, for instance, the use of hello messages for neighbor detection and a more efficient exchange of topology information. Although these modifications improve the quality of link detection, several drawbacks remain.

4.3.2 Clustering Protocols

Topology information is needed for network clustering, which divides a network into groups of nodes. As the amount of topology information needed for this task varies, clustering protocols (see [3] for a survey, and Chap. 9) usually incorporate functionality for topology detection. Furthermore, this functionally is restricted to the discovery of communication topologies, as interference and sensing topologies are not considered for clustering.

To establish clusterings that are optimal w.r.t. specified clustering criteria, complete global communication topologies would be needed. As this would yield high communication and time complexity, most clustering algorithms only use partial topologies. For instance, to form a d-hop dominating set, i.e., a set $V' \subseteq V$ where all nodes $v_i \in V$ are in d-hop neighborhood of some node $v_k \in V'$, knowledge about the d-hop neighborhood may be sufficient.

A well-known clustering protocol for wireless ad hoc networks is *MaxMinD* cluster formation [4]. Clustering objectives are connectivity achieved by intra- and inter-cluster routing, and balanced cluster sizes. In the *floodmax* phase, cluster heads are predetermined. For this purpose, nodes determine the largest node id in their d-hop neighborhood. In the *floodmin* phase, cluster sizes are balanced, based on additional information from the $(2 \cdot d)$-hop neighborhood.

Compared to ATDP, topology detection of MaxMinD has a number of drawbacks. First, link detection is based on single observations only. *d*-hop neighbors and potential cluster heads are recorded already after a single successful message reception. This implies that weak links are recorded, too, if a message has been received. Furthermore, it is assumed that links are symmetrical.

MaxMinD strives for a small number of clusters, heuristically maximizing cluster sizes defined by *d*-hop neighborhoods. This is achieved by considering communication links covering long distances between nodes. However, in wireless networks, such links tend to be weak, resulting in unreliable routes for intra- and inter-cluster communication.

Other clustering protocols for wireless ad hoc networks, such as *Algorithm for Cluster Establishment* (*ACE*) [8], work in similar ways as MaxMinD. Differences are, for instance, the use of hello messages for neighbor detection and different metrics for cluster formation. As in MaxMinD, weak links are recorded, too, to achieve a small number of clusters.

4.3.3 Detection of Interference Topologies

To perform SDMA, information about the interference topology is required. With SDMA, nodes outside each other's interference range may use the medium simultaneously. This can be exploited to increase network-wide throughput, based on exclusive reservation of time slots.

In [11], the authors report on interference experiments with IEEE 802.11 nodes to predict interference among pairs of links, with rather accurate results. However, given a network of n nodes, $O(n^2)$ experiments are required to determine the interference topology. For a test bed of 22 nodes, the authors report an experiment duration of about 28 h.

In [12], the *radio interference detection* (*RID*) protocol is presented. RID assesses interference based on the transmission of detection sequences, consisting of two frames each. The first frame is transmitted with high transmission power, whereas the second frame is sent with reduced transmission power. For this to work, the assumption is that the communication range of high power transmissions is equal to or larger than the interference range of transmissions with reduced power. Thus, by receiving the first frame, a receiving node acquires knowledge about the sender id and can then assess the link type by measuring the received signal strength of the second frame and comparing it to a threshold value. A major drawback of this approach is that only reduced transmission power can be used during network operation.

4.4 Conclusions

In this chapter, we have explained the concepts and requirements of topology detection, have presented our protocol ATDP for the automatic detection of communication, interference, and sensing topologies, and have surveyed and compared related work. We have argued that solid information about communication topologies is vital for other network functionalities, in particular for routing and clustering. Furthermore, knowledge of interference topologies enables SDMA to increase network-wide throughput. Finally, knowledge about sensing topologies can help to tackle the hidden station problem and to transmit black burst encoded bit sequences.

With its functionality and reliability, ATDP exceeds existing protocols for automatic topology detection. To our knowledge, ATDP is the only protocol that detects sensing topologies and is one of very few protocols that address interference topologies. Furthermore, unlike many other protocols, ATDP does not rely on single or a few observations, but continues link observation until stability criteria are met. In particular, only reliable communication and sensing links are considered for information exchange.

Regarding the detection of interference and sensing topologies, we state that this requires prior knowledge about transmission activities of nodes, i.e., transmission schedules. In ATDP, this is achieved by virtual time slotting combined with exclusive and globally known slot reservations. Exclusive slot reservations also prevent the collision of management messages, which is a problem in other link-state protocols.

We have already noted that link stabilization in ATDP is rather strict, as links only stabilize if the same link type is observed during a sequence of observations. Although this has proven feasible in our experiments, the notion of link stabilization may be relaxed, for instance, by allowing for a small percentage of deviating observations.

Literature

Chair for Networked Systems

1. Gotzhein R (2019) ProNet 4.0—a wireless real-time communication system for Industry 4.0. White Paper, Networked Systems Group, Department of Computer Science, University of Kaiserslautern. http://vs.informatik.uni-kl.de/publications/2014/Go14/whitePaperEN-ProNet4.0.pdf. Last accessed: 27 Aug 2019
2. Kramer C, Christmann D, Gotzhein R (2015) Automatic topology discovery in TDMA-based ad hoc networks. In: Proceedings of the 11th international wireless communications & mobile computing conference (IWCMC 2015), Dubrovnik, Croatia, 24–28 Aug 2015, pp 634–639

Further References

3. Abbasi AA, Younis M (2007) A survey on clustering algorithms for wireless sensor networks. Comput Commun 30(14):2826–2841
4. Amis AD, Prakash R, Vuong THP, Huynh DT (2000) Max-Min D-cluster formation in wireless ad hoc networks. In: Proceedings of INFOCOM 2000, vol 1, Tel Aviv, Israel, pp 32–41
5. Boukerche A, Turgut B, Aydin N, Ahmad MZ, Bölöni L, Turgut D (2011) Routing protocols in ad hoc networks: a survey, computer networks 55, pp 3032–3080
6. Chipcon AS (2019) CC2420 ZigBee-ready RF transceiver. http://www-inst.eecs.berkeley.edu/~cs150/Documents/CC2420.pdf. Last accessed: 27 Aug 2019
7. Chen TW, Gerla M (1998) Global state routing: a new routing scheme for ad hoc wireless networks. In: Proceedings of the IEEE international conference on communications (ICC 98), pp 171–175
8. Chan H, Perrig A (2004) ACE—an emergent algorithm for highly uniform cluster formation. In: Karl H, Wolisz A (eds) European conference on wireless sensor networks (EWSN 2004), LNCS 2920. Springer, pp 154–171
9. MEMSIC Inc (2019) Imote 2 datasheet. https://vs.cs.uni-kl.de/downloads/Imote2NET_ED_Datasheet.pdf. Last accessed: 27 Aug 2019
10. Jacquet P, Mühletaler P, Clausen T, Laouiti A, Qayyum A, Viennot L (2001) Optimized link state routing protocol for ad hoc networks. In: Proceedings of the IEEE international multitopic conference (INMIC 2001), Lahore, Pakistan, Dec 2001, pp 62–68
11. Padhye J, Agarwal S, Padmanabhan VN, Qiu L, Rao A, Zill B (2005) Estimation of link interference in static multi-hop wireless networks. In: Proceedings of the 5th ACM SIGCOMM conference on internet measurement (IMC 2005), Berkeley, CA, USA, USENIX Association, pp 305–310
12. Zhou G, He T, Stankovic JA, Abdelzaher T (2005) RID—radio interference detection in wireless sensor networks. In: 24th annual joint conference of the IEEE computer and communications societies (INFOCOM), proceedings, vol 2, IEEE, pp 891–901
13. Sobrinho JL, Krishnakumar AS (1999) Quality of Service in Ad Hoc Carrier Sense Multiple Access Networks, IEEE Journal on Selected Areas in Communications, 17(8):1353–1368

Chapter 5
Medium Access Schemes

The ultimate task of a communication system is the exchange of payload messages among a set of nodes via a physical medium. If the medium is shared among these nodes, methods controlling medium occupancy called *medium access schemes* are applied. In this chapter, we explain foundations of medium access, present medium access schemes of ProNet 4.0 [1] including our novel *mode access scheme* [2–5], survey and compare related work, and draw conclusions.

5.1 Foundations

In this chapter, we establish the context of accessing a wireless medium, explain concepts and measures, and address areas of operation.

5.1.1 Context

In wireless communication systems, nodes transmit and receive frames via radio channels that are shared among all network nodes. Since the medium is shared, collisions occur if transmissions overlap in time, frequency, and space. Depending on positions of senders and receivers, received signal strength, and other factors, frame collisions can be destructive, resulting in frame corruption or frame loss. Therefore, measures to reduce the probability of collisions or to avoid them entirely are to be taken, located in a Medium Access Control (MAC) protocol on MAC layer (see Fig. 1.1).

MAC protocols provide several protocol functionalities. They define the format of frames, e.g., data frames and management frames, and support addressing schemes, e.g., unicast, multicast, and broadcast. In the core of a MAC protocol is the medium access scheme to control medium occupancy. Furthermore, a MAC

R. Gotzhein, *Real-time Communication Protocols for Multi-hop Ad-hoc Networks*, Computer Communications and Networks,
https://doi.org/10.1007/978-3-030-33319-5_5

protocol provides error detection, for instance, detection of transmission errors such as bit errors and frame loss, and error handling, e.g., by bit error correction and retransmission of frames. In wireless networks, further measures to tackle the hidden and exposed station problems may be taken. On the other hand, routing is usually not addressed by MAC protocols. In this chapter, our focus is on medium access schemes in wireless networks.

5.1.2 Concepts and Measures

We distinguish three basic types of medium access. Exclusive access means that at any point in time, only one node has the right of using the medium for transmission. With shared access, all nodes may use the medium at any time. Mode access reduces the right of medium usage to subsets of nodes.

- *Exclusive medium access* is usually based on network-wide reservation of time intervals assigned to single nodes. In single-hop networks, token passing is a feasible method to provide exclusive access. In networks with global time slotting, exclusive reservation of time slots is a suitable approach.
- *Shared medium access* bears the risk of destructive collisions, if several nodes use the medium at the same time. To reduce the probability of collisions, the following measures are applied:
  - A straightforward measure is to listen on the medium before starting transmission, called Listen Before Talk (LBT) or Carrier Sense Multiple Access (CSMA). Only if the medium is detected as idle, transmission begins.
  - If a node detects the medium as busy, which may be due to an ongoing transmission, it waits until the medium becomes idle again. However, during this period, other nodes may do the same. To prevent these nodes from accessing the medium at the same time, some form of medium contention, e.g., random backoff (explained below) is performed.
- *Mode medium access* defines, for specific time intervals, subsets of nodes that may contend for the medium. Furthermore, messages that may have to be sent during the same time interval are assigned unique priorities. By enforcing these priorities, medium access becomes deterministic.

We distinguish two types of medium contention schemes. Passive contention is performed by listening on the medium for a random duration. Active contention is done by transmitting contention patterns on the medium. Passive and active contention can also be combined.

Passive contention is usually associated with random backoff. As soon as a node detects a previously busy medium as idle, it determines a random time interval during which it refrains from sending. If the medium is still idle at the end of this interval, the node has won contention and starts transmission. If the medium becomes busy during this interval, the node has lost contention and has to contend again. By configuring value ranges of random backoff intervals, strict and statistical frame priorities can be supported.

Active contention can be performed by transmitting busy tones or binary countdown sequences:

- A *busy tone* is a jamming signal of defined length. Contending nodes transmit busy tones of different length after detecting the medium as idle and then listen on the medium. If the medium is still busy, another node in range is sending a longer busy tone and thereby gains preference. The node sending the longest busy tone wins contention.
- A *binary countdown sequence* is a bit sequence encoding a unique priority. When the medium is idle or at aligned points in time, nodes contend by bitwise transmission of their sequences. Bits are encoded as dominant and recessive signals, with the medium being monitored instead of transmitting recessive signals. If during a binary countdown sequence, a node contends with a recessive signal, but monitors a dominant signal, it loses contention and resigns. At the end of the sequence, the node signaling the highest priority wins.

In wireless networks, passive contention and active contention are effective in single-hop range only. For instance, a node may not detect busy tones or binary countdown sequences, as it may be out of range. Furthermore, a node may detect the medium as idle even if another node in interference range of the intended receiver is transmitting. In some MAC protocols, this problem is tackled by a sequence of short reservation frames between sender and receiver. However, collisions cannot be ruled out entirely.

5.1.3 Areas of Operation

Medium access schemes support the operation and efficient usage of shared communication media. The decision about the access scheme(s) depends on the type of traffic (periodic, sporadic), availability of synchronization, required guarantees, and efficiency of bandwidth usage.

In case of periodic traffic, communication is time-triggered, i.e., messages are to be sent at determined points in time. For timeliness and reliability of transmissions, exclusive medium access schemes, e.g., based on reserved time slots, are required. Thereby, collision-free medium access at defined points in time can be guaranteed. Furthermore, due to the periodicity of transmissions, bandwidth is used efficiently.

In case of sporadic traffic, communication is event-triggered, i.e., messages are to be sent when events occur. For short reaction delays, shared medium access schemes are the first choice. Thereby, short-term, possibly priority-driven medium access is basically possible. In addition, no reservations need to be made in advance; therefore, bandwidth is used efficiently.

In case of a mix of periodic and sporadic traffic, both types of medium access schemes should be used together, as none of them supports both types of traffic without significant drawbacks. For instance, in case of sporadic traffic with short reaction delays, exclusive access schemes have bad bandwidth usage, as worst-case reservations of time slots are required. In case of periodic traffic, shared access schemes cannot guarantee timeliness and reliability of transmissions. A solution could be to alternate between exclusive and shared medium access. However, this implies that during a phase of exclusive access, event-triggered messages are delayed until the next contention phase. Here, mode medium access with restricted contention is a proper approach.

5.2 Medium Access Schemes in ProNet 4.0

The protocol stack ProNet 4.0 [1] supports several medium access schemes: exclusive access, shared access with priorities, and mode access. These access schemes are implemented by protocol modules ProRes, ProCont, and ProMod, respectively, and activated during dedicated virtual time regions (see Fig. 5.1 and Chap. 3). ProRes, ProCont, and ProMod are self-contained MAC protocols, i.e., they cover frame formats, addressing schemes, and error control, too. However, in this chapter, we will focus on medium access schemes.

Figure 5.1 shows an example of virtual time slotting for medium access and synchronization. Exclusive regions host time-triggered traffic with strict deadlines and access guarantees. Event-triggered communication is placed into shared regions. Mode regions support a mix of periodic and sporadic traffic, with access guarantees for high-priority messages.

Virtual time regions can be freely placed and dimensioned, so, for instance, exclusive regions may prevail in case of purely periodic traffic. Sporadic messages have to wait for the next shared or mode region. Therefore, in the case of short reaction delays, these regions should be placed in short distances. The global time slotting scheme of ProSlot (see Chap. 3) provides all necessary flexibility.

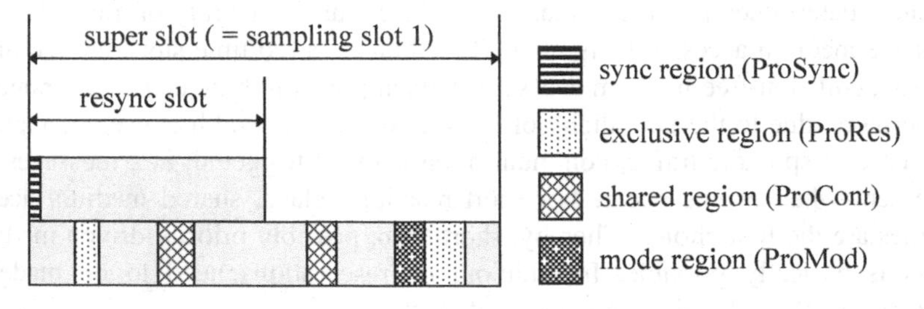

Fig. 5.1 Virtual time slotting—medium access regions (for explanations, see also Chap. 3)

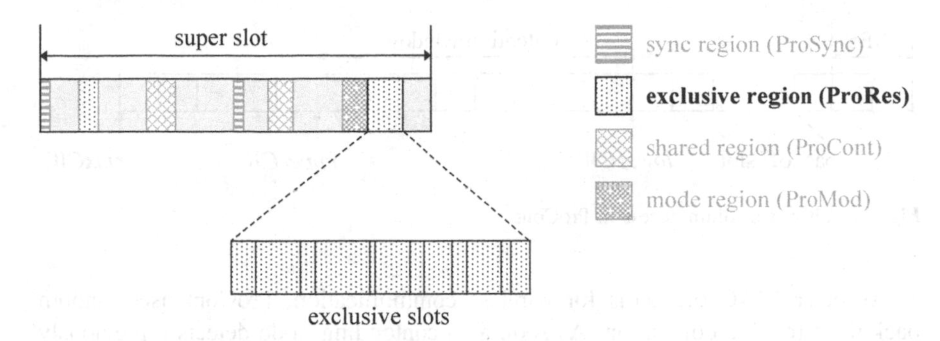

Fig. 5.2 Virtual time slotting—exclusive regions and exclusive slots

5.2.1 Exclusive Medium Access—ProRes

Exclusive medium access is granted in exclusive regions. As illustrated in Fig. 5.2, these regions are decomposed into sequences of exclusive time slots, which may differ in length. Global time slotting is provided by ProSlot (see Chap. 3) and highly flexible.

For exclusive medium access, it is necessary that a time slot be assigned to a single node, either before deployment or at runtime. We assume that dynamic slot assignments are handled by a resource reservation protocol during route selection (see Chap. 8). Additional flexibility can be achieved by configuring size and placement of exclusive slots at runtime.

As an exclusive slot is assigned to a single sending node, medium access is straightforward. After an initial delay defined by the maximum tick offset (see Chap. 2), the sender starts its frame transmission. Since slot assignments are exclusive, no contention is required, and transmissions are collision-free. Therefore, reliability depends on link quality only, which is checked during topology detection (see Chap. 4).

5.2.2 Shared Medium Access—ProCont

Shared medium access is granted in shared regions. No prior assignment of time slots to sending nodes is performed. Instead, nodes with frames to be sent contend for the medium and start transmission only after winning contention.

In ProNet 4.0, the protocol module ProCont controls medium access in shared regions (see Fig. 5.1). To reduce the risk of destructive collisions, ProCont takes several measures. First, nodes perform Carrier Sense Multiple Access (CSMA) before sending and access the medium only if it was detected as idle. Second, if the medium was busy, a passive contention scheme is applied.

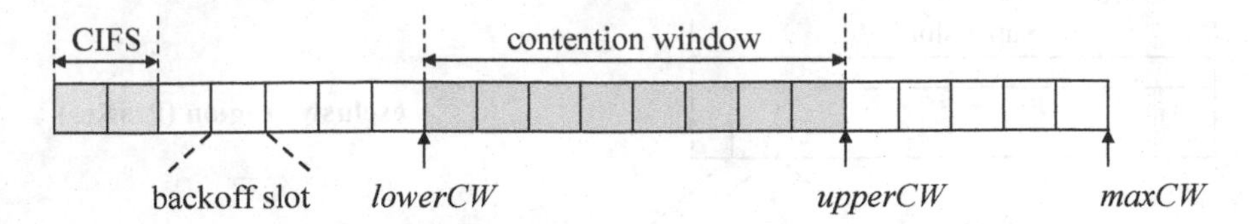

Fig. 5.3 Shared medium access in ProCont

As other MAC protocols for wireless communication, ProCont uses random backoff to resolve contention. As soon as a contending node detects a previously busy medium as idle, it continues to listen on the medium for a random backoff interval. If the medium stays idle, the node wins contention and starts its frame transmission. Otherwise, it loses contention and has to compete again.

To detect a medium as idle, a node listens for a fixed duration d_{CIFS} of the contention inter frame space (CIFS, see Fig. 5.3). It then draws a random value n_{random} from the current contention window, which is an interval of natural numbers. The duration of the random backoff interval d_{random} is obtained by multiplying n_{random} with the fixed duration of a backoff slot $d_{\mathrm{backoffSlot}}$. Random backoff is realized by a backoff counter c_{backoff}, which is set to the initial value n_{random} and then decremented after each backoff slot, if the medium is still idle. Medium access is performed as soon as c_{backoff} is counted to zero.

Compared to existing MAC protocols, there is one aspect of ProCont that deserves special attention. For shared medium access, both strict and statistical frame priorities are supported. For each frame, a frame-specific start of the contention window $frame_{\mathrm{lowerBound}}$ and a frame priority $frame_{\mathrm{priority}}$ are defined. From these parameters and the maximum size *maxCW* of the contention window, the frame-specific contention window is derived as follows (see Fig. 5.3):

- $lowerCW =_{\mathrm{df}} \min\{frame_{\mathrm{lowerBound}}, maxCW\}$
- $upperCW =_{\mathrm{df}} \min\{frame_{\mathrm{priority}}, maxCW\}$

By choosing $frame_{\mathrm{lowerBound}} \leq frame_{\mathrm{priority}}$, we can realize frame priorities. Higher priorities are achieved by shifting the frame-specific contention window to the beginning of the interval [0…*maxCW*]; for lower priorities, it can be shifted toward the end. For frames with $frame_{\mathrm{priority}} > maxCW$, the contention window always ends at *maxCW*. Yet, frames can be scheduled locally according to their frame priority before starting contention.

To realize statistical priorities, frame-specific contention windows may overlap. In addition, after losing contention, the backoff counter value is kept, and countdown continues during the next contention phase. For strict priorities, non-overlapping contention windows are required. Furthermore, a new random value n_{random} from the current contention window is drawn, and the backoff counter is set to n_{random} at the start of the next contention phase.

Of course, the passive priority-driven contention scheme of ProCont is only effective in the single-hop range. To address the hidden station problem, ProCont also supports the RTS/CTS mechanism (e.g., see [11]) for unicast transfers.

5.2.3 *Mode Medium Access—ProMod*

In this chapter, we present our novel concept of *mode-triggered scheduling with mode medium access* [2–5]. Our approach can substantially improve bandwidth usage in many scenarios of industrial relevance, while preserving timeliness. With mode-triggered scheduling, a well-controlled amount of dynamic contention for time slots can be defined. With some technologies, mode medium access reaches consensus among contending nodes extremely fast and deterministically.

5.2.3.1 Motivation

To motivate mode medium access, we consider a scenario from the control domain, where four control systems are operating via a shared wireless communication system:

- Two control systems of type "inverted pendulum" (see Chap. 1), each equipped with three sensors to measure position, velocity, and angle, one actuator to apply a voltage to a motor, and a controller.
- Two control systems of type "flow rate system", each equipped with two sensors to measure flow rate, one actuator to operate a valve, and a controller.
- For each control system, there is an emergency shutdown to move it to a fail-safe state. This also happens if steering values are not received for a defined period of time, to protect against failure of the communication system.
- To synchronize value sampling, the sensor nodes of each control system are to be tick-synchronized.

For convenience, we assume that all nodes are in single-hop range, with stable communication links. During system analysis, we have identified functionalities, message types, and communication requirements (see Table 5.1). Most messages are of periodic nature, with the exception of emergency messages, which are to be sent in case of very rare emergency events, and therefore sporadic. All messages require high or even very high transfer reliability and carry only light payload. Furthermore, all messages have tight timing constraints regarding maximal total transfer delay, which is the sum of medium access delay and transmission delay.

Periodic messages have transmission intervals that correspond to sampling intervals, ranging from preferred (short) intervals to maximal (long) intervals. By convention, maximal intervals are necessary and sufficient to keep a controlled system stable. For instance, if every 72 ms, the inverted pendulum can complete a

Table 5.1 Control system scenario—results of system analysis

Functionalities	Message types	Message characteristics	Communication requirements
Tick sync	m_{tick}	periodic data	very high reliability, small data volume, fixed transmission interval of 72 ms, maximal total transfer delay of 4 ms
Inverted pendulum 1	$m_{s\_1,1\ldots3}$, m_{a_1}	periodic data	high reliability, small data volume, transmission interval of 36–72 ms, maximal total transfer delay of 12 ms
Inverted pendulum 2	$m_{s\_2,1\ldots3}$, m_{a_2}	periodic data	high reliability, small data volume, transmission interval of 36–72 ms, maximal total transfer delay of 12 ms
Flow rate control A	$m_{s\_A,1\ldots2}$, m_{a_A}	periodic data	high reliability, small data volume, transmission interval of 72–144 ms, maximal total transfer delay of 20 ms
Flow rate control B	$m_{s\_B,1\ldots2}$, m_{a_B}	periodic data	high reliability, small data volume, transmission interval of 72–144 ms, maximal total transfer delay of 20 ms
Emergency shutdown	$m_{e,1..4}$	very rare sporadic events	very high reliability, small data volume, maximal total transfer delay of 100 ms

control cycle, consisting of sampling and steering, the control objective is achieved. Thus, at least 4 messages are to be exchanged every 72 ms. In case of sporadic disturbances, however, stabilization delay may be high. To shorten this delay and thereby increase quality of control, the sampling interval may be temporarily reduced. This also has an impact on the transmission interval. By dividing the maximum transmission interval of the inverted pendulum in half, 4 messages are to be exchanged every 36 ms. However, as 4 messages every 72 ms are still sufficient to reestablish stability, it is not mandatory that the additional 4 messages can actually be sent.

Table 5.2 Event-triggered scheduling—message priorities

Functionalities	Message types	Message priorities
Tick sync	m_{tick}	0
Inverted pendulum 1	$m_{s\_1,1\ldots3}$, m_{a_1}	2
Inverted pendulum 2	$m_{s\_2,1\ldots3}$, m_{a_2}	2
Flow rate control A	$m_{s\_A,1\ldots2}$, m_{a_A}	3
Flow rate control B	$m_{s\_B,1\ldots2}$, m_{a_B}	3
Emergency shutdown	$m_{e,1\ldots4}$	1

From the application requirements, we derive the number of messages to be sent during each transmission interval:

- For tick synchronization, 1 message per 72 ms is required.
- Each inverted pendulum exchanges 4 to 8 messages per maximal sampling period of 72 ms.
- For flow rate control, 3–6 messages per maximal sampling period of 144 ms and control system are to be exchanged.
- For emergency shutdown, a maximum number of 4 messages are to be exchanged during a period of 100 ms.

If a range of messages per maximal sampling period is given, the smaller number of transmissions is mandatory. It is expected, however, that the higher number of transmissions can be temporarily supported, too.

Solution 1: Event-triggered priority-driven scheduling

In a first solution, we assume that a wireless communication system with event-triggered strict priority-driven scheduling based on shared medium access is available (e.g., see Sect. 5.2.2). Table 5.2 shows a possible mapping of communication requirements to an assignment of static message priorities. As wireless communication systems usually only support a small number of priorities, several messages are assigned the same priority.

- As tick synchronization has the most stringent requirements regarding total transfer delay, we assign the highest priority 0 to messages of type m_{tick}.
- We assign high priority 1 to emergency event messages $m_{e,1\ldots4}$. Given the maximal total transfer delay of 100 ms, this does not seem mandatory; however, we want to rule out that sporadic emergency messages are delayed indefinitely by frequent periodic messages of higher priority.
- As inverted pendulums have shorter transmission intervals than flow rate control systems, their messages are assigned higher priority.
- Finally, we assign a low priority to messages of flow rate control systems.

Table 5.3 Time-triggered scheduling—time slots per cycle

Functionalities	Message types	Slots/cycle
Tick sync	m_{tick}	2
Inverted pendulum 1	$m_{s\_1,1\ldots3}$, m_{a_1}	8–16
Inverted pendulum 2	$m_{s\_2,1\ldots3}$, m_{a_2}	8–16
Flow rate control A	$m_{s\_A,1\ldots2}$, m_{a_A}	3–6
Flow rate control B	$m_{s\_B,1\ldots2}$, m_{a_B}	3–6
Emergency shutdown	$m_{e,1\ldots4}$	8

While this priority assignment is systematical and coherent, it bears a number of potential problems. In event-triggered priority-driven communication systems, high-priority messages may delay messages of lower priority indefinitely. Thus, the inverted pendulums may dominate and subsequently destabilize the flow rate control systems if they exchange messages at maximal rates, leading to a violation of maximal total transmission delays. Furthermore, messages of different types are assigned the same priority. This implies that in addition, a random-based contention scheme is to be applied, which may result in destructive collisions and thereby reduces transfer reliability. In summary, we can state that shared medium access is not adequate to support the deterministic communication requirements of this scenario.

Solution 2: Time-triggered scheduling

The second solution builds on time-triggered scheduling with exclusive medium access. For each message type, a fixed number of time slots per sampling interval are exclusively assigned, thereby avoiding collisions entirely, which achieves very high reliability and timeliness. Table 5.3 shows the number of required time slots per cycle of 144 ms.

The length of a cycle is determined as the least common multiple of the transmission intervals of tick resynchronization, inverted pendulums, and flow rate control systems, which is 144 ms. As this is two times the length of the resynchronization interval, two sync slots per cycle are assigned. Similarly, we obtain a demand of 8–16 (3–6) time slots per cycle for each inverted pendulum (flow rate control system). As the requirements call for a maximum total transfer delay of 100 ms for emergency shutdowns, 2 time slots per control system and cycle are required, yielding a total of 8 time slots per cycle. Adding up, there is a need for up to 54 time slots per cycle. If we assume a slot size of 4 ms, only 36 time slots per cycle are available. Thus, to implement the communication requirements, two wireless communication systems would have to be deployed. In summary, we can state that exclusive medium access yields timeliness and high reliability; however, it lacks bandwidth efficiency, as exclusively reserved slots may not be needed in each cycle.

5.2.3.2 Mode-Triggered Scheduling

To overcome the problems of event-triggered and time-triggered scheduling, we now present a third solution that is based on our novel concept of *mode-triggered scheduling with mode medium access* [2–5]. The main idea is to assign time slots to subsets of message types and to apply a deterministic priority-driven scheduling strategy implemented by a highly efficient distributed arbitration scheme. Mode medium access supports exclusive access as well as high bandwidth usage.

In a first step, we uniquely associate *transmission modes* (or *modes* for short) with message types. A mode is a high-level design concept and may be derived, for instance, from system operation phases or transmission purposes. In the control scenario, we use modes *management*, $safety_{\{1,2,A,B\}}$, and *emergency*.

Definition 5.1 (*Modes*) *Modes* are modeled as a non-empty, finite set $M = \{m_1,\ldots, m_r\}$.

Definition 5.2 (*Mode assignment*) Let M be a set of modes, MT be set of message types. Then ma: $MT \rightarrow M$ is a function uniquely associating a mode with each message type, called *mode assignment*.

Table 5.4 shows a mode assignment ma_c for the control scenario, where 6 modes are associated with 19 message types. As we will show later, the number of modes influences the degree of restricted contention, with more modes permitting better medium usage. For this reason, we distinguish 4 kinds of safety modes instead of having only one such mode.

For mode-triggered traffic, virtual time regions called *mode regions* are formed (see Fig. 5.4). Mode regions are then substructured into scheduling units called *mode slots*.

Definition 5.3 (*Global time slotting*) Time is divided into an infinite set of consecutively numbered *super slots* $S = \{S_1, S_2, \ldots\}$ of equal length. Super slots contain one or more *mode regions*. All mode regions of a super slot are divided into a finite set $S = \{s_1, \ldots, s_n\}$ of consecutively numbered *mode slots*, which may have different lengths. Each mode slot s_j has one occurrence per super slot S_i, referred to as $s_{i,j}$.

Table 5.4 Mode-triggered scheduling—mode assignment ma_c

Functionalities	Message types	Mode
Tick sync	m_{tick}	*management*
Inverted pendulum 1	$m_{s\_1,1\ldots3}$, m_{a_1}	$safety_1$
Inverted pendulum 2	$m_{s\_2,1\ldots3}$, m_{a_2}	$safety_2$
Flow rate control A	$m_{s\_A,1\ldots2}$, m_{a_A}	$safety_A$
Flow rate control B	$m_{s\_B,1\ldots2}$, m_{a_B}	$safety_B$
Emergency shutdown	$m_{e,1\ldots4}$	*emergency*

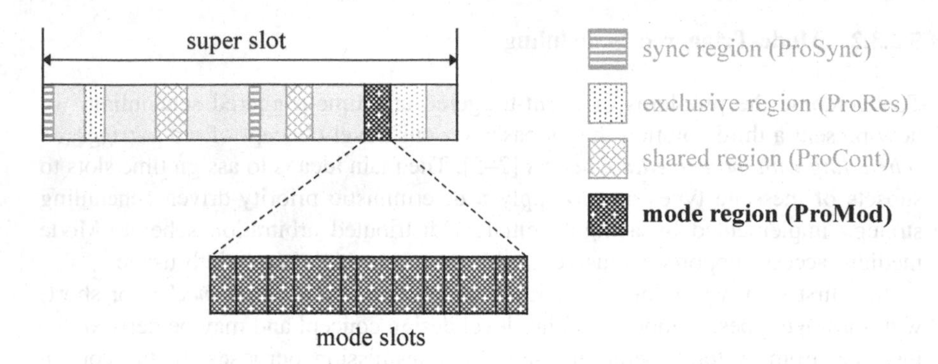

Fig. 5.4 Virtual time slotting—mode regions and mode slots

In the following, we assume that addressing of message instances is based on message types: A message type identifier is uniquely associated with one sending node and one or more receiving nodes.

For each mode $m_k \in M$, a mode slot $s_j \in S$ can be assigned to a message type $mt \in MT$ of that mode. This assignment implicitly applies to all mode slots $s_{i,j}$ of super slots $S_i \in$ S. We note that with this stipulation, a mode slot can be assigned to up to $n_{\text{mode}} = |M|$ message types.

Definition 5.4 (*Slot assignment*) Let M be a set of modes, S be a set of mode slots, MT be a set of message types, and ma be a mode assignment. Then sa: $M \times S \rightarrow_{\text{p}} MT$ is a partial function called *slot assignment*. We stipulate that sa respects mode assignment, i.e., $\forall m \in M, s \in S, mt \in MT. (sa(m, s) = mt \Rightarrow ma(mt) = m)$.

The function sa is partial, i.e., it is possible to have empty or incomplete slot assignments. In case of an empty assignment, a slot remains unused. If there is a single slot assignment, the slot is reserved exclusively for messages of the assigned message type. In case of multiple assignments, messages of the assigned message types have to contend for slot usage at runtime.

For mode-triggered scheduling, it is important that message types assigned to the same slot have different modes, which is captured by the constraint in Definition 5.4. This limits the number of message types per slot and thus the maximum degree of contention to $n_{\text{mode}} = |M|$. For finer distinctions, further modes can be introduced.

Table 5.5 shows a slot assignment sa_{c} for the control scenario (see Sect. 5.2.3.1) respecting mode assignment ma_{c} in Table 5.4. In most cases, slots are assigned to one message type only, i.e., exclusively. Exclusive assignments are appropriate in case of strictly periodic messages, where assigned slots are always used. In case of sporadic messages, or when messages are not strictly periodic, assigned slots may not always be needed. In the control scenario, this is the case during periods where sufficient quality of control is already achieved using maximal intervals. In this situation, it is appropriate to assign the same slot to several message types. In Table 5.5, some slots are assigned 3 or 4 times. This raises the question of how these slots are arbitrated in case of dynamic contention of messages. We note that different from the time-triggered solution with a need of 54 time slots per cycle (see Sect. 5.2.3.1), 32 slots per cycle are now sufficient.

For mode-triggered scheduling, we introduce, for each slot and mode, unique mode preferences, which determine the priority of messages that are scheduled for the same slot.

Definition 5.5 (*Mode preference assignment*) Let *M* be a set of modes, *S* be a set of mode slots, *ma* and *sa* be mode and slot assignments, respectively. Then *mpa*: $M \times S \rightarrow_p \mathbb{N}_0$ is a partial function called *mode preference assignment*, assigning, to each tuple of mode and slot for which *sa* is defined, a unique preference, with lower values representing higher preferences.

Table 5.6 shows a mode preference assignment mpa_c for the control scenario satisfying the constraints stated in Definition 5.5. Preferences are chosen such that message types get the highest preference 0 in case of exclusive mode slot assignment. Furthermore, message types associated with mode emergency always get the highest preference.

The possibility to assign mode preferences to message types on a per slot basis instead of having the same mode preference for all slots is a crucial and outstanding aspect of mode-triggered scheduling. Compared to static message type-based priority assignments as, for instance, in Controller Area Network (CAN) [6], this opens up a new dimension of scheduling. In Table 5.6, mode preferences 0, 1, and 2 are assigned to mode $safety_1$ in different slots. Thus, messages with this mode can compete with different priorities, which support temporal message urgencies. In slot s_{14}, up to 4 messages[1] may contend for medium access, with the message instance winning contention being uniquely determined by the mode preference assigned to the message type's mode in that slot.

In Table 5.6, we have assigned minimal consecutive integer values as mode preferences, starting with 0. As we will show later, this supports a more efficient medium arbitration, as compared to using arbitrary integer values.

Definition 5.6 (*Message instance assignment*) Let S be a set of super slots, *S* be a set of mode slots, and *M* be a set of modes. Furthermore, let *MI* be a set of *message instances* (*messages* for short) to be sent at runtime. Then *mia*: $\mathrm{S} \times S \times M \rightarrow_p MI$ is a partial function defining a *message instance assignment*.

A message instance assignment defines which messages are scheduled for which slot instance and mode. As it is not known offline which messages are to be sent at what point of execution, this assignment is determined dynamically. For instance, additional messages of an inverted pendulum may have to be scheduled in case of sporadic disturbances, to shorten the stabilization delay. Furthermore, messages

[1]Even if a message type is assigned to a mode and a slot, this does not imply that in every such slot, a message of this type is to be sent. Especially sporadic messages are only sent from time to time. In particular, in some occurrences of slot s_{14}, there may be no message to be sent.

Table 5.5 Mode-triggered scheduling—slot assignment sa_c

Mode	Slot																	
	s_1	s_2	s_3	s_4	s_5	s_6	s_7	s_8	s_9	s_{10}	s_{11}	s_{12}	s_{13}	s_{14}	s_{15}	s_{16}	s_{17}	s_{18}
management	m_{tick}																	
$safety_1$			$m_{s\_1,1}$	$m_{s\_1,2}$	$m_{s\_1,3}$		m_{a_1}							$m_{s\_1,1}$	$m_{s\_1,2}$	$m_{s\_1,3}$		m_{a_1}
$safety_2$								$m_{s\_2,1}$	$m_{s\_2,2}$	$m_{s\_2,3}$		m_{a_2}		$m_{s\_2,1}$	$m_{s\_2,2}$	$m_{s\_2,3}$		m_{a_2}
$safety_A$						$m_{s\_A,1}$					$m_{s\_A,2}$		m_{a_A}					
$safety_B$														$m_{s\_B,1}$	$m_{s\_B,2}$			m_{a_B}
emergency														m_{e_1}	m_{e_2}	m_{e_3}		m_{e_4}
Mode	Slot																	
	s_{19}	s_{20}	s_{21}	s_{22}	s_{23}	s_{24}	s_{25}	s_{26}	s_{27}	s_{28}	s_{29}	s_{30}	s_{31}	s_{32}	s_{33}	s_{34}	s_{35}	s_{36}
management	m_{tick}																	
$safety_1$			$m_{s\_1,1}$	$m_{s\_1,2}$	$m_{s\_1,3}$		m_{a_1}							$m_{s\_1,1}$	$m_{s\_1,2}$	$m_{s\_1,3}$		m_{a_1}
$safety_2$								$m_{s\_2,1}$	$m_{s\_2,2}$	$m_{s\_2,3}$		m_{a_2}		$m_{s\_2,1}$	$m_{s\_2,2}$	$m_{s\_2,3}$		m_{a_2}
$safety_A$														$m_{s\_A,1}$	$m_{s\_A,2}$			m_{a_A}
$safety_B$						$m_{s\_B,1}$					$m_{s\_B,2}$		m_{a_B}					
emergency														m_{e_1}	m_{e_2}	m_{e_3}		m_{e_4}

Table 5.6 Mode-triggered scheduling—mode preference assignment mpa_c

Mode	Slot																	
	s_1	s_2	s_3	s_4	s_5	s_6	s_7	s_8	s_9	s_{10}	s_{11}	s_{12}	s_{13}	s_{14}	s_{15}	s_{16}	s_{17}	s_{18}
management	0																	
safety$_1$			0	0	0		0							1	1	1		1
safety$_2$								0	0	0		0		2	2	2		2
safety$_A$						0					0		0					
safety$_B$														3	3			3
emergency														0	0	0		0
Mode	Slot																	
	s_{19}	s_{20}	s_{21}	s_{22}	s_{23}	s_{24}	s_{25}	s_{26}	s_{27}	s_{28}	s_{29}	s_{30}	s_{31}	s_{32}	s_{33}	s_{34}	s_{35}	s_{36}
management	0																	
safety$_1$			0	0	0		0							2	2	2		2
safety$_2$								0	0	0		0		1	1	1		1
safety$_A$														3	3			3
safety$_B$						0					0		0					
emergency														0	0	0		0

may have to be scheduled more than once, if they compete with other messages of higher preference in the assigned slot instance, or be discarded.

As a constraint, a (dynamic) message instance assignment has to respect the (static) slot assignment:

Definition 5.7 (*Consistency*) Let *sa* be a slot assignment, *mia* be a message instance assignment. Let mt: $MI \rightarrow MT$ be a function associating, with each message instance, its message type. The function *mia is consistent with sa* iff $\forall S \in \mathrm{S}, s \in S, m \in M, mi \in MI, mt \in MT. (mia(S, s, m) = mi \wedge mt(mi) = mt \Rightarrow sa(m, s) = mt)$.

With these preparations, we can now formally define mode-triggered scheduling:

Definition 5.8 (*Mode-triggered scheduling*) Let mia: $\mathrm{S} \times S \times M \rightarrow_{\mathrm{p}} MI$ be a message instance assignment, *mpa* be a mode preference assignment, and $tx \subseteq \mathrm{S} \times S \times MI$ be a relation defining whether a message is sent in a given slot instance. Then *mode-triggered scheduling* is the strategy where, for each slot instance, the scheduled message with the highest mode preference is sent. Formally: $\forall S \in \mathrm{S}, s \in S. \exists m \in M, mi \in MI. (mia(S, s, m) = mi \wedge \forall m' \in M, mi' \in MI. (m \neq m' \wedge mia(S, s, m') = mi' \Rightarrow mpa(m, s) < mpa(m', s)) \Rightarrow \mathrm{tx}(S, s, mi))$.

From the definitions, it follows that if there is dynamic contention among nodes, because two or more message instances are dynamically assigned to the same slot instance, the contention is deterministically resolved.

In the control scenario, for instance, sporadic emergency message instances $m_{e,1\ldots4}$ always win a possible contention due to their highest mode preference. If no such message instance is assigned, other assigned message instances may use the slot, in the order of mode preferences. Thus, mode-triggered scheduling improves bandwidth usage compared to time-triggered scheduling significantly.

Other than solutions 1 and 2 (see Sect. 5.2.3.2), mode-triggered scheduling with mode medium access efficiently realizes the requirements of the control scenario. Exclusive medium access is granted when necessary. Due to deterministic medium access, no collisions occur, therefore, communication is highly reliable. Furthermore, real-time traffic is delayed only within specified time bounds. Finally, compared to solution 2, no additional wireless communication system needs to be deployed, as the required bandwidth is reduced from 54 (exclusive) slots to 32 (mode) slots, i.e., by 40%.

5.2.3.3 Mode Medium Access with Fast Mode Signaling

By fast mode signaling, we refer to methods to propagate the mode of all message instances assigned to a slot instance that has the highest preference in that slot fast and reliably across the network. Fast means that at the beginning of the slot instance or with a very short delay, all nodes are aware whether they may or may not transmit an assigned message instance. Fast mode signaling is crucial for the efficient implementation of mode medium access.

Conceptually, fast mode signaling can be realized in different ways, related to the scope of modes. For instance, a mode can be valid globally, i.e., for a given slot instance, the same mode applies to all nodes (see, e.g., TTP/C [14]). It follows that a slot instance is used only if there is an assigned message instance associated with the valid mode. Global modes can be signaled, for instance, at the beginning of a super slot, or at the beginning of each mode slot. An advantage is that a message instance associated with the global mode can be sent without further contention. However, if for a given slot instance, no message instance of that mode is assigned, the slot instance remains unused. Therefore, global modes are inappropriate to realize mode-triggered scheduling.

Another approach of fast mode signaling is to reduce the scope of modes to individual nodes and single slot instances. For a given slot instance, each node determines the set of its locally assigned message instances. The current local mode is then defined to be the mode with the highest mode preference of the associated message types of that set or undefined if the set is empty. At the beginning of each slot instance, all nodes with defined local mode signal their mode preference, with the node signaling the highest preference winning contention. Here, it is important that mode signaling is very fast, and that it can be performed in a collision-protected way. Since modes and mode preferences of a slot are unique, the winning node is uniquely determined.

5.2.3.4 Implementation of Mode Medium Access

Mode medium access can be implemented by passive and active contention (see Sect. 5.1.2). For passive contention, each node can perform a deterministic backoff with a duration corresponding to its current local mode preference. To realize higher preferences, shorter backoffs are applied. Since mode preferences of a slot instance are unique, the node that signals the highest preference wins. For active contention, nodes can send busy tones of different length, associated with the current local mode preference. For higher preferences, longer busy tones are transmitted.

Contention by deterministic backoffs or busy tones is feasible in single-hop networks. For multi-hop networks, other mechanisms are required, and until recently, it has been an open question whether deterministic solutions for wireless networks actually exist. In Chap. 6, we will present a binary countdown protocol for deterministic arbitration in multi-hop wireless networks, which solves the problem. However, this kind of arbitration comes at some expense and therefore may not be sufficiently efficient for fast mode signaling.

Mode medium access has been implemented for single-hop scenarios in ProMod [4], a module of the ProNet 4.0 protocol stack [1], applying a passive contention scheme. The implementation is based on virtual time slotting with mode regions

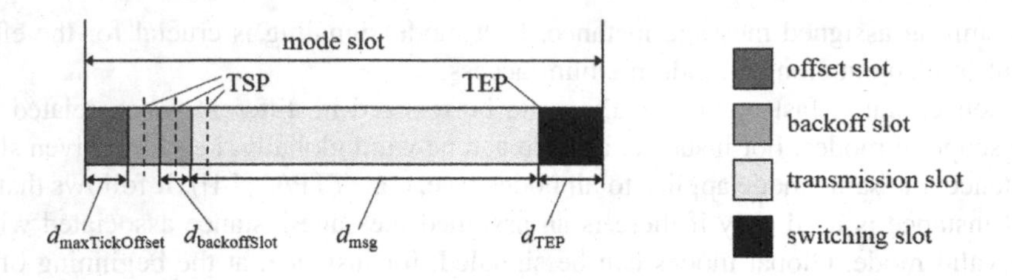

Fig. 5.5 Temporal structure of a mode slot with deterministic backoff

subdivided into mode slots, as shown in Fig. 5.4. In each mode slot, mode medium access with fast mode signaling is performed.

To achieve deterministic contention, the timing of mode medium access is crucial [4]. Figure 5.5 shows the temporal structure of a mode slot *s* supporting 3 mode preferences. To avoid overlap with a transmission in the previous mode slot, contending nodes, i.e., nodes with a message assigned to *s*, have to wait for the maximum tick offset $d_{\text{maxTickOffset}}$ (see Chap. 2) before starting arbitration. Each node then listens on the medium for $mpa(m, s) \geq 0$ backoff slots of length $d_{\text{backoffSlot}}$, where *m* is the current local mode associated with the assigned message instance, and *mpa* is the mode preference assignment. If the medium is idle for this listening period, the node wins contention and starts transmission at the transmission start point (TSP) following the end of the associated backoff slot. Transmission ends at transmission end point (TEP) to ensure that overlap with the next mode slot or slot region is avoided.

The structure of a backoff slot is determined by two parameters: the setting of TSP with a delay of d_{TSP} relative to the start of the backoff slot, and the backoff slot duration $d_{\text{backoffSlot}}$. Both parameters have to be chosen such that all contending nodes are able to associate the start of a message transmission with the same backoff slot. TSP has to consider the maximum tick offset $d_{\text{maxTickOffset}}$ and the delay d_{switch} of the sender to switch from receive mode to transmit mode:

$$d_{\text{TSP}} \geq \max(d_{\text{maxTickOffset}}, d_{\text{switch}}) \tag{5.1}$$

This is to ensure that transmission only starts after the corresponding backoff slot of a node with a slow clock has begun, i.e., after $d_{\text{maxTickOffset}}$. Furthermore, switching can only start at the beginning of the current backoff slot, yielding a minimum delay of d_{switch} before transmission can start.

The duration $d_{\text{backoffSlot}}$ is determined as follows:

$$d_{\text{backoffSlot}} \geq d_{\text{maxTickOffset}} + d_{\text{TSP}} + d_{\text{maxProp}} + d_{\text{maxCCA}} \tag{5.2}$$

This is to ensure that all nodes detect the start of transmission in the same backoff slot. The delay $d_{\text{maxTickOffset}}$ is taken into account once more to ensure that a node with a fast clock listens long enough. It then takes up to d_{TSP} before transmission starts, up to d_{maxProp} for it to arrive, and up to d_{maxCCA} to be detected.

After transmission ends, the sending node switches its transceiver from transmit mode back to receive mode. Furthermore, to avoid overlap with a transmission in the next mode slot, the transmission interval has to end at least $d_{\text{maxTickOffset}}$ before the mode slot ends. This is captured by the following constraint:

$$d_{\text{TEP}} \geq \max(d_{\text{maxTickOffset}}, d_{\text{switch}}) \quad (5.3)$$

For the implementation of mode medium access in ProMod, Imote2 motes [13] with the Chipcon CC2420 radio transceiver [7] have been used. Tick synchronization is performed by Black Burst Synchronization (BBS, see Chap. 2), implemented in ProSync of the ProNet 4.0 protocol stack [1]. Inserting corresponding values from data sheets and the analysis of BBS for a single-hop scenario with a resynchronization interval of 1 s into constraints (5.1) and (5.2), we obtain $d_{\text{TSP}} \geq 192$ μs and $d_{\text{backoffSlot}} \geq 529$ μs.

5.2.3.5 Experimental Results

In this chapter, we report on real-world experiments that were conducted to show feasibility and reliability of the implementation of mode medium access [4]. Figure 5.6 shows the topology of the experiment setup, with 4 nodes in single-hop range. Nodes v_1, v_2, and v_3 take the role of senders, and node v_s is synchronization master and records receptions from senders.

Figure 5.6 also shows the virtual time structure, with a super slot hosting a resync slot and a sampling slot with the same length of 1000 ms (see Chap. 3). In the middle of the super slot, a mode region consisting of one mode slot s is configured.

In the scenario, modes $M = \{regular, safety, emergency\}$ are defined. Three message types mt_1, mt_2, and mt_3 assigned to these modes in mode slot s are configured, with node v_i sending message instances of type mt_i. Furthermore, mode preferences in mode slot s are 0 (*emergency*), 1 (*safety*), and 2 (*regular*). To distinguish between these mode preferences, a mode slot contains 2 backoff slots (see Fig. 5.5). Message instances of mode *regular* are assigned in every mode slot, of type *safety* in every 4th mode slot, and of type *emergency* in every 8th mode slot.

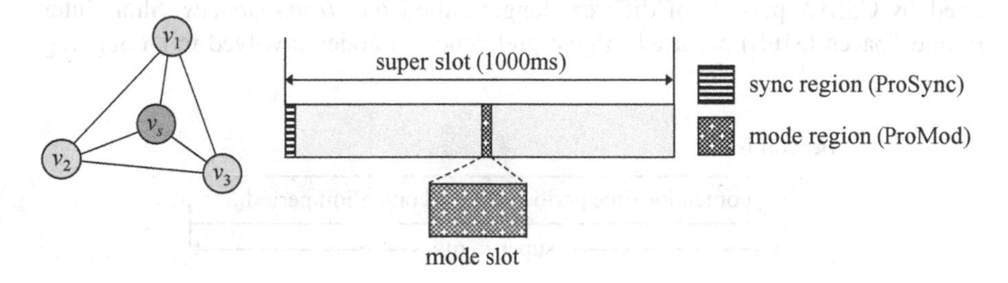

Fig. 5.6 Topology and virtual time structure of the experiment

With this strictly periodical traffic pattern, 2 (3) nodes contend in every 4th (8th) mode slot.

To assess feasibility and reliability of the implementation of mode medium access, experiments with a duration of 5,000 super slots each have been conducted. All experiments were carried out in a controlled environment, where the influence of other senders could be ruled out, i.e., with the single-network property (see Sect. 1.3) being satisfied. In all experiments and all mode slots, the node with the assigned message instance and associated highest mode preference has won contention, and the transmitted message instance was correctly received. This shows that the implementation of mode medium access is indeed feasible and highly reliable.

5.3 Related Work

In this chapter, we survey and compare medium access schemes used in wireless communication technologies. Our survey is not intended to be comprehensive, but to address some state-of-the-practice approaches, and to point out similarities and differences to medium access schemes of ProRes, ProCont, and ProMod.

5.3.1 IEEE 802.11—Wi-Fi

Wi-Fi (IEEE 802.11 [11]) is a widespread wireless technology incorporated by virtually all mobile devices, such as smartphones, notebooks, and tablets. In single-hop networks also called *cells*, medium access can be supervised by a base station called *point coordinator*. Here, time is structured into super frames started by a beacon frame of the point coordinator and decomposed into contention-free period and contention period (see Fig. 5.7). In ad hoc mode, i.e., without point coordinator, time is a continuous contention period.

Medium access is by contention, which is partially deterministic, partially random. Before starting a transmission, nodes are required to perform Carrier Sense Multiple Access (CSMA) and to reduce potential collisions. Contention is prioritized by CSMA periods of different length called *interframe spaces*. Short Inter Frame Spaces (SIFS) are used to give preference to nodes involved in an ongoing

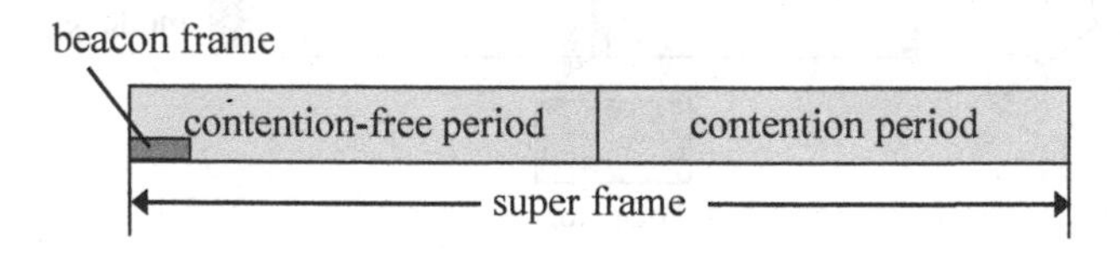

Fig. 5.7 Wi-Fi with point coordinator—time structure

frame sequence, e.g., RTS-CTS-DATA-ACK. The base station applies PIFS (Point Coordinator Function (PCF) Inter Frame Spaces), which are longer than SIFS. In both cases, contention is resolved deterministically, as at any point in time during execution, there is at most one contending node per interframe space and strict preference between contending nodes. This scheme has some similarities with fast mode signaling with priorities 0 (SIFS) and 1 (PIFS), as provided by ProMod. A difference is that nodes applying priority 0 are not predetermined, but have to win general contention first. Furthermore, the points in time when contention is performed are not determined as in ProMod.

In legacy Wi-Fi[2], the point coordinator contends for medium access during the contention free period. By applying PIFS, it gains exclusive medium access. It can then send data to nodes of the same cell, or poll them. Polled nodes apply SIFS to get exclusive medium access once. It is up to the point coordinator which nodes to poll in which order. Therefore, this form of exclusive access is not deterministic as in ProRes.

Shared medium access is provided via general contention, where DIFS (Distributed Coordinator Function (DCF) Inter Frame Spaces), which are longer than SIFS and PIFS, are used. Different from SIFS and DIFS, this does not resolve contention deterministically. Therefore, the CSMA period is extended by a random backoff interval controlled by a backoff counter. This further reduces the risk of collisions and, however, does not rule them out.[3]

Different from ProCont, IEEE 802.11 does not support frame priorities. Nodes losing general contention have to contend again. However, they keep their backoff counter value and continue countdown after the next CSMA period of length DIFS, which improves their chances of winning contention. In a variant of IEEE 802.11, frame priorities can be supported by configuring up to 8 traffic categories with different interframe spaces and contention window sizes. As in legacy IEEE 802.11, backoff counter values are kept after losing contention, which means that frame priorities are statistical.

5.3.2 ZigBee—IEEE 802.15.4

ZigBee [17] is a wireless technology for short distance communication in home and industrial automation, with low transmission rates and low energy consumption. It incorporates IEEE 802.15.4 [12] and adds routing, security, and application functionalities.

---

[2]In a variant of IEEE 802.11, the point coordinator can also contend for medium access during the contention period. This extension is referred to as Hybrid Coordination Function (HCF).

[3]The risk of collisions is further reduced by exchanging an RTS-CTS sequence before the actual data transmission. Since we are focusing on medium access schemes, we will not discuss this further.

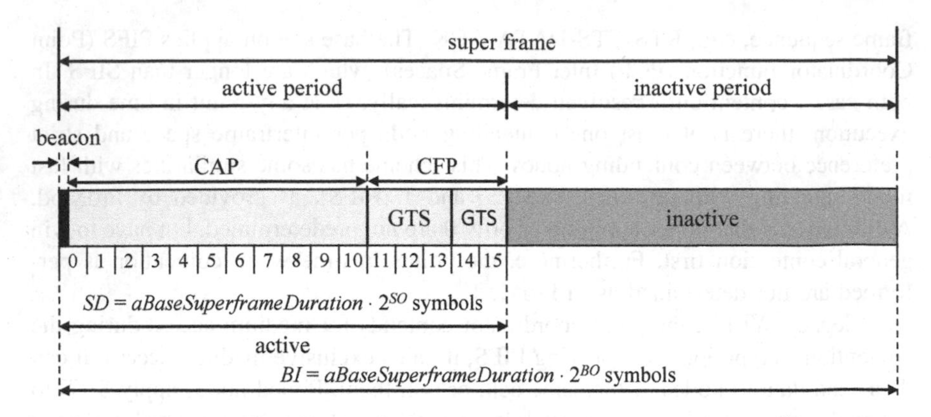

Fig. 5.8 ZigBee with PAN coordinator—time structure in beaconed mode (see [12], Fig. 8)

When operated in beaconed mode with a personal area network (PAN) coordinator synchronizing device nodes in single-hop distance via beacon frames, time is structured into super frames (see Fig. 5.8). Each super frame consists of active and inactive period. The active period is decomposed into 16 super frame slots of equal length, which are subdivided into dynamic intervals referred to as contention access period (CAP) and contention-free period (CFP).

In beaconed mode, ZigBee supports shared and exclusive medium access. Shared medium access is provided during the CAP and based on slotted CSMA/CA. To contend for the medium, a node locates the next backoff slot boundary. It then waits for a random number of backoff slots, without listening on the medium, which reduces energy consumption. If the medium is idle at the beginning of the two slots following this backoff interval, the node wins contention and starts transmission. Different from ProCont, no frame priorities are supported.

ZigBee provides exclusive medium access during the CFP, based on exclusively reserved slots called guaranteed time slots (GTSs). Reservations can be requested by device nodes during the CAP and are granted on a first-come-first-served basis by the PAN coordinator, subject to availability. Compared to ProRes and ProSlot, the time structure of CFPs and GTSs is quite inflexible, which strongly limits the number and placement of reserved slots. However, reservations can be handled dynamically.

5.3.3 WirelessHART

WirelessHART [8, 16] is a communication technology for wireless networking of intelligent field devices and applied in the production domain. It builds on IEEE 802.15.4 PHY [12] and adds functionality for medium access control, multi-channel operation, routing, end-to-end communication, and application support.

Communication in WirelessHART is based on global time slotting (see Sect. 3.3.2). First, time is physically structured into slots of 10 ms, which are identified by their Absolute Slot Number (ASN). Then, slots are composed into virtual time regions called super frames.

WirelessHART provides exclusive medium access, based on time-triggered scheduling. A centralized network manager handles time synchronization, channel management, scheduling, reservation, and route discovery. Time slots are exclusively assigned to nodes and used to transmit several kinds of unicast and broadcast frames. Exclusive medium access is based on reservations and therefore comparable to ProRes.

5.3.4 TTP/C

Time-Triggered Protocol Class C (*TTP/C*) [14] is a wired time-triggered field bus that explicitly defines and applies the concept of mode. As mode medium access, TTP/C supports dynamic mode changes; however, it lacks efficiency and flexibility.

TTP/C structures time into a sequence of cluster cycles, which are substructured into a fixed number of TDMA rounds consisting of a fixed number of slots. In a TDMA round, slots are exclusively assigned to nodes, with at most one slot assignment per node. It is possible that in different TDMA rounds, a slot is exclusively assigned to different nodes.

Modes in TTP/C are introduced as global system operating modes. Global means that the current mode applies to all nodes, which may only send messages associated with the current operating mode in assigned slots. As in mode-triggered scheduling with mode medium access, TTP/C supports the configuration of the set of modes and mode changes at runtime, which are requested by authorized nodes. Mode change requests are signaled in a dedicated control field of regular messages. Mode changes become effective only at the beginning of the TDMA round following their reception (deferred mode change) and are performed by all nodes. If more than one request is received during a TDMA round, the last request is granted.

Compared to mode medium access, the use of TTP/C modes lacks efficiency and flexibility. In TTP/C, mode changes are only performed at the beginning of TDMA rounds. During a TDMA round, the current mode applies to all nodes. Furthermore, slots are always assigned exclusively. Thus, if nodes have no message to send for the mode in place, slots remain unused. Furthermore, mode changes are very slow. In addition, the time structure of TTP/C is quite inflexible.

5.4 Conclusions

In this chapter, we have explained concepts and measures of medium access, have presented medium access schemes of ProNet 4.0, and have surveyed and compared medium access schemes used in state-of-the-practice wireless communication technologies. We have distinguished three basic types of medium access. Exclusive access means that at any point in time, only one node has the right of using the medium for transmission. With shared access, several nodes may access the medium simultaneously, with potential collisions. Mode access restricts shared access to defined subsets of nodes, renders contention deterministic, and thereby avoids collisions.

Among the medium access schemes supported by ProNet 4.0, we have focused on the novel mode medium access with mode-triggered scheduling. This way, a well-controlled amount of dynamic contention for time slots is permitted. We have introduced fast mode signaling to agree on the current transmission mode of highest preference extremely fast and reliably. Furthermore, we have outlined the implementation of mode medium access and have reported on the results of real experiments showing feasibility and reliability in a wireless network.

While mode medium access as presented in this chapter supports deterministic contention with a predictable winner, less rigid forms are perceivable, too. For instance, it is possible to assign, for a given mode slot, the same mode preference to more than one message type. In this case, nodes with assigned message instances associated with the same mode preference would have to contend, e.g., by performing an additional random backoff in their backoff slot. A similar scheme is used by Elimination by Sieving/Deadline Bursting with DCF (ES/DB-DCF) [15] in the elimination phase. Another option is to assign mode preference 0 uniquely to one message type per slot and to assign mode preference 1 collectively to several other message types in the same slot. This enforces exclusive access for assigned message instances associated with the highest mode preference and permits other message instances to contend randomly, which improves bandwidth usage if message instances of highest preference are not always assigned while preserving real-time capability.

While the focus of this book is on wireless communication technologies, we would like to point out that mode medium access is also applicable in wired technologies. In [2], we have presented a solution based on Controller Area Network (CAN) [6], with a working prototype for TTCAN [2, 4], a Time-Triggered Protocol in the automotive domain [10]. Solutions for FlexRay [9] are described in [3].

Literature

Chair for Networked Systems

1. Gotzhein R (2014) ProNet 4.0—a wireless real-time communication system for Industry 4.0. White Paper, Networked Systems Group, Department of Computer Science, University of Kaiserslautern. http://vs.informatik.uni-kl.de/publications/2014/Go14/whitePaperEN-ProNet4.0.pdf. Last accessed: Aug 27 2019
2. Braun T, Gotzhein R, Kuhn T (2014) Mode-based scheduling with fast mode-signaling—a method for efficient usage of network time slots. In: Proceedings of the 6th international conference on computer science and information technology (ICCSIT 2013), Paris, France, 20–21 Dec 2013. Journal of Advances in Computer Networks (JACN) 2(1):48–57
3. Gotzhein R, Kuhn T (2012) Zeit-und Prioritäts-gesteuerter Sende/Empfangsknoten, Deutsches Patent-und Markenamt, Az 10 2010 039 488, June 6, 2012 (date of granting)
4. Braun T (2016) Reliable mode-based communication and virtual prototyping in the development of distributed real-time systems, PhD Thesis (in German), Computer Science Department, University of Kaiserslautern
5. Braun T, Gotzhein R (2017) Zeit-und Prioritäts-gesteuerter Sende/Empfangsknoten—Technische Realisierungen mit FlexRay und LIN, Deutsches Patent-und Markenamt, Az 10 2012 200 475, Oct 17, 2013 (date of granting); European Patent Register, EP2614996 B1, Nov 11, 2017 (date of granting)

Further References

6. Etschberger K (2001) Controller area network—basics, protocols, chips, and applications. IXXAT Press
7. Chipcon AS (2019) CC2420 ZigBee-ready RF transceiver. http://www-inst.eecs.berkeley.edu/~cs150/Documents/CC2420.pdf. Last accessed: 27 Aug 2019
8. Chen D, Nixon M, Mok A (2010) WirelessHART™—real-time mesh network for industrial automation. Springer
9. FlexRay Consortium (2011) FlexRay communications system—protocol specification v3.0.1
10. Führer T, Müller B, Dieterle W, Hartwich F, Hugel R, Walther M (2000) Time triggered communication on CAN (Time Triggered CAN—TTCAN). In: 7th international CAN conference, Amsterdam
11. Institute of Electrical and Electronics Engineers (2012) IEEE Standard 802 part 11: wireless LAN medium access control (MAC) and physical layer (PHY) specifications. IEEE Computer Society, New York, NY, USA, Feb 2012
12. Institute of Electrical and Electronics Engineers (2011) IEEE Standard 802 part 15.4: low-rate wireless personal area networks (LR-WPANs). IEEE Computer Society, New York, USA, Sep 2011
13. MEMSIC Inc (2019) Imote 2 datasheet. https://vs.cs.uni-kl.de/downloads/Imote2NET_ED_Datasheet.pdf. Last accessed: 27 Aug 2019
14. Kopetz H, Grünsteidl G (1993) TTP—a time-triggered protocol for fault-tolerant real-time systems, Digest of papers, 23rd international symposium on fault-tolerant computing (FTCS-23), Toulouse, France, 22–24 June 1993, pp 524–533
15. Pal A, Dogan A, Özgüner F (2002) MAC layer protocols for real-time traffic in ad-hoc wireless networks. In: Proceedings of the IEEE international conference on parallel processing (ICPP'02), Washington DC, USA, 2002, pp 539–546

16. International Electrotechnical Commission (IEC) (2010) Industrial communication networks—wireless communication network and communication profiles—WirelessHART (IEC 62591 ed 1.0), Geneva, Switzerland, April 2010
17. ZigBee™ Alliance (2005) ZigBee specification, version 1.0, June 2005, 378p. www.zigbee.org. Last accessed 27 Aug 2019

Chapter 6
Deterministic Arbitration

Predictable behavior is of paramount importance in distributed real-time computer systems. In particular, this concerns predictability of timing and of arbitration procedures. In this chapter, we explain the foundations of deterministic arbitration in wireless ad hoc networks, present *Arbitration and Cooperative Transfer Protocol* (*ACTP*) [2, 3], survey and compare related work, and draw conclusions.

6.1 Foundations

In this chapter, we provide the context of deterministic arbitration in wireless ad hoc networks, explain concepts and measures, and address areas of operation.

6.1.1 Context

In distributed systems, nodes exchange messages to share data and to coordinate their behavior. To achieve predictable behavior, deterministic protocols are required. In Chap. 2, we have presented Black Burst Synchronization (BBS), our deterministic protocol for tick and time synchronization, with upper bounds for tick and time offset, and for convergence delay. In Chap. 5, we have addressed exclusive medium access and the novel mode medium access with an implementation that provides deterministic outcomes for medium arbitration in single-hop range.

In distributed real-time systems, there are further functionalities requiring deterministic arbitration:

- *Leader election*: A leader is needed for centralized coordination of activities, such as provision of a service registry, route discovery, and resource reservation.

R. Gotzhein, *Real-time Communication Protocols for Multi-hop Ad-hoc Networks*, Computer Communications and Networks,
https://doi.org/10.1007/978-3-030-33319-5_6

- *Consensus on a common data value*: Consensus is needed for agreement on the identity of a leader, on a global clock value, and on termination of a distributed algorithm.
- *Signaling of mode changes*: Real-time systems may run in different operating modes, with different settings and functionalities. For instance, Internet of Things (IoT) devices may be executed in operational mode, fail-safe mode, diagnosis mode, or configuration mode. To enable mode-driven operation, the current mode has to be signaled efficiently and reliably.

In this chapter, we identify and present protocol mechanisms that can be applied to tackle these problems in wireless ad hoc networks.

6.1.2 Concepts and Measures

For deterministic arbitration in wireless ad hoc networks, we distinguish the following general objectives:

- *n-hop deterministic arbitration*: Nodes contend with all nodes in their n-hop neighborhood in order to determine a unique winner. Special cases are $n = 2$, which solves the hidden station problem, or $n = n_{\text{maxHops}}$, where n_{maxHops} denotes the maximum network diameter, which means that contention is network-wide.
- *n-hop deterministic value transfer*: A single node communicates a data value to all nodes in n-hop neighborhood. For instance, a master node may propagate its clock value of the most recent reference tick to all other nodes, with n set to n_{maxHops}.
- *n-hop deterministic arbitration and value transfer*: This combines the previous objectives. For instance, a leader may be elected (arbitration) with its identity made public to all nodes participating in the election (value transfer).

In networks in general and in wireless ad hoc networks in particular, several problems make it difficult to achieve these objectives. Communication may be impeded by weak links and by destructive collisions due to interference of concurrent transmissions, particularly in wireless networks. Furthermore, if broadcasts are used, collisions may go undetected. Finally, the multi-hop nature of networks poses an additional challenge.

To resolve contention, passive and active contention schemes can be considered. Passive contention is performed by listening on the medium. In Sect. 5.2.3.4, we have presented an implementation of mode medium access, using a deterministic passive contention scheme for wireless ad hoc networks based on individual backoffs of unique length. While this is feasible in single-hop networks, it is unclear how it can be extended to multi-hop topologies.

Active contention is done by transmitting contention patterns. If several nodes in range contend, collisions are unavoidable. Therefore, for deterministic outcomes, it is mandatory that active contention schemes are collision-resistant. In the literature, two active schemes with this property have been proposed:

- *Busy tone protocols*: When the medium is idle or at predefined points in time, nodes contend by transmitting jamming signals of different length. After finishing transmission, nodes listen on the medium. If the medium is detected idle, meaning that the listening node has transmitted the longest jamming signal, the node wins contention; otherwise, it loses.
- *Binary countdown protocols* (dominance protocols): When the medium is idle or at predefined points in time, nodes contend by bitwise transmission of unique arbitration bit sequences of fixed length representing a priority. Bits are encoded as dominant and recessive signals, with the medium being monitored instead of transmitting recessive signals. If during an arbitration bit sequence, a node contends with a recessive signal, but monitors a dominant signal, it loses contention and resigns. At the end of the arbitration bit sequence, the node signaling the highest priority remains and wins contention.

As passive contention schemes, active schemes can be applied in single-hop networks. In the literature, there have been attempts to extend these schemes to multi-hop topologies. However, until recently, it has been an open question whether deterministic solutions for multi-hop wireless networks actually exist. In Sect. 6.2, we will present a solution.

6.1.3 Areas of Operation

Deterministic arbitration is an essential functionality of distributed real-time computer systems, enabling applications on user and system level.

6.1.3.1 *Leader Election*

A well-known and well-studied problem requiring deterministic arbitration is the leader election problem, where a node is to be uniquely chosen out of a set of competitors [7]. Leaders—also called coordinators—are required in a variety of applications. For instance, to manage the operation of a distributed system, a centralized scheduler could be elected. To determine routes with exclusive slot assignments, a node acting as routing and reservation manager could be chosen. In both cases, the election is among all nodes, i.e., network-wide.

There are also applications where elections occur in n-hop neighborhood, for instance, in case of clustering, where the objective is to divide a network into groups of nodes (see Chap. 9). Here, a set of leaders called cluster heads and their followers are chosen.

Many leader election algorithms assume communication to be reliable. However, when applying wireless technologies for ad hoc networks, this cannot be taken for granted. For instance, links may be weak and therefore lossy, and destructive collisions may occur, yielding unbounded election delays.

6.1.3.2 *Consensus*

Another well-known and well-studied problem requiring deterministic arbitration is the consensus problem, where agreement on a common data value is to be reached among a set of nodes [6, 9]. For instance, it may not be sufficient to elect an anonymous leader, but also be necessary to share its node address.

In Chap. 2, we have presented Black Burst Synchronization (BBS), a deterministic protocol for tick and time synchronization in wireless ad hoc networks. The exchange of clock values of the last reference point in time is another scenario requiring distributed agreement. For instance, the master node's clock value could be propagated across the network. Alternatively, nodes could agree on the clock value of the fastest clock, which, in addition, requires network-wide arbitration.

In Chap. 4, we have presented Automatic Topology Discovery Protocol (ATDP), which detects and disseminates communication, interference, and sensing topologies. As soon as all nodes have a consistent view, consensus on termination is to be reached.

6.1.3.3 *Quality of Service Routing*

In multi-hop networks, routing is required to provide end-to-end communication. In distributed real-time systems, routes have to satisfy specific quality of service (QoS) requirements regarding, for instance, performance and reliability. In [4], we have introduced Black Burst-based QoS routing (BBQR). The objective of BBQR is to discover routes satisfying a specified QoS requirement. In the process of route discovery, BBQR applies n-hop deterministic arbitration and value transfer several times, with different settings for the arbitration radius n.

- In Phase 1, BBQR selects a single node to continue its route search, thereby serializing route requests to avoid the problem of mutual blocking of resources. This is achieved by setting $n = n_{\mathrm{maxHops}}$. At the end of this phase, there is a unique winner, and all nodes know its node id.
- In Phase 2, BBQR detects feasible routes, i.e., routes satisfying the QoS requirement, with $n = 1$. At the end of this phase, all nodes have learned about their hop distance to the destination node.
- In Phase 3, BBQR selects one feasible route with minimal hop count, with $n = 2$ to resolve potential conflicts. Furthermore, nodes along the selected route learn about their predecessor.

- In Phase 4, BBQR informs nodes along the selected route about their successor by regular unicast frames.

BBQR has been devised for multi-hop wireless ad hoc networks and qualifies as a deterministic protocol. By using a collision-resistant active contention scheme, route discovery is reliable and predictable.

6.2 Arbitrating and Cooperative Transfer Protocol (ACTP)

In this chapter, we present *Arbitrating and Cooperative Transfer Protocol* (*ACTP*) [2, 3], a protocol for *n*-hop deterministic arbitration and value transfer in multi-hop ad hoc networks. ACTP is implemented in protocol module ProArb of the ProNet 4.0 protocol stack [1] and activated during dedicated virtual time regions.

6.2.1 Encoding of Bit Sequences with Black Bursts

To encode bit sequences used for arbitration and value transfer, ACTP uses the concept of black bursts [12] (see Sect. 2.2.2). A black burst is a period of transmission energy of defined length. Transmission of black bursts starts at locally determined points in time, without prior medium arbitration. If black bursts collide, a receiving node can still detect a period of energy on the medium, which renders collisions non-destructive provided certain timing constraints are observed.

Bit sequences are encoded with black bursts as follows: a logical 1 is encoded by transmitting a black burst of defined duration $d_{BB} > 0$ µs; a logical 0 is encoded by not transmitting a black burst, i.e., $d_{BB} = 0$ µs. If several nodes in range transmit a bit, a logical 1 will dominate a logical 0, which therefore is called recessive bit, while 1 is dominant. This feature can be used to encode a bitwise logical OR by sending bit sequences in an aligned way.

6.2.2 Time Structure of ACTP

The time structure of ACTP builds on global time slotting (see Chap. 3). Arbitration and value transfer occur in virtual time regions called *arbitrated regions*. As illustrated in Fig. 6.1, these regions are decomposed into sequences of *arbitrated slots*. Slot lengths depend on the required arbitration range (in hops), and on the size of values and data to be exchanged, and are therefore not fixed.

Arbitrated slots are further decomposed into a *bit sequence phase* for arbitration and/or value transfer and a possibly empty *data phase* for transmission of one or

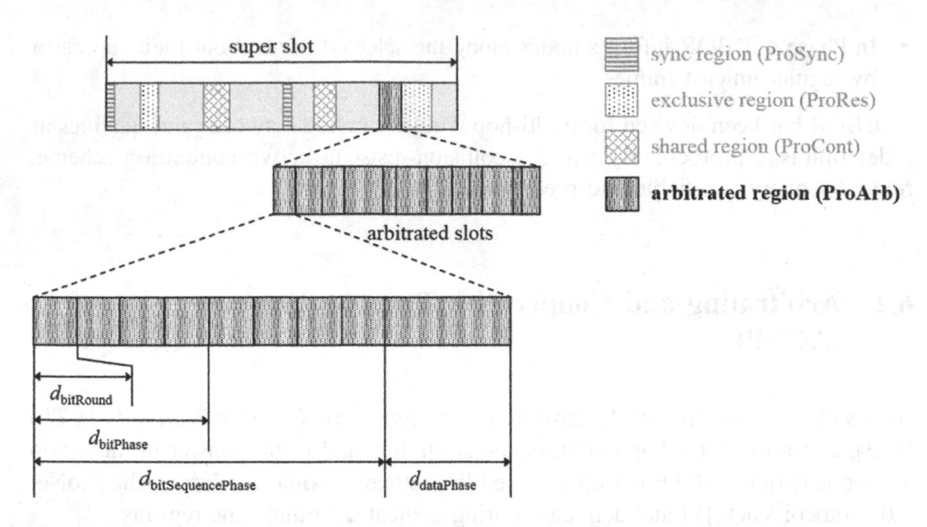

Fig. 6.1 Virtual time slotting—arbitrated region (for explanations, see also Sect. 3.2)

more regular frames. Depending on the arbitration range, this could be exploited to send a data frame over several hops, without further arbitration and free of collisions. A bit sequence phase consists of n_{bits} *bit phases*, where n_{bits} is the length of the arbitration bit sequence of that arbitrated slot. A bit phase is composed of n_{hops} *bit rounds*, where n_{hops} is the arbitration range. In Fig. 6.1, $n_{bits} = 2$, and $n_{hops} = 4$. If $d_{bitRound}$ is the (constant) duration of a bit round, we have $d_{bitPhase} = n_{hops} \cdot d_{bitRound}$ and $d_{bitSequencePhase} = n_{bits} \cdot d_{bitPhase}$ for the duration of bit phase and bit sequence phase, respectively.

6.2.3 Arbitration and Cooperative Transfer in Single-hop Networks

When performing ACTP in single-hop networks, i.e., $n_{hops} = n_{maxHops} = 1$, bit phases consist of a single bit round only. Since this simplifies the operation of ACTP, we start by considering the single-hop algorithm, which can be classified as a binary countdown protocol (see Sect. 6.1.2).

At the beginning of an arbitrated slot, all contending nodes are active; all other nodes are passive, i.e., they listen only.

- Active nodes transmit their arbitration bit sequence bit by bit as follows:
  - In (bit round 1 of) the current bit phase i, $1 \leq i \leq n_{bits}$, an active node "transmits" the ith bit of its arbitration bit sequence.
  - If an active node has transmitted a logical 1, i.e., a dominant bit, it remains active.

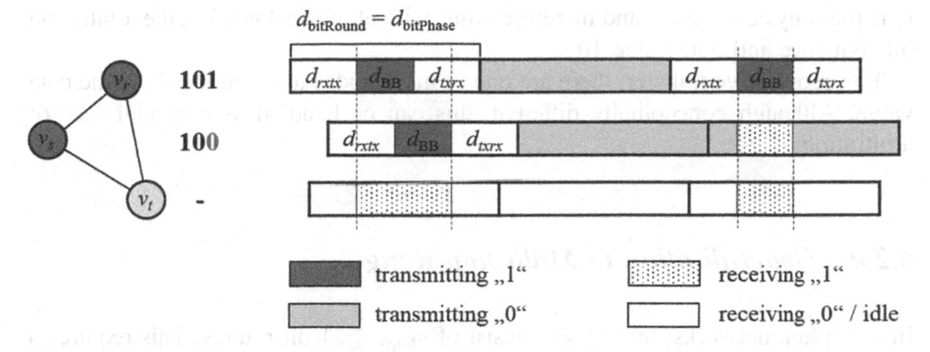

Fig. 6.2 Arbitration and cooperative transfer—single-hop scenario (see [14], Fig. 2)

  - If an active node has "transmitted" a logical 0, i.e., a recessive bit, it remains active only if it does not receive a logical 1 in this bit round. Otherwise, it becomes passive, i.e., it stops contending.

- Data values are recorded as follows:
  - If an active node has sent a logical 1 in (bit round 1 of) the current bit phase i, $1 \leq i \leq n_{\text{bits}}$, it records a logical 1 as the ith bit.
  - If a passive node has received a logical 1 in (bit round 1 of) the current bit phase i, it records a logical 1 as the ith bit.
  - In all other cases, a node records a logical 0 as the ith bit.

After n_{bits} bit phases, only one node is still active and therefore winner if the initial set of contending nodes has not been empty, and if all bit sequences have been unique. Furthermore, all nodes know the bit sequence of the winner. Thereby, 1-hop arbitration and transfer of a data value as given by the arbitration bit sequence are achieved. In addition, the winner can send a regular data frame in the data phase if configured.

Figure 6.2 illustrates the single-hop case of ACTP. Here, nodes v_r, v_s, and v_t form a single-hop network. Nodes v_r and v_s are contending with arbitration bit sequences 101 and 100 of length $n_{\text{bits}} = 3$; node v_t is not contending, but listening in order to learn about the data value of the winner.

In the first and only bit round of bit phase 1, nodes v_r and v_s transmit a logical 1, encoded as black burst of duration d_{BB}. Before doing so, they switch their transceivers from receive (*rx*) mode to transmit (*tx*) mode; afterward, they switch back. Figure 6.2 also shows that there is a tick offset between nodes, leading to slightly shifted arbitrated slots of participating nodes. Consequently, node v_t observes an extended energy interval when receiving the logical 1. In bit round 1 of bit phase 2, both v_r and v_s transmit a logical 0. However, when listening on the medium, they do not receive a logical 1; therefore, they both remain active. Finally, in bit round 1 of bit phase 3, nodes v_r and v_s transmit 1 and 0, respectively. As node v_s receives a logical 1, it loses and becomes passive. Thus, at the end of the bit sequence phase,

v_r is the only active node and therefore wins, with all nodes knowing the arbitration bit sequence and data value 101.

For cooperative transfer, there are one or more nodes transmitting the same data value. Although conceptually different, this can be handled as a special case of arbitration.

6.2.4 Generalization to Multi-hop Range

In multi-hop networks, bit phases consist of $n_{\text{hops}} \geq 1$ bit rounds. This requires a generalization of the operation of ACTP, i.e., an extension of the binary countdown protocol approach. As in the single-hop case, all contending nodes are active at the beginning of an arbitrated slot; all other nodes are passive. In addition, passive nodes act as repeaters.

- Active nodes transmit their arbitration bit sequence bit by bit as follows:
  - In bit round 1 of the current bit phase i, $1 \leq i \leq n_{\text{bits}}$, an active node "transmits" the ith bit of its arbitration bit sequence.
  - If an active node has transmitted a logical 1 in bit round 1, it remains active and silent throughout the remaining bit rounds of this bit phase.
  - If an active node has "transmitted" a logical 0 in bit round 1, it remains active throughout this bit phase only if it does not receive a logical 1 in some bit round j, $1 \leq j \leq n_{\text{hops}}$. Otherwise, it becomes passive in the bit round in which it receives a logical 1 and immediately starts acting as repeater.
- Passive nodes act as repeaters, by forwarding dominant bits:
  - If a passive node receives a logical 1 in bit round j for the first time in this bit phase, $1 \leq j < n_{\text{hops}}$, it transmits a logical 1 in bit round $j + 1$. This especially holds for nodes that have just become passive in bit round j.
  - If a passive node has repeated a logical 1 in bit round $j < n_{\text{hops}}$, it stops repeating for the rest of this bit phase.
- Data values are recorded as follows:
  - If an active node has sent a logical 1 in bit round 1 of the current bit phase i, $1 \leq i \leq n_{\text{bits}}$, it records a logical 1 as the ith bit.
  - If a node has received a logical 1 in some bit round j, $1 \leq j \leq n_{\text{hops}}$, of bit phase i, $1 \leq i \leq n_{\text{bits}}$, it records a logical 1 as the ith bit.
  - In all other cases, a node records a logical 0 as the ith bit.

Figure 6.3 illustrates the multi-hop bit sequence phase of ACTP. Here, nodes v_r, v_s, v_t, v_u, and v_m form a multi-hop network, with the topology shown in the figure. At the beginning of the bit sequence phase, all nodes are active and contend with unique bit sequences of length $n_{\text{bits}} = 4$, with $n_{\text{hops}} = n_{\text{maxHops}} = 3$. These settings also determine the time structure of the bit sequence phase.

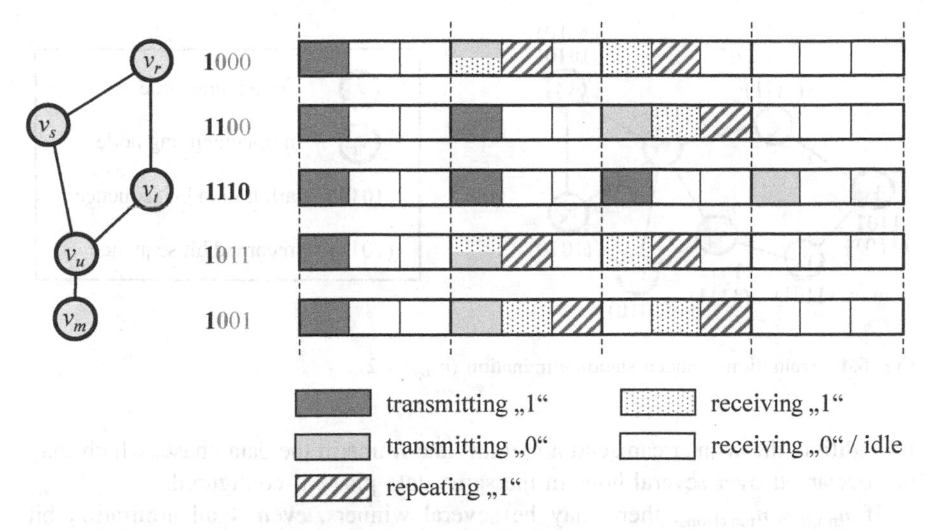

Fig. 6.3 Arbitration and cooperative transfer—multi-hop scenario

- Bit phase 1: In the first bit round, all nodes are active and transmit the first bit of their arbitration bit sequence, which is a logical 1. Therefore, they record a 1 as the first bit and remain silent for the rest of bit phase 1.
- Bit phase 2: In the first bit round, all nodes are still active and transmit their second bit. Here, only v_s and v_t send a logical 1, while all other nodes "transmit" a logical 0. As v_r and v_u are in range of v_s and v_t, they receive a logical 1 in bit round 2, and therefore, become passive taking immediate effect. For this reason, they repeat the logical 1 in bit round 2 of bit phase 2. Node v_m has not received a logical 1 in bit round 1 yet, and therefore, is still active at the beginning of bit round 2. However, in bit round 2, it receives a logical 1 repeated by v_u, and therefore, turns passive, repeating this logical 1 in bit round 3.
- Bit phase 3: In the first bit round, only nodes v_s and v_t are still active and transmit their third bit. The logical 1 of v_t is received by v_r and v_u and repeated in bit round 2. Here, v_s receives it and becomes passive, while the receiver v_m has already been passive before. In bit round 3, the logical 1 is repeated by v_s and v_m.
- Bit phase 4: In the first bit round, node v_t is the only remaining active node. By transmitting a logical 0, it finishes transmission of its arbitration bit sequence. Since v_t does not receive a logical 1 during bit phase 4, it remains active and wins arbitration. By then, all nodes have recorded the complete winner arbitration bit sequence and data value, which is 1110.

After n_{bits} bit phases, exactly one node is still active and therefore winner if the initial set of contending nodes has not been empty, if $n_{\text{hops}} = n_{\text{maxHops}}$, and if all arbitration bit sequences have been unique. Furthermore, all nodes know the arbitration bit sequence of the winner. Thereby, n_{maxHops}-hop arbitration and transfer of a data value are achieved. This is illustrated by the example in Fig. 6.3.

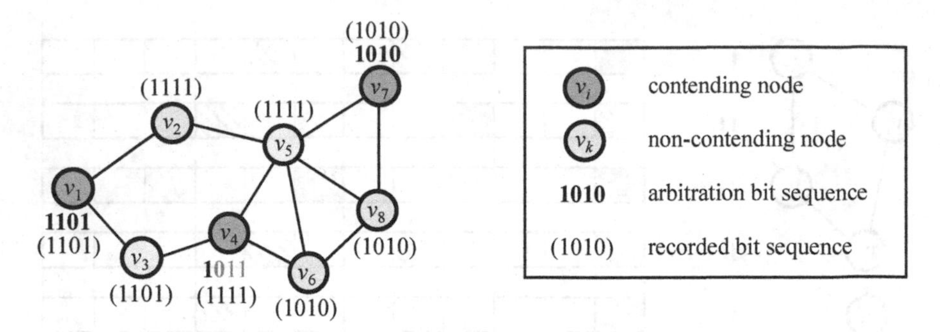

Fig. 6.4 Arbitration—hidden station elimination ($n_{\text{hops}} = 2$)

In addition, the winner can send a regular data frame in the data phase, which may be forwarded over several hops in the same data phase if configured.

If $n_{\text{hops}} < n_{\text{maxHops}}$, there may be several winners, even if all arbitration bit sequences have been unique. Furthermore, it can happen that nodes in $n \leq n_{\text{hops}}$ hop distance of a winner v have contended with a bit sequence representing a higher priority, however, have lost because there is another contending node v' in their n_{hops} range dominating their arbitration bit sequence, which is not in n_{hops} range of v. In such a case, these nodes record the logical OR of all arbitrating bit sequences of winners in their n_{hops} range. Thus, while arbitration still works as specified, value transfer is impaired.

Figure 6.4 illustrates this behavior for $n_{\text{hops}} = 2$, which is a suitable setting for hidden station elimination. In the scenario, nodes v_1, v_4, and v_7 contend with arbitration bit sequences 1101, 1011, and 1010, respectively. Node v_4 loses in bit phase 2 when receiving a logical 1 from repeating node v_3 in bit round 2 and becomes passive, acting as repeater. Since $n_{\text{hops}} = 2$, node v_4, which is in 2-hop distance of active nodes v_1 and v_7, only records bit values received in bit round 2 of bit phases 2–4, but does not repeat them. At the end of the bit sequence phase, nodes v_1 and v_7 remain as winners. Here, node v_7 wins although v_4 has a dominating arbitration bit sequence, which, however, is in turn dominated by the arbitration bit sequence of v_1. Nodes in two-hop range of either v_1 or v_7 record the corresponding arbitration bit sequence 1101 or 1010. Nodes in two-hop range of both v_1 and v_7 record the logical OR, i.e., 1111, as shown in the figure.

For cooperative transfer, there are one or more nodes transmitting the same data value. Although conceptually different, this can be handled as a special case of arbitration, with only one active node. Alternatively, the entire bit sequence could be propagated hop by hop, which requires a different time structure of an arbitrated slot. Here, a bit sequence phase is decomposed into n_{hops} bit sequence rounds, each consisting of n_{bits} bit rounds.

6.2.5 *Abstract Analysis of ACTP*

Performance and complexity of ACTP can be analyzed based on configuration and hardware parameters. For a given hardware platform, concrete values can be determined (see Sect. 6.2.5).

To determine arbitration delay d_{arbPhase}, we have to analyze the minimal duration d_{bitRound} of a single bit round. Together with the parameters n_{hops} and n_{bits} of an arbitrated slot, we obtain

$$d_{\text{arbPhase}} = d_{\text{bitSequencePhase}} = n_{\text{bits}} \cdot d_{\text{bitPhase}}, \text{ with } d_{\text{bitPhase}} = n_{\text{hops}} \cdot d_{\text{bitRound}} \quad (6.1)$$

To determine d_{bitRound}, we abstract from propagation delay, delays caused by internal hardware commands, and processing delays, and derive the following constraints:

$$d_{bitRound} \geq d_{\text{BB}} + 2 \cdot d_{\text{switch}} \quad (6.2)$$

This constraint results from the observation that a node sending a logical 1 must be able to complete this within one bit round, which consists of the transmission of a black burst and of switching delays.

$$d_{\text{BB}} \geq d_{\text{maxCCA}} \quad (6.3)$$

A black burst has to be long enough such that nodes in range can detect its energy with the clear channel assessment (CCA) mechanism.

$$d_{\text{TSP}} \geq \max(d_{\text{maxTickOffset}}, d_{\text{switch}}) \quad (6.4)$$

The transmission start point (TSP) of a bit round has to be placed at least d_{TSP} after the beginning of a bit round such that all nodes locate the start of black burst transmission in the same bit round (parameter $d_{\text{maxTickOffset}}$), and the transmitting node has enough time to switch from receive to transmit mode (parameter d_{switch}).

$$d_{\text{TEP}} \geq \max(d_{\text{switch}}, d_{\text{maxCCA}} + d_{\text{maxTickOffset}}) \quad (6.5)$$

The transmission end point (TEP) has to be placed at least d_{TEP} before the end of a bit round, to ensure that all nodes locate the end of black burst transmission in the same bit round (parameter d_{maxCCA}), taking the maximum tick offset into account (parameter $d_{\text{maxTickOffset}}$), and the transmitting node has enough time to switch back to receive mode (parameter d_{switch}).

$$d_{\text{bitRound}} \geq d_{\text{TSP}} + d_{\text{BB}} + d_{\text{TEP}} \quad (6.6)$$

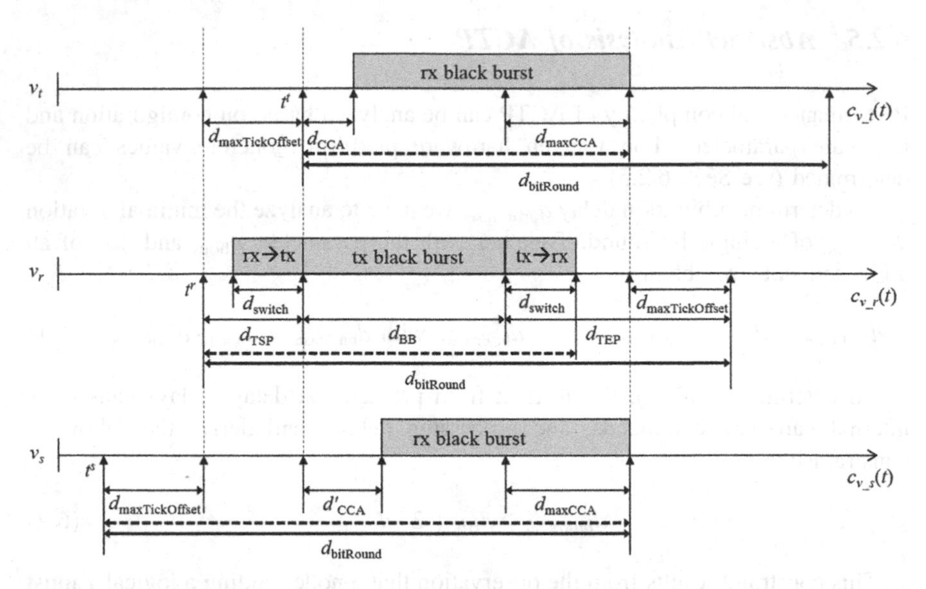

Fig. 6.5 Arbitration and cooperative transfer—timing of bit rounds

A bit round has to be configured such that the constraints (6.2) to (6.5) are considered. This is captured by constraint (6.6). Since $d_{TEP} \geq d_{TSP} \geq d_{switch}$, this also covers constraint (6.2).

Figure 6.5 illustrates the timing constraints of bit rounds. Node v_r is sending a black burst, which is received by nodes v_t and v_s in range of v_r. The local start time of the current bit round of v_r is at t^r. Node v_t (v_s) has the slowest (fastest) clock; therefore, the local start time of this bit round of v_t (v_s) is shifted forward (backward) by $d_{maxTickOffset}$, yielding a local start time of t^t (t^s).

As argued before, the sender v_r has to wait at least $d_{maxTickOffset}$ before starting its transmission. However, during this time span, it can already start switching to transmit (*tx*) mode. In Fig. 6.5, $d_{switch} \leq d_{maxTickOffset}$, therefore, $d_{TSP} = d_{maxTickOffset}$ (see constraint 6.4). This is followed by the transmission of a black burst and switching back to receive (*rx*) mode.

Nodes v_s and v_t start detecting the black burst with delays d_{CCA} and d'_{CCA}, respectively, during the same bit round. Black burst reception ends at most d_{maxCCA} after v_r has finished its transmission. In Fig. 6.4, $d_{switch} \leq d_{maxCCA} + d_{maxTickOffset}$, therefore, $d_{TEP} = d_{maxCCA} + d_{maxTickOffset}$ (see constraint 6.5).

The thick dashed arrows indicate, for each node, the minimal bit round duration. As it turns out, this duration is smaller for node v_t, which has the slowest clock, longer for node v_s with the fastest clock, and in-between for the sender v_r. To cover the worst case, the bit round duration required by v_s is the duration to be considered, as expressed in constraint (6.6) and indicated by the thick arrows.

If a data phase is configured in the arbitrating slot, its duration $d_{\text{dataPhase}}$ has to take n_{hops} (number of hops), d_{frame} (frame transmission duration), and $d_{\text{maxTickOffset}}$ into account. As duration of an arbitrating slot, we get $d_{\text{arbSlot}} = d_{\text{bitSequencePhase}} + d_{\text{dataPhase}}$.

With $d_{\text{maxTickOffset}}$, d_{maxCCA}, and d_{switch} being constant, time complexity of ACTP is in the order of $n_{\text{hops}} \cdot n_{\text{bits}}$, and therefore, independent of the number of nodes. For communication complexity, we have an order of $n_{\text{nodes}} \cdot n_{\text{bits}}$. Space complexity is constant, and there is no structural complexity, as ACTP is topology-independent and therefore robust against node movements and node failure.

6.2.6 Concrete Analysis of ACTP

In this chapter, we extend the abstract analysis of ACTP, based on a real hardware platform, by inserting parameters from data sheets and results of the concrete analysis of tick synchronization (see Chap. 2).

To analyze ACTP, we have chosen the Chipcon CC2420 radio transceiver [5]. Furthermore, we use results of the concrete analysis of Black Burst Synchronization (BBS) from Chap. 2.

Table 6.1 shows the results of the evaluation of the arbitration performance of ACTP, for several parameter settings. We consider topologies with a maximum network diameter n_{maxHops} of 1, 4, and 10 hops and resynchronization intervals d_{resInt} of 1 s and 5 s, yielding maximum tick offsets $d_{\text{maxTickOffset}}$ as shown in the table. In our implementation, d_{BB} is 160 μs, and values for d_{switch} and d_{maxCCA} are taken from the CC2420 data sheet.

Table 6.1 Arbitration performance of ACTP

n_{maxHops}	1	4			10		
d_{resInt} (s)	1	1			5		
$d_{\text{maxTickOffset}}$ (μs)	208	592			1680		
d_{switch} (μs)	192						
d_{BB} (μs)	160						
d_{maxCCA} (μs)	128						
d_{TSP} (μs)	208	592			1680		
d_{TEP} (μs)	336	720			1808		
d_{bitRound} (ms)	0.70	1.47			3.65		
n_{hops}	1	1	2	4	1	2	10
d_{bitPhase} (ms)	0.70	1.47	2.94	5.89	3.65	7.30	36.48
n_{bits}	4						
$d_{\text{bitSequencePhase}}$ (ms)	2.82	5.89	11.78	23.55	14.59	29.18	145.92

In a single-hop network, minimum bit round duration d_{bitRound} is 0.70 ms (see constraint 6.6). This value increases in topologies with higher diameter, due to higher maximal offsets. For instance, in a 10-hop network, d_{bitRound} is 3.65 ms. For bit phase duration d_{bitPhase}, we consider n_{maxHops}-hop arbitration as well as 1-hop and 2-hop arbitration, where applicable. In all settings, n_{bits} is 4, yielding the bit sequence phase durations $d_{\text{bitSequencePhase}}$ shown in the table.

In the settings listed in Table 6.1, bit sequence phase durations are in the order of milliseconds, i.e., relatively high. These durations even increase when using longer arbitration bit sequences. This leads us to the conclusion that this overhead should in general not be spent for a single frame transmission, but rather in applications that require deterministic arbitration less frequently, such as leader election and consensus, or to transmit several frames without further arbitration. In Chap. 4, we have presented Automatic Topology Detection Protocol (ATDP), which uses ACTP to reach deterministic consensus on termination, a decision to be taken only once.

6.2.7 Experimental Assessment of ACTP

In this chapter, we report on real-world experiments that were conducted to show feasibility and reliability of ACTP [2, 3]. Experiments were performed in testbeds of Imote2 motes [8], using the CC2420 radio transceiver [5].

Feasibility Experiments

To show feasibility, ACTP was implemented and executed in a representative small-scale topology that covers scenarios found in larger networks [2]. Due to weather conditions, experiments were conducted indoors, with other networks operating nearby. Nevertheless, operation of ACTP has been correct and stable.

Figure 6.6 shows topology and virtual slot timing of the experiments. Time is divided into super slots with a duration of 1 s, which start with a sync region. In each super slot, ten arbitrated regions are placed, consisting of a single arbitrated slot of 20 ms duration. Black Burst Synchronization (BBS, see Chap. 2) establishes tick synchronization, with a worst-case offset $d_{\text{maxTickOffset}}$ of 336 μs.

The topology consists of four nodes including the timing master v_m, with a maximum network diameter n_{maxHops} of 2. Arbitration range n_{hops} is set to n_{maxHops},

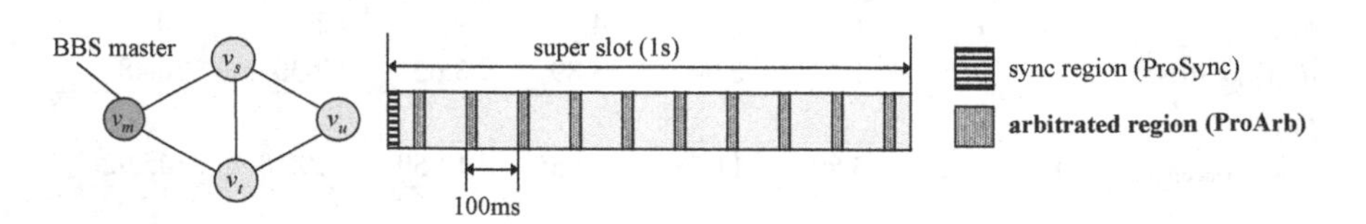

Fig. 6.6 Feasibility of ACTP—topology and virtual slot timing

Table 6.2 Assignment of arbitration bit sequences (in decimal notation) (see [3], Table 5.3)

	Slot 1	Slot 2	Slot 3	Slot 4	Slot 5	Slot 6	Slot 7	Slot 8	Slot 9	Slot 10
v_m	**213**	156	76	103	**178**	45	91	57	27	30
v_s	22	**189**	99	71	110	**155**	209	59	**219**	73
v_t	187	188	**155**	66	44	79	**210**	23	51	**201**
v_u	89	14	103	**245**	3	100	104	**198**	113	33

and arbitration bit sequences have length $n_{\text{bits}} = 8$. For each arbitrated slot, unique arbitration bit sequences are assigned to nodes such that each node wins 2 or 3 arbitrations per super slot (see Table 6.2).

The results of the experiments show feasibility of ACTP. In particular, the results of contention, conversion of active nodes to passive nodes and operation as repeaters, collision-resistant transmission of black bursts, and timing behavior were as expected. This also provides some evidence that the timing constraints have been correctly derived.

Reliability experiments

To assess reliability of ACTP, experiments were conducted over a long period of time, to determine success rates and false observation rates. With the same settings and topology as in the feasibility experiments (see Fig. 6.6), experiments continued for 5 h, i.e., for 18,000 super slots, with a total of 180,000 arbitrations.

Figure 6.7 shows node-specific and overall success rates. An arbitration counts as node-specific success if the node wins arbitration and is the intended winner, or if the node loses arbitration and receives the arbitration bit sequence of the intended winner correctly. An arbitration counts as overall success if it is a node-specific success for all nodes, which implies that there is exactly one winner. Although there were other networks operating nearby, success rates were very high (>99.9%). The lowest overall success rates were observed in arbitrated slots 1 and 5, where 13 out

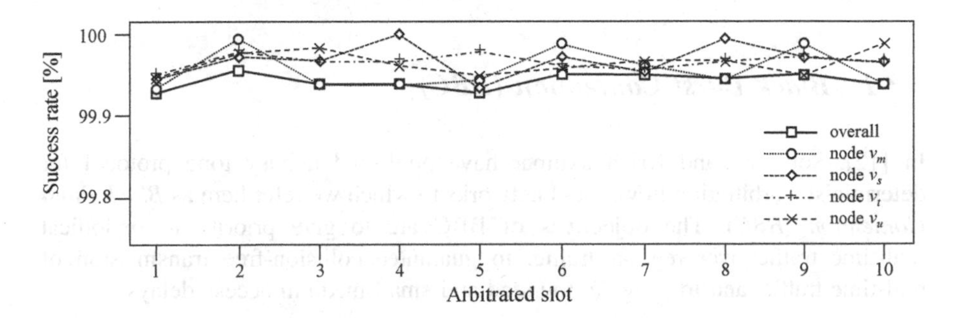

Fig. 6.7 Reliability of ACTP—success rates per node and overall (see [2], Fig. 10)

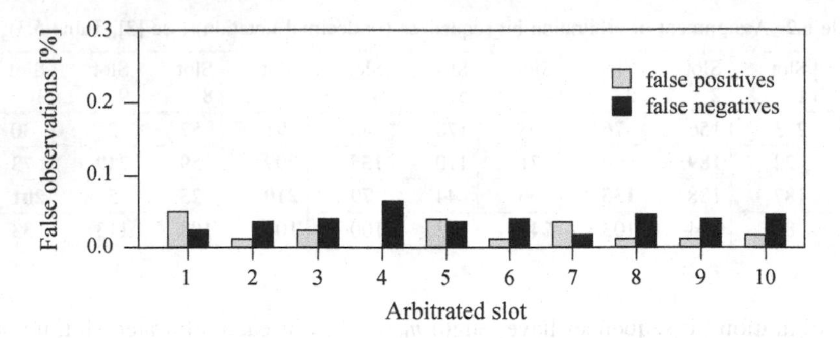

Fig. 6.8 Reliability of ACTP—false positives and false negatives (see [2], Fig. 11)

of 18,000 arbitrations (~0.07%) fail. Node-specific success rates exhibit no degradation between nodes v_s and v_t in the center of the topology and nodes v_m and v_u at the boundary, which shows that ACTP enforces the specified arbitration range of $n_{\text{hops}} = 2$.

To further investigate unsuccessful arbitrations, the role of false positives, i.e., the faulty detection of a logical 1, and false negatives, i.e., the faulty detection of a logical 0 during an arbitration bit sequence was investigated (see Fig. 6.8). In 37 out of 180,000 arbitrations (~0.02%), false positives have led to wrong or multiple winners; in 67 arbitrations (~0.037%), false negatives have caused wrong recordings of the arbitration bit sequence. A straightforward explanation is the interference of other networks operating nearby. In this particular case, problems could be traced back to interfering Bluetooth and Wi-Fi networks.

6.3 Related Work

In this chapter, we survey and compare deterministic arbitration protocols devised for wireless communication technologies. Our survey is not intended to be comprehensive, but to identify related protocols, and to point out differences to ACTP.

6.3.1 Black Burst Contention (BBC)

In [12], Sobrinho and Krishnakumar have published a busy tone protocol for deterministic arbitration in wireless networks to which we refer here as *Black Burst Contention* (*BBC*). The objectives of BBC are to give priority to periodical real-time traffic over regular traffic, to guarantee collision-free transmission of real-time traffic, and to provide bounded and small medium access delays.

The communication model of BBC distinguishes between communication, interference, and sensing links (see Chap. 4):

- Node v_j is in communication range of v_i if v_j can successfully receive from v_i.
- Node v_j is in interference range of v_i if v_i can interfere with a reception at v_j.
- Node v_j is in sensing range of v_i if v_j can detect whether v_i is transmitting.

It is assumed that if a node is in communication (interference) range, it is also in interference (sensing) range. Though not explicitly stated, BBC builds on the validity of the single-network property (see Sect. 1.3). Furthermore, it requires that there are no hidden stations: If a node v_i is in communication range of v_j, and v_k is in interference range of v_j, then v_k is in sensing range of v_i. This implies that carrier sensing before sending is sufficient to (almost) avoid destructive frame collisions. Due to the active contention scheme of BBC, this implication holds despite switching delay, propagation delay, and sensing delay if timing constraints are observed. Figure 4.1 in Chap. 4 shows a multi-hop topology that satisfies the assumption.

For the busy tone algorithm in [12], the authors further strengthen the "no hidden station" requirement by considering only topologies where all nodes are in mutual sensing range.[1] Furthermore, they classify nodes into regular nodes using a random backoff contention scheme and real-time nodes using the busy tone contention scheme. Busy tones—called black bursts—are composed of an integral number of black slots. BBC guarantees that black bursts sent by contending real-time nodes differ in length by at least one black slot, which yields unique winners.

Real-time nodes execute real-time sessions, which are long time intervals where a node requires periodical medium access to transmit real-time frames. To start a real-time session, a real-time node has to contend by random backoff, i.e., just like regular nodes. Once it has successfully transferred the first real-time frame, the rules of busy tone contention apply.

During a real-time session, real-time nodes start contending for medium access after the medium has been found idle for an interframe spacing of duration d_{med}, thereby gaining preference over regular nodes, which wait for $d_{long} > d_{med}$. After d_{med}, a contending real-time node sends a black burst. The length d_{BB} of this black burst is a function f_{BB} of the current contention delay d_{cont}, i.e., the length of the time interval since the real-time frame was scheduled until the start of contention. BBC ensures that contending nodes use black bursts of different length.

After a contending real-time node finishes its black burst transmission, it listens on the medium. If the medium is busy, i.e., another node is sending a longer black burst, the node loses contention. The node sending the longest black burst is the only node that senses the medium as idle. It wins contention and sends its real-time frame. Without going into further details, we note that timing in the contention phase is very critical.

[1] Otherwise, bounded access delays cannot be assured.

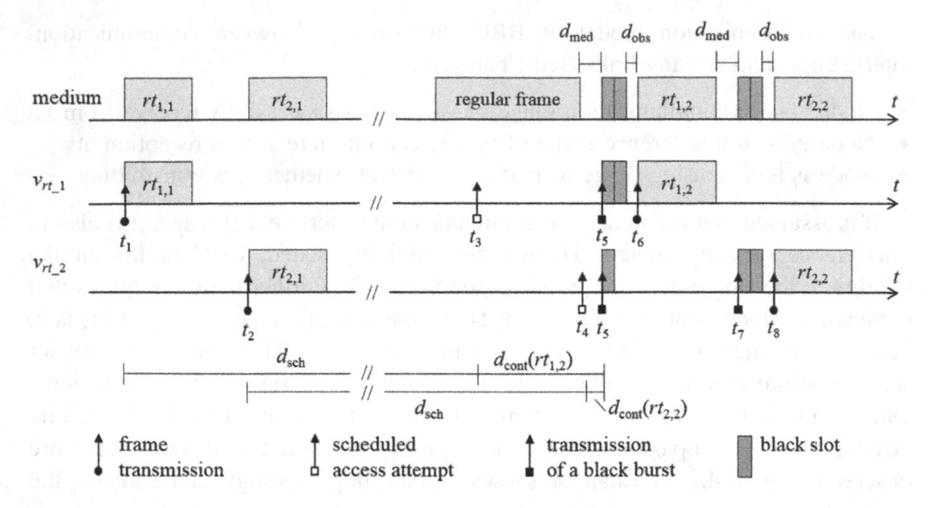

Fig. 6.9 Operation of BBC—execution scenario (see [12], Fig. 2)

Figure 6.9 shows an execution scenario of BBC, with real-time nodes v_{rt_1} and v_{rt_2} starting their real-time sessions at t_1 and t_2, respectively, by transmitting real-time frames $rt_{1,1}$ and $rt_{2,1}$. The diagram shows the activity of v_{rt_1}, v_{rt_2}, and the resulting medium occupancy. Subsequent real-time frames are scheduled with a transmission interval of d_{sch}, which is identical for all real-time nodes. In the example, v_{rt_1} and v_{rt_2} schedule frames $rt_{1,2}$ and $rt_{2,2}$ at $t_3 = t_1 + d_{sch}$ and $t_4 = t_2 + d_{sch}$, respectively. However, since the medium is detected as busy due to an ongoing transmission of a regular frame, both nodes have to wait until the medium becomes idle. After an interframe space d_{med}, they both start contending by transmitting black bursts at t_5. As the contention delay $d_{cont}(rt_{1,2}) = t_5–t_3$ is longer than $d_{cont}(rt_{2,2}) = t_5–t_4$, the function f_{BB} determines, for each node black bursts of different length, consisting of two and one black slots, respectively. Thus, v_{rt_1} wins contention and starts transmitting $rt_{1,2}$ at t_6. Node v_{rt_2} contends again at t_7, now with a black burst consisting of two black slots, wins, and transmits $rt_{2,2}$ at t_8.

One strength of BBC is that contention of real-time frames is deterministic once real-time sessions have been established. Furthermore, real-time frames collide neither with regular frames nor with other real-time frames.

A drawback of BBC is the topology constraints, requiring that nodes in interference range of a receiver be always in sensing range of the sender. This strongly limits the applicability of BBC in multi-hop networks. In particular, BBC does not support network-wide leader election or network-wide consensus. Furthermore, there is no upper bound for the transmission delay of frames that start a real-time session. To assess the performance, the authors have simulated BBC; however, no implementation on a real hardware platform exists.

A general disadvantage of busy tone protocols, which therefore also applies to BBC, is that busy tones can get quite long in situations of high contention. Since the number of black slots forming a busy tone grows linearly with the number of priorities, and because black slots tend to be relatively long, the efficiency of contention deteriorates significantly.

6.3.2 *SYN-MAC*

In [13], Wu et al. have presented a binary countdown protocol for potentially deterministic arbitration in wireless networks called SYNchronized MAC (SYN-MAC). The objective of SYN-MAC is to provide unicast frame transfers with very low collision probabilities.

In SYN-MAC, time is structured into slots (called frames) of fixed length, which are aligned network-wide, using external time synchronization. Slots consist of three phases (see Fig. 6.10). During the *contention phase*, 1-hop random value-based medium arbitration using binary countdown is performed. During the *hidden station elimination phase*, the reservation achieved during the contention phase is extended to neighbors of the receiving node. Data and acknowledgment are exchanged in the *data transfer phase*. The contention phase consists of k contention slots of fixed length, which are composed of switching interval and signaling interval. In the signaling interval, short MAC frames carrying the intended receiver's MAC address are sent.

At the beginning of the contention phase, active nodes, i.e., nodes intending to send data, generate a random bit sequence b of length k. Then, for contention slot i, $1 \leq i \leq k$, they send a contention signal CS_i, if $b[i] = 1$; otherwise, i.e., if $b[i] = 0$, they listen on the medium. If an active node perceives the medium as busy while listening, it stops contending and continues as passive node.

Passive nodes, i.e., nodes not intending to send data or nodes having lost contention, listen on the medium in contention slots $i \leq k$. As soon as they receive an uncorrupted contention signal CS_i, they stop listening. If CS_i carries its own MAC address, a node marks itself as the intended receiver and records the contention slot number i. For correct results, communication, interference, and sensing range must be identical.

Figure 6.11 illustrates the contention phase for a sample topology, with nodes B, C, G, and H as active nodes when contention starts. Node B resigns during contention slot 1, perceiving the medium as busy due to CS_1 transmitted by node C.

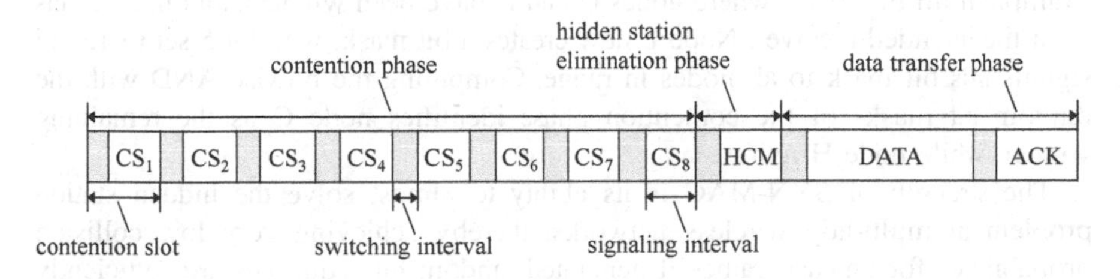

Fig. 6.10 SYN-MAC—time structure (see [13], Fig. 1)

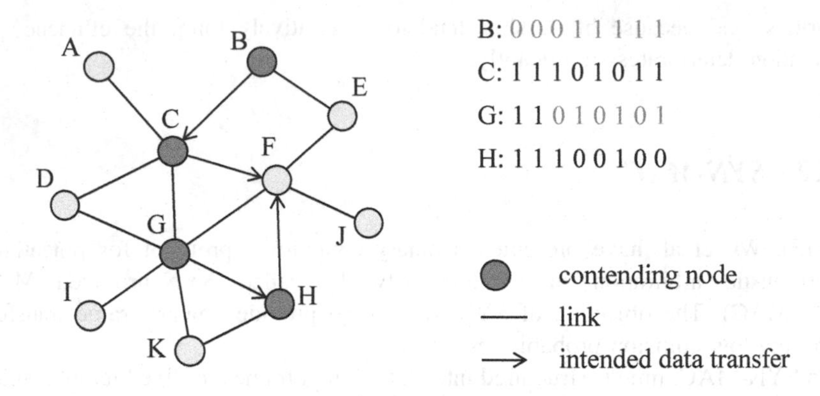

Fig. 6.11 SYN-MAC—contention phase (see [13], Fig. 2)

Furthermore, node B receives CS_1 uncorrupted, however, is not the intended receiver of C. In contention slot 3, node G resigns too. Here, CS_3 of nodes C and H collide; therefore, G cannot yet find out whether it is an intended receiver and has to continue listening until CS_5. At the end of the contention phase, nodes C and H remain as winners, although their random bit sequences are different, a problem to be resolved in the subsequent hidden station elimination phase. Moreover, F has received an uncorrupted contention signal CS_5 of node C in contention slot 5, and therefore, has recorded contention slot number 5.

At the beginning of the hidden station elimination phase, there is at most one winner per collision domain, unless two or more active nodes have generated identical random numbers. A collision domain is defined for each node v and consists of v and its neighbors in sensing range. However, intended receivers may be in range of more than one winner. To determine a unique winner, an intended receiver creates a bit mask of length k, with only the ith bit set to 1, where i is the contention slot number recorded during the contention phase. Then, it sends an Hidden station Clear Message (HCM), carrying this bit mask.

Winner nodes receiving an HCM frame compute the bitwise "and" of their random bit sequence and received bit mask. If the result is different from 0, the node remains as winner and can continue with the data phase. Otherwise, it loses and stops.

Figure 6.12 illustrates the hidden station elimination phase, continuing the example from Fig. 6.11, where nodes C and H have been winners, and node F has been the intended receiver. Node F now creates a bit mask, with bit 5 set to 1, and signals this bit mask to all nodes in range. Computing the bitwise AND with the random bit masks of the contention phase identifies node C as the remaining winner, while node H retires.

The strength of SYN-MAC is its ability to almost solve the hidden station problem in multi-hop wireless networks, thereby achieving very low collision probabilities for unicast frames if generated random bit sequences are sufficiently long. Another advantage that is common to all binary countdown protocols is their

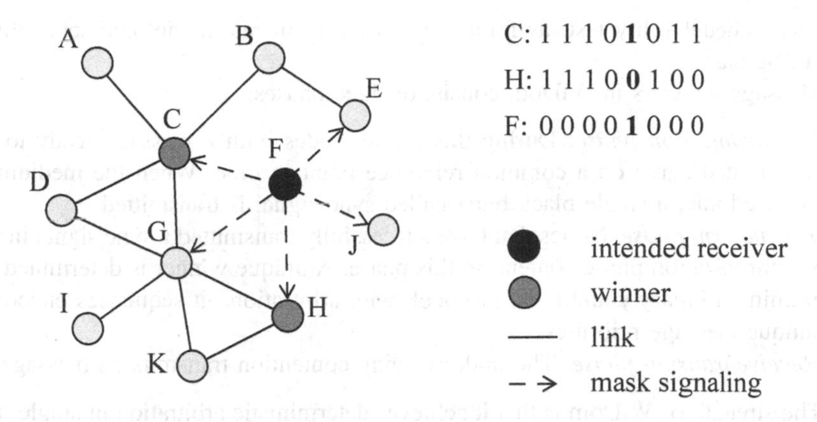

Fig. 6.12 SYN-MAC—hidden station elimination phase

arbitration efficiency, due to binary encoding of arbitration bit sequences. Furthermore, by associating non-overlapping random number intervals with priorities, it is possible to support frame priorities.

A drawback of SYN-MAC is that the operation is not entirely deterministic, as nodes of the same collision domain may generate identical random numbers. Furthermore, the arbitration range is limited to 2 hops, which implies that SYN-MAC does not support network-wide leader election or network-wide consensus in larger networks. Problematic is the assumption that communication range and sensing range are identical, and the fact that timing constraints, which are crucial for the operation of the protocol, have not been worked out. To assess performance, the authors have simulated SYN-MAC; however, no implementation on a real hardware platform exists.

6.3.3 Dominance Protocol for Wireless Medium Access (WiDom)

In [10], Pereira, Andersson, and Tovar have presented a binary countdown protocol for deterministic arbitration in single-hop wireless networks[2] called Wireless Dominance (WiDom) protocol. The objective of WiDom is to provide collision-free transfer of sporadic messages with deadlines.

As a general assumption, the traffic load consists of streams of sporadic messages, with a lower bound on message interarrival times, an upper message length, and transmission deadlines. To ensure that all transmission deadlines are met, a

[2]This is later extended to multi-hop networks [11], which, however, may lead to concurrent arbitrations that are not aligned temporarily and therefore may fail.

suitable schedulability test, as found, for instance, in rate-monotonic scheduling, has to be passed.

Message transfers in WiDom consist of three phases:

- *Synchronization phase*: During this phase, nodes with a message ready to be transmitted agree on a common reference point in time. When the medium is detected idle, a single black burst called sync signal is transmitted.
- *Tournament phase*: Nodes that have successfully transmitted a sync signal in the synchronization phase contend in this phase. A unique winner is determined by running a binary countdown protocol, with arbitration bit sequences encoding unique message priorities.
- *Receive/transmit phase*: The node winning contention transmits its message.

The strength of WiDom is that it achieves deterministic arbitration in single-hop wireless networks, using collision-resistant schemes for tick synchronization and arbitration. Another benefit is that WiDom has been implemented on a real hardware platform, and that it has been experimentally assessed in a testbed consisting of up to ten motes.

A drawback of WiDom is its very high overhead, as for each message transmission, synchronization, and tournament are required. Another problem is that carrier pulses used for synchronization are not protected against corruption, which makes WiDom prone to false positives and false negatives. Finally, WiDom is restricted to single-hop networks. In [11], the authors report on an extension for 2-hop arbitrations. However, this does not solve the problem of overlapping sync phases in different parts of the network.

6.4 Conclusions

In this chapter, we have explained concepts and measures of deterministic arbitration in wireless ad hoc networks, have presented Arbitration and Cooperative Transfer Protocol (ACTP), and have surveyed and compared related work. ACTP is a binary countdown protocol enabling deterministic arbitration and value transfer within a configurable hop radius, which supports applications such as leader election and distributed consensus. The protocol has low time and space complexity and is robust against node movements and node failure. ACTP has been implemented on Imote2 motes and has been assessed in real experiments that have shown its feasibility and reliability. Furthermore, ACTP has been implemented in ProArb, a protocol module of our protocol stack ProNet 4.0 [1].

Protocols for deterministic arbitration require reference points in time where arbitration starts. In single-hop networks, these reference points can be established by having some node starting an arbitration, with others joining when they notice this start. This is, however, only feasible with active contention schemes, i.e., with busy tones or binary countdowns. In multi-hop networks, network-wide tick synchronization of nodes is needed. In our protocol stack ProNet 4.0, we have incorporated the protocol module ProSync that implements Black Burst Synchronization (BBS, see Chap. 2) for this purpose. The protocols addressed in Sect. 6.3 either operate in single-hop environments only (BBC, WiDom), or assume the availability of a suitable synchronization protocol (SYN-MAC).

Literature

Chair for Networked Systems

1. Gotzhein R (2019) ProNet 4.0—a wireless real-time communication system for industry 4.0. White Paper, Networked Systems Group, Department of Computer Science, University of Kaiserslautern, 2014. http://vs.informatik.uni-kl.de/publications/2014/Go14/whitePaperEN-ProNet4.0.pdf. Accessed 27 Aug 2019
2. Christmann D, Gotzhein R, Rohr S (2012) The arbitrating value transfer protocol (AVTP)—deterministic binary countdown in wireless multi-hop networks. In: Proceedings of the 21st international conference on computer communication networks (ICCCN 2012), Munich, Germany, Jul 30–Aug 2, 2012, pp 1–9
3. Christmann D (2015) On the development of a wireless binary countdown protocol and the applicability of SDL to such time-critical systems. PhD Thesis, Computer Science Department, University of Kaiserslautern
4. Becker P, Birtel M, Christmann D, Gotzhein R (2011) Black burst-based quality-of-service routing (BBQR) for wireless ad-hoc networks. In: Proceedings of the 11th international conference on new technologies in distributed systems (NOTERE 2011), Paris, pp 1–8

Further References

5. Chipcon AS (2019) CC2420 ZigBee-ready RF transceiver. http://www-inst.eecs.berkeley.edu/~cs150/Documents/CC2420.pdf. Accessed 27 Aug 2019
6. Dwork C, Lynch N, Stockmeyer L (1988) Consensus in the presence of partial synchrony. J ACM 35:288–323
7. Garcia-Molina H (1982) Elections in a distributed computing system. IEEE Trans Comput C-31(1):48–59
8. MEMSIC Inc (2019) Imote 2 datasheet. https://vs.cs.uni-kl.de/downloads/Imote2NET_ED_Datasheet.pdf. Accessed 27 Aug 2019
9. Ren W, Beard RW, Atkins EM (2005) A survey of consensus problems in multi-agent coordination, american control conference, Portland, OR, USA, June 8–10, 2005, pp. 1859-1864
10. Pereira N, Andersson B, Tovar E (2007) WiDom: a dominance protocol for wireless medium access. IEEE Trans Ind Inform 3(2):120–130
11. Pereira N, Andersson B, Tovar E, Rowe A (2007) Static-priority scheduling over wireless networks with multiple broadcast domains. In: 28th IEEE international real-time systems symposium (RTSS 2007), Tucson, Arizona, 3–6 Dec 2007
12. Sobrinho JL, Krishnakumar AS (1999) Quality of service in ad hoc carrier sense multiple access networks. IEEE J Sel Areas Commun 17(8):1353–1368

13. Wu H, Utgikar A, Tzeng N (2005) SYN-MAC: A distributed medium access protocol for synchronized wireless networks. Mobile Netw Appl 10:627–637 (Springer)
14. Gotzhein R, Kuhn T (2011) Black burst synchronization (BBS)—a protocol for deterministic tick and time synchronization in wireless networks. Comput Netw, Elsevier 55(13):3015–3031

Chapter 7
Duty Cycling

When using battery-powered nodes, energy consumption is a major concern. Duty cycling is a functionality to save energy and thereby extend network lifetime. In this chapter, we explain foundations of duty cycling in wireless ad hoc networks, present our duty cycling approach in ProNet 4.0 [1] that has been originally developed for MacZ [2], survey and compare related work, and draw conclusions.

7.1 Foundations

In this chapter, we provide the context of duty cycling in wireless ad-hoc networks, explain concepts, measures, and requirements, and address areas of operation.

7.1.1 Context

In systems consisting of battery-powered nodes, such as wireless sensor networks and wireless networked control systems, energy consumption is a major concern, constraining the uptime of nodes and therefore the network's lifetime. As a rule, energy consumption should be kept as low as possible. This can be achieved, for instance, by using low-power hardware platforms such as microcontrollers and motes, low-power communication technologies such as ZigBee [15] and Bluetooth Low Energy [5], and by devising energy-efficient algorithms, and protocols. Furthermore, hardware components can be temporarily shut off or switched to an energy-saving mode when not needed.

While it is desirable to reduce energy consumption, it is mandatory to consider computation and communication requirements. For instance, in networked control systems consisting of sensors, actuators, and controllers communicating over a wireless medium, the main objective is achieving optimal system behavior such as

R. Gotzhein, *Real-time Communication Protocols for Multi-hop Ad-hoc Networks*, Computer Communications and Networks,
https://doi.org/10.1007/978-3-030-33319-5_7

stability of the controlled system. Therefore, system components can only be switched to energy-saving modes if this objective is not put in jeopardy. In some systems, always-on nodes are required, which leaves little room for energy saving. However, in scenarios where the full computation power is only needed sporadically or where communication events are rare, the potential for energy saving is substantial.

7.1.2 Concepts, Measures, and Requirements

Duty cycling is a functionality to reduce energy consumption of nodes. To save energy, hardware components such as CPU, memory, transceiver, LEDs, and sensing devices can be switched to active mode (e.g., receive, transmit) when needed, and to non-active mode (e.g., sleep, idle) mode when not needed. The *duty schedule* defines the duration of alternating intervals of active and non-active modes. A duty schedule may be static, consisting of predetermined strictly or weakly periodic active and non-active mode intervals, dynamic, where active and non-active mode intervals are determined at runtime or a combination of both. The *duty cycle* denotes the ratio of time a hardware component is in active mode.

Figure 7.1 shows a strictly periodical duty schedule of duration $d_{\text{dutySchedule}}$ and an active interval of length d_{active}, yielding duty cycle $r_{\text{dutyCycle}} = d_{\text{active}}/d_{\text{dutySchedule}}$. In addition to these common notions, we introduce the *waste ratio* $r_{\text{waste}} = (d_{\text{active}} - d_{\text{dataTransfer}})/d_{\text{dataTransfer}}$, where $d_{\text{dataTransfer}}$ sums up the duration of payload transmissions and receptions of a duty schedule. Thus, waste ratio of 0.5 means that 50% overhead are produced.

To achieve minimal energy consumption, all hardware components of a node are to be considered together. In this chapter, we focus on wireless communication, and therefore consider the transceiver only. In the literature (see, for instance, Ye et al. [13]), there are four sources of energy waste associated with the transceiver:

- *Idle-listening* occurs when a node listens on the medium without detecting any transmissions. Without specific knowledge about the communication behavior of other nodes, a node has to listen whenever not sending itself.

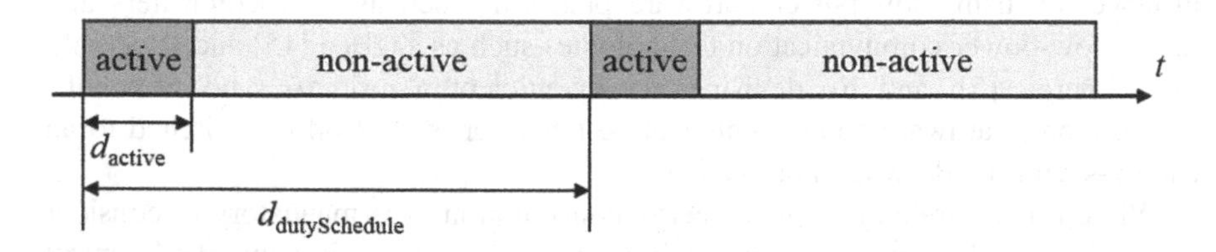

Fig. 7.1 Strictly periodical duty schedule

- *Overhearing* occurs when a node receives a frame with different destination. Without specific knowledge, a node cannot determine in advance whether an incoming frame can be disregarded.
- *Collisions* occur when a receiving node is in interference range of two or more nodes with transmissions overlapping in time. If the incoming frame is corrupted, it has to be retransmitted, thereby increasing the overall energy consumption.
- *Overhead* occurs when non-payload data (such as preamble and control fields of DATA frames) or management frames (e.g., Request-To-Send (RTS), Clear-To-Send (CTS), ACKnowledgement (ACK), and beacons) are transmitted or received.

Figure 7.2 illustrates these sources of energy waste for a Multiple Access with Collision Avoidance for Wireless (MACAW) protocol, i.e., a MAC protocol with passive medium contention and an RTS-CTS-DATA-ACK sequence for single-hop unicast transmission. In the example, nodes v_1, v_2, and v_3 form a line topology. After waking up, they start listening on the medium, with node v_2 contending and winning. During this phase, all nodes waste energy due to idle-listening, as the medium remains idle. Furthermore, idle-listening periods may differ in length, as nodes may wake up at different points in time due to clock offsets.

When winning contention, node v_2 switches from receive mode to transmit mode, which is another source of energy waste (not mentioned in Ye et al. [13]). Then, there is overhead due to the exchange of an RTS-CTS sequence for v_2 and v_3. In the scenario, the RTS frame is also received by v_1. However, because v_1 is not the destination, this is energy waste due to overhearing. In fact, the information contained in the RTS frame helps v_1 to set its network allocation vector (NAV), and to switch its transceiver to sleep mode, thereby saving energy during the subsequent data exchange of v_2 and v_3. Data exchange happens after the successful reception of the CTS frame by v_2 and is followed by an ACK frame. In the scenario, no collisions occur.

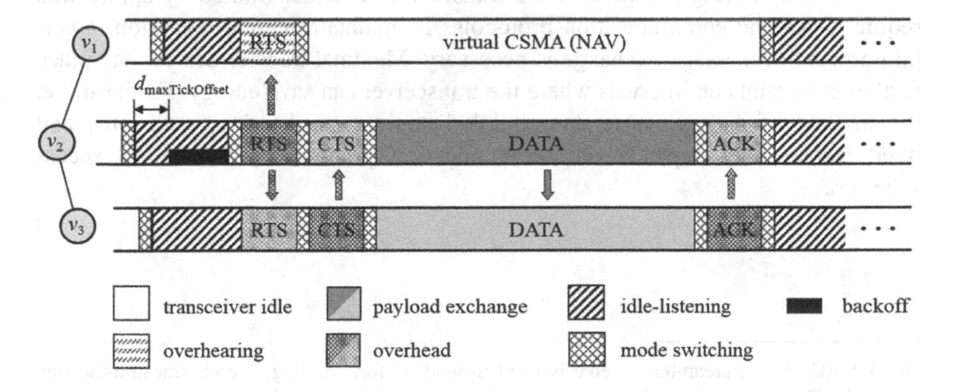

Fig. 7.2 Sources of energy waste for a MACAW protocol (see [2], Fig. 4)

Operational modes	Energy consumption
powerDown	0.02 mA
idle	0.426 mA
receive	18.8mA
send	8.5-17.4 mA

Current mode	Next mode	Duration
powerDown	idle	1000 μs
idle	receive	192 μs
	transmit	192 μs
receive	powerDown	-
	idle	-
	transmit	192 μs
send	receive	192 μs

Fig. 7.3 Energy model of the transceiver CC2420—tabular notation

To determine absolute energy consumption, an energy model of the transceiver and information about transmission rates and frame sizes are required. The energy model determines operational modes, energy consumed in each mode, and transitions between modes with energy consumption and/or duration. Figure 7.3 shows the energy model of the CC2420 transceiver [6] in a tabular notation, which can also be represented as a graph. The transceiver has four operational modes. Interestingly, energy consumption in receive mode is higher than in transmit mode even when sending with maximum transmission power. Figure 7.3 also shows state transitions and their durations. The energy model is incomplete, as some transition durations and the energy consumption associated with transitions are missing.

To exchange messages, the transceivers of communicating nodes have to be in active mode. Therefore, duty cycling requires distributed coordination. In particular, nodes have to agree on wake-up times and duty schedules to perform synchronized mode switching. For this reason, duty cycling protocols usually[1] build on tick or time synchronization, preferably with low and bounded offsets to maximize energy savings.

The energy-saving potential of the transceiver is further limited by application requirements and communication protocols. To maintain regular operation, a certain amount of message exchange is necessary. Maximal transfer delays may place further constraints on intervals where the transceiver can save energy. Furthermore, the amount of energy waste due to idle-listening, overhearing, collisions, and overhead strongly depends on communication protocols and medium access schemes.

[1]In B-MAC [11], a preamble scheme is used instead. Before sending, a node transmits a long preamble. The other nodes wake up in regular intervals and check for the preamble. If they detect the preamble, they wait for the message, otherwise, they change to sleep mode.

7.1.3 *Areas of Operation*

Duty cycling extends the lifetime of nodes with scarce energy resources, and thereby of networks. This is of particular importance in harsh environments where nodes are not easily accessible for battery replacement and in industrial applications where the cost of maintenance is a factor.

7.1.3.1 Monitoring Slope Stability in Permafrost Areas

PermaSense [12] is a consortium of Swiss research institutions and companies, developing and deploying wireless sensor networks customized for long-term operation in high-mountain environments. A project run by the consortium uses wireless sensor networks in cryosphere research, to monitor the stability of high-alpine mountain slopes in permafrost areas of the Alps. Slope instability can cause massive landslides, which can result in the deaths of people, and therefore has to be detected as early and reliably as possible. In a period of global warming, the threat of slope instability is increasing.

To monitor slope stability, the consortium has developed and deployed wireless sensor networks to measure microseismic activity in remote high-mountain regions. Sensor networks consist of sensor nodes, which sample data, and a base station collecting data values. Deploying these nodes has been a big, expensive, and dangerous effort, involving teams of mountaineers and even helicopters. It is self-evident that deployed sensor networks have to be long living, which, among other things, requires low-energy consumption (to avoid battery replacements) or the use of energy harvesting techniques. Among the measures taken to achieve this objective is the development of a communication stack called low-power wireless bus (LWB) [9] supporting time-triggered message scheduling and duty cycling.

7.1.3.2 Monitoring Wagons of Freight Trains

In the Asset Monitoring for Rail Applications (AMRA) project at Bosch BEG [4], wagons of freight trains are equipped with a variety of sensors. Sensors detect shocks, flats, open cargo doors, cargo status, and location. Furthermore, an local data collector (LDC) is placed on each wagon, to gather and aggregate sensor data. LDCs forward their status data to a global data collector, which is located on the railroad engine, and forwarded via global system for mobile communications (GSM) to a remote data collector in a stationary headquarter where all trains are monitored. The system is used to detect malfunctions and to plan maintenance activities.

An important concern of the AMRA project is energy saving, as sensor nodes and LDCs are battery-driven. For this reason, Bluetooth Low Energy [5] has been chosen as technology for on-wagon and inter-wagon communication. Furthermore, duty cycling is applied using a duty schedule of weakly periodic active mode intervals. With these measures, a node lifetime of up to 5 years is to be reached.

7.2 Duty Cycling in ProNet 4.0

In this chapter, we present our duty cycling protocol devised for ProNet 4.0 [1]. Duty cycling is based on network-wide tick synchronization with BBS (Black Burst Synchronization, see Chap. 2). Originally, the protocol has been conceived for MacZ, a predecessor of ProNet 4.0 with reduced functionality implemented on the MICAz [10] hardware platform [2]. Compared to this original version, we have made extensions to cover arbitrated regions.

7.2.1 Virtual Time Slotting

Global time slotting is a basic functionality of ProNet 4.0 (see Chap. 3). Following an analysis of timing requirements, time is structured into physical time slots, which are composed into virtual time regions. We have argued that virtual time slotting provides a high degree of flexibility, while keeping unusable slot portions small.

Figure 7.4 shows an example of virtual time slotting, with virtual time regions of different types and lengths formed and placed into super slots in a non-overlapping, but otherwise flexible way. Time intervals in-between active regions, i.e., regions where messages may be exchanged, form passive regions called *idle regions*, where transceivers can be switched to an energy-saving mode.

Virtual time slotting is node-specific and can be performed statically and/or dynamically. In case of static virtual slotting, the time structure of a super slot is repeated, yielding a weakly period pattern of active and passive virtual time regions. Compared to other duty cycling approaches (see Sect. 7.3), which assume a strictly periodic pattern of sync, data, and sleep phases, this provides considerable flexibility. When combined with dynamic virtual time slotting, even more flexibility is achieved. For instance, depending on the varying need for exclusive time slots,

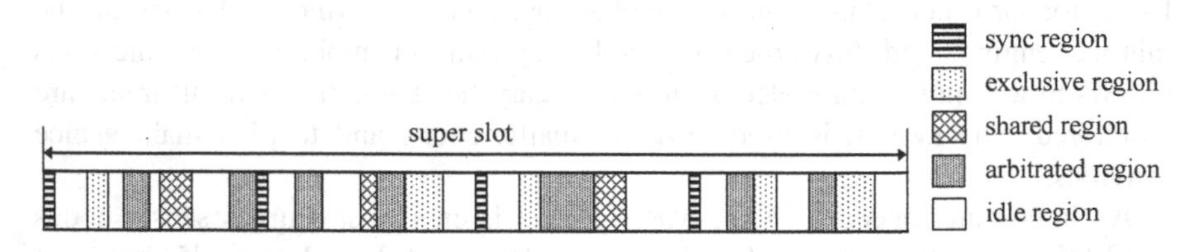

Fig. 7.4 Virtual time slotting (for explanations, see Chap. 3)

exclusive regions may be extended or shortened during execution, with an effect on adjacent idle regions, which, as a result, may shrink and grow.

7.2.2 Duty Cycling in Idle Regions

Virtual time slotting structures time into a sequence of active (non-idle) and passive (idle) regions. While active regions are used for message exchange, it may still be possible to switch the transceivers of some or all nodes to energy-saving mode, depending on medium access scheme and traffic pattern. During idle regions, message exchange is disabled network-wide. Therefore, by default, all transceivers can change to sleep or idle mode, to minimize energy consumption.

Figure 7.5 shows a super slot structured into non-idle and idle regions. At the beginning of an idle region, the transceiver is switched to energy-saving mode, and returns to active mode right before the idle region ends. It follows that during an idle region, switching delays are the only source of energy waste.

The full potential of duty cycling in idle regions is achieved, if three constraints are satisfied. First, the length of an idle region has to exceed switching delays. Second, at the local start of an idle region, all other nodes have finished message exchanges of the previous active region. Third, all other nodes do not start medium occupancy before the local end of an idle region.

7.2.3 Duty Cycling in Exclusive Regions

Exclusive regions host deterministically scheduled, contention-free traffic, and are substructured into exclusive slots of possibly different length. In each exclusive slot, one or more messages can be exchanged, with slot boundaries being respected. In particular, this means that messages transferred in an exclusive slot are associated with the same slot by receiving nodes and that switching delays are observed.

Although exclusive regions are active regions, there is high potential for energy saving. Since nodes involved in message exchanges of exclusive slots are determined in advance, nodes not involved in an exchange can switch their transceivers

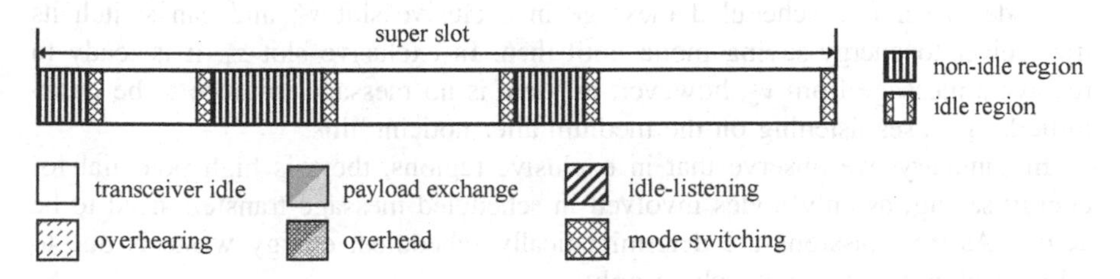

Fig. 7.5 Duty cycling in idle regions

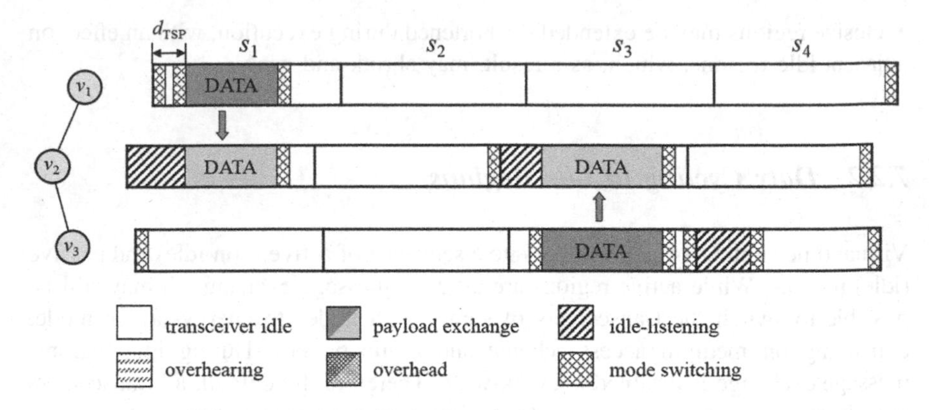

Fig. 7.6 Duty cycling in exclusive slots (see [2], Fig. 3)

to energy-saving mode. This also holds for nodes after they have finished a scheduled message transfer in a slot, and, in particular, if there is no message instance to be transmitted in an assigned slot.

Figure 7.6 shows an exclusive region consisting of four exclusive slots. We stipulate that at the beginning and the end of an exclusive region, all transceivers are in receive mode. This may require switching from idle or transmit mode to receive mode at the end of the preceding virtual time region, and switching to receive mode when the exclusive region ends.

In exclusive slot s_1, a transfer from node v_1 to v_2 is scheduled. Before starting transmission, v_1 has to wait until the transmission start point (TSP), i.e., for the maximum tick offset. If this interval is sufficiently long, there is some potential for energy saving, as shown in the figure. After transmission, v_1 can switch its transceiver to energy-saving mode for the remainder of the exclusive region, as it does not participate in further scheduled message exchanges. Thus, energy waste of v_1 is due to mode switching and overhead (preamble and control fields of the data frame) only.

To receive the scheduled message from v_1, node v_2 has to start listening on the medium at the local start of exclusive slot s_1, which constitutes idle-listening. The length of this period depends on the current and maximum tick offsets between both nodes. After receiving the data frame, node v_2 can switch to energy-saving mode until exclusive slot s_3, where this scheme is repeated.

Node v_3 sends a scheduled message in exclusive slot s_3, and can switch its transceiver to energy-saving mode until then. In exclusive slot s_4, it is ready to receive a message from v_2, however, as there is no message instance to be transmitted, v_3 ceases listening on the medium after noticing this.

In summary, we observe that in exclusive regions, there is high potential for energy saving, as only nodes involved in scheduled message transfers need to be active. As transmissions are deterministically scheduled, energy waste is due to idle-listening and mode switching only.

7.2.4 Duty Cycling in Shared Regions

Shared regions host unscheduled, contention-based traffic. As the occurrence of message transfers is not predictable, nodes have to stay active and listen on the medium even if they are not contending for medium access. Thus, energy waste due to idle-listening is high in case of low traffic. With increasing traffic, idle-listening decreases, whereas energy waste due to overhearing and collisions increases. Furthermore, when running a Multiple Access with Collision Avoidance for Wireless (MACAW) protocol, there is energy waste due to overhead. This is further explained in Sect. 7.1.2, and illustrated in Fig. 7.2.

Duty cycling in shared regions can be improved by distinguishing between a reservation interval, where all nodes are active and nodes intending to communicate exchange a reservation sequence consisting of Request-To-Send (RTS) and Clear-To-Send (CTS), and by having the actual data exchange in a subsequent time interval where all nodes not involved in this exchange switch their transceiver to energy-saving mode. To some degree, this effect is also achieved by applying the network allocation vector (NAV) to save energy.

7.2.5 Duty Cycling in Arbitrated Regions

Arbitrated regions host deterministic n-hop arbitrations and value transfers. They are decomposed into arbitrated slots, with slot lengths depending on arbitration range and on the size of values and data to be exchanged. For a detailed treatment, we refer to Chap. 6. Since all nodes are potentially involved in an arbitration process, as either contending or repeating nodes, they have to stay in active mode for almost the entire duration of arbitrated slots.

Figure 7.7 illustrates the potential of duty cycling during an arbitrated slot, for a single-hop network, i.e., $n_{\mathrm{hops}} = n_{\mathrm{maxHops}} = 1$, and an arbitration bit sequence of length $n_{\mathrm{bits}} = 2$. In the example, v_2 contends with the bit sequence 10. We stipulate that at the beginning of a bit round, all transceivers are in receive mode. As v_2 sends a logical 1 with a delay of d_{TSP} relative to the local start of its first bit round, it may switch to energy-saving mode provided the interval d_{TSP} is long enough for two switching delays. It then sends its black burst and may switch to energy-saving mode for the rest of this bit round. In the second bit round, v_2 has to listen on the medium in case another node sends a logical 1. After winning contention, v_2 sends a data frame during the data phase of the arbitrated slot.

Nodes v_1 and v_3 do not contend, but listen on the medium to record the current arbitration bit sequence. Afterward, v_3 receives the data frame, while v_1 may switch its transceiver to energy-saving mode.

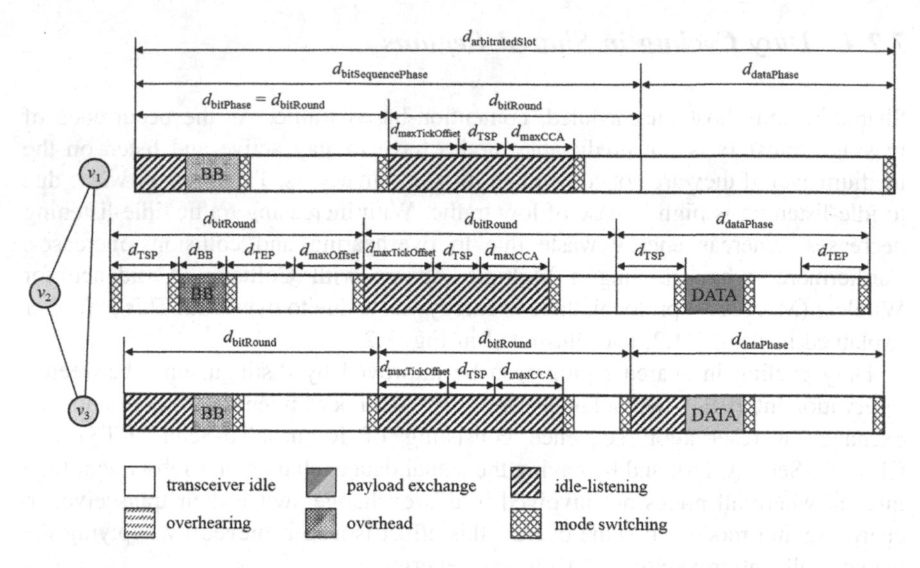

Fig. 7.7 Duty cycling in arbitrated slots

In summary, we observe that in arbitrated regions, there is not much potential for energy saving, as both active (contending) and passive (repeating) nodes are involved in the arbitration process. Since arbitration is deterministic and occurring collisions are non-destructive, energy waste is due to idle-listening, overhead (exchange of the arbitration sequence), and mode switching.

7.2.6 Analytical Assessment

To assess the energy efficiency of our duty cycling protocol [2], we analytically determine duty cycles and waste factors for a real-life application scenario operated on MICAz motes [10] equipped with the CC2420 transceiver [6]. At the University of Kaiserslautern, an ambient intelligence system for the improvement of the training effects of a group of racing cyclists called assisted bicycle trainer (ABT) has been developed [3]. During training sessions, ABT collects status data (e.g., velocity, headwind, heart rate, and pedal power output) of each cyclist, periodically forwards the data to the trainer node, and shows an aggregation of the data on the trainer display. Based on these data and his experience, the trainer can modify training parameters (e.g., velocity, position) by sending commands to cyclists.

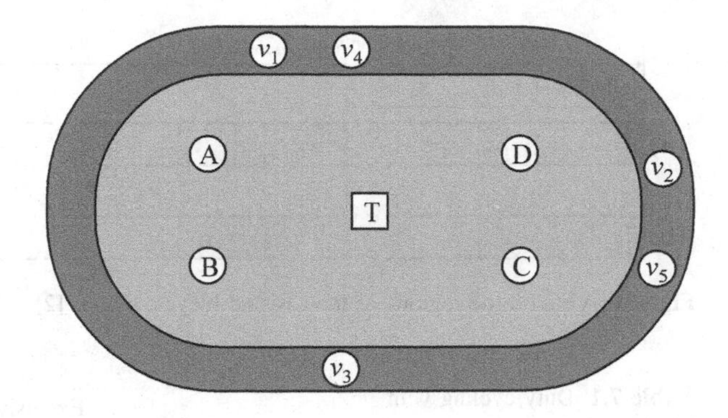

Fig. 7.8 Assisted bicycle trainer—network topology (see [2], Fig. 6 and 7)

7.2.6.1 Application Scenario

In a typical training session, a number of cyclists (nodes v_1 to v_5) are pedaling an oval cycling track,[2] with the trainer (node T) standing in the center of the track (see Fig. 7.8). To improve connectivity, stationary repeater nodes A to D may be placed. Individual status data are collected on each bicycle once per second, i.e., $d_{\mathrm{appInt}} = 1$ s, and communicated to the trainer, possibly via some repeater node, depending on the current position of a cyclist on the track. Given data rate r_{tx} and frame size $l_{\mathrm{dataFrame}}$, the duration $d_{\mathrm{dataTransfer}}$ for a single data frame transmission, i.e., the minimally required time of a transceiver in active mode per application interval (duty schedule duration) is

$$d_{\mathrm{dataTransfer}} = l_{\mathrm{dataFrame}} / r_{\mathrm{tx}} \tag{7.1}$$

This yields a minimal duty cycle of $d_{\mathrm{dataTransfer}}/d_{\mathrm{appInt}}$. However, depending on the type of communication protocol, further factors may increase the required time d_{active} of a transceiver in active mode, which together determines waste ratio and duty cycle.

$$\begin{aligned} d_{\mathrm{active}} = {} & d_{\mathrm{dataTransfer}} + d_{\mathrm{idleListening}} + d_{\mathrm{overhead}} + d_{\mathrm{collision}} \\ & + d_{\mathrm{overhearing}} + d_{\mathrm{switching}} \end{aligned} \tag{7.2}$$

$$r_{\mathrm{waste}} = (d_{\mathrm{active}} - d_{\mathrm{dataTransfer}}) / d_{\mathrm{dataTransfer}} \tag{7.3}$$

$$r_{\mathrm{dutyCycle}} = d_{\mathrm{active}} / d_{\mathrm{appInt}} \tag{7.4}$$

[2]The assisted bicycle trainer can also be used for road training.

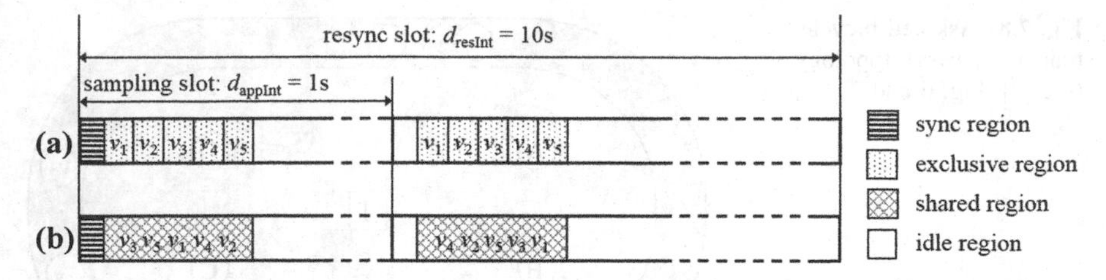

Fig. 7.9 Virtual time regions of the assisted bicycle trainer [2]

Table 7.1 Duty cycling with MacZ—cyclist nodes

	Exclusive		Shared	
	1-hop	2-hop	1-hop	2-hop
$d_{\text{dataTransfer}}$ [ms] (7.1)	0.70	0.70	0.70	0.70
$d_{\text{idleListening}}$ [ms]	–	–	5.33	7.20
d_{overhead} [ms]	0.23	0.38	1.19	1.34
$d_{\text{collision}}$ [ms]	–	–	–	–
$d_{\text{overhearing}}$ [ms]	–	–	0.34	1.44
$d_{\text{switching}}$ [ms]	0.38	0.38	2.30	3.84
d_{active} [ms] (7.2)	1.31	1.46	10.16	14.52
r_{waste} (%) (7.3)	0.86	1.07	13.43	19.63
$r_{\text{dutyCycle}}$ (%) (7.4)	0.13	0.15	1.02	1.45

7.2.6.2 Exclusive Regions

In the first communication solution, we configure virtual time regions as shown in Fig. 7.9a. Resynchronization is performed every 10 s, i.e., $d_{\text{resInt}} = 10$ s, sampling occurs once per second, i.e., $d_{\text{appInt}} = 1$ s. Exclusive regions are placed with an offset of d_{conv}—the time needed for resynchronization—relative to the beginning of each sampling slot, and subdivided into five exclusive slots assigned to cyclist nodes v_1 to v_5. Exclusive slots are sized such that data frame transfer (10 bytes preamble and control fields, 12 bytes payload), switching delays, and maximum synchronization offset fit.

Table 7.1 (columns "exclusive") shows the results of our analytical assessment of duty cycling with exclusive reservations. To determine values for the sources of energy waste, we have used parameter values of the CC2420 transceiver and Black Burst Synchronization (BBS), see Chap. 2. Due to exclusive slot assignments and because cyclist nodes only act as senders,[3] there is no energy waste due to idle-listening, collisions, and overhearing. Energy waste occurs due to mode switching, which is $2 \cdot d_{\text{switch}}$ per data transfer, and due to overhead caused by

[3]For simplicity, we omit exclusive slot assignments (once per super slot) for the trainer, to communicate commands to cyclists. During these slots, cyclist nodes act as receivers and therefore perform idle-listening, too.

resynchronization, i.e., $d_{overhead} = (d_{conv}/d_{resInt}) \cdot d_{appInt}$. In the single-hop case, cyclist nodes are active for 1.31 ms per application interval, yielding a very low duty cycle $r_{dutyCycle}$ of 0.13%. The waste ratio r_{waste} is 0.86 and mainly caused by switching delays that are hardware-specific and therefore cannot be avoided. Results of the two-hop case are similar.

7.2.6.3 Shared Regions

In the second communication solution, we configure virtual time regions as shown in Fig. 7.9b. As in the first solution, d_{resInt} = 10 s and d_{appInt} = 1 s. Shared regions are placed with an offset of d_{conv} relative to the beginning of each sampling slot and are collectively used by cyclist nodes v_1 to v_5.

Table 7.1 (columns "shared") shows the results of our analytical assessment of duty cycling with medium contention. We have simplified the analysis by assuming that communication is collision-free. Because of passive medium contention, there is a substantial amount of idle-listening.

As soon as a cyclist node notices that it has lost contention, it switches its transceiver to energy-saving mode for the duration of the following CTS-DATA-ACK frame sequence, which increases switching delays. We assume that on average, a node has to contend $n_{cyclist}/2$ times to win contention, and then switches its transceiver to energy-saving mode.

Compared to duty cycling with exclusive regions, duty cycles are about 10 times higher, waste ratios about 20 times. This clearly shows that communication based on reservations has far more potential for energy saving compared to passive medium contention.

7.3 Related Work

In this chapter, we survey and compare duty cycling protocols devised for wireless communication technologies. Our survey is not intended to be comprehensive, but to identify related protocols, and to point out differences to our duty cycling approach.

7.3.1 Sensor-MAC (S-MAC)

In [13], Ye, Heidemann, and Estrin have presented *S-MAC* (*Sensor-MAC*), a duty cycling protocol for wireless sensor networks. Objectives of S-MAC are energy saving, scalability and collision avoidance in multi-hop networks. S-MAC is devised for applications with long idle periods and networks with light traffic load.

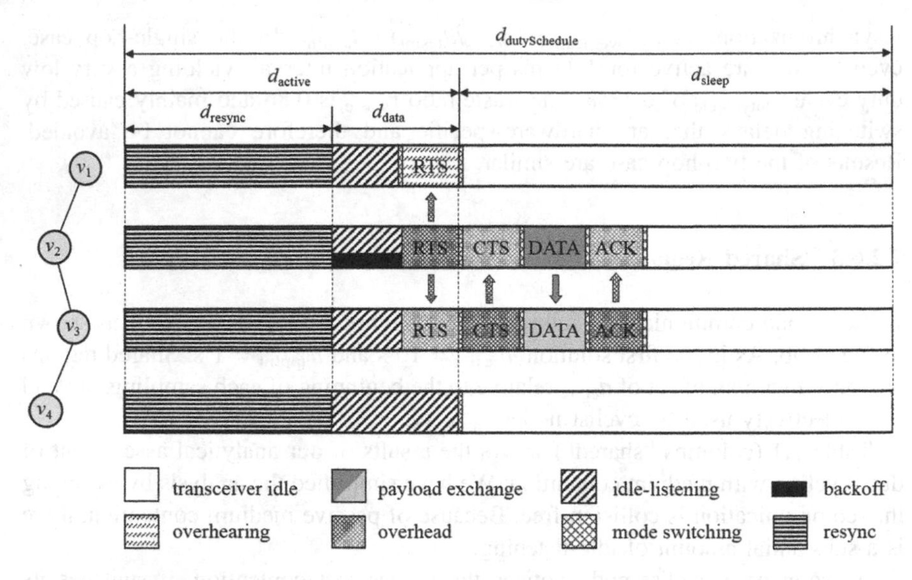

Fig. 7.10 Operation of S-MAC

S-MAC divides time into duty schedules of fixed duration, consisting of (short) active and (long) sleep phases (see Fig. 7.10). Active phases are further decomposed into sync and data phase. During the sync phase, S-MAC runs a synchronization protocol that establishes synchronization clusters, each consisting of a cluster head (called synchronizer) and followers. Nodes belonging to the same synchronization cluster synchronize their active phases. To establish network-wide connectivity, suitable cluster overlaps are required, which cannot be guaranteed by the approach. Furthermore, there is the risk of overlapping active phases of neighboring clusters, which may result in collisions. As it is straightforward to replace the synchronization protocol of S-MAC by an algorithm that avoids these and further problems, we will not go into further detail.

During the data phase, nodes passively contend for the medium, using a random back-off scheme. Nodes winning contention start a unicast transfer sequence by sending an Request-To-Send (RTS) frame. This already ends the data phase, and the sleep phase begins. However, nodes involved in an RTS exchange, i.e., senders and receivers, stay active to complete the transfer sequence continued by Clear-To-Send (CTS) frame, DATA frame, and ACK frame before switching their transceivers to energy-saving mode.

Figure 7.10 shows a scenario with a line topology. For simplicity, all nodes are aligned in time. Node v_2 wins contention and starts its transfer sequence. Node v_3 is the intended receiver, and therefore responds with a CTS frame in the sleep phase, followed by DATA and ACK frames. Node v_1 also receives the RTS frame; however, as it is not the intended receiver, it can switch its transceiver to energy-saving mode immediately. Node v_4 is not involved, but has to listen for possible RTS frames.

Figure 7.10 also shows where energy is wasted. Although not detailed, we can assume that the synchronization protocol in the sync phase is rather expensive in terms of energy, as several broadcasts have to be exchanged, and because nodes have to listen on the medium during the entire sync phase. During the data phase, nodes are active, too, either contending for the medium or listening for RTS frames. This causes energy waste due to idle-listening, overhead, overhearing, and mode switching. Further overhead is produced by the exchange of CTS and ACK frames in the sleeping interval.

In single-hop networks, throughput of S-MAC is limited to one data frame per duty schedule, which is the reason why the protocol is applicable in networks with light traffic load only. A major drawback of S-MAC in multi-hop networks is that messages can cover a distance of only one hop per duty schedule, which leads to high end-to-end delays, if sleep intervals are long (as intended, to achieve high energy savings). In [14], the authors of S-MAC propose an improvement that works for a distance of up to 2 hops per duty schedule, called adaptive listening.

Another drawback of S-MAC is the high synchronization overhead at the beginning of every active phase, especially when using the synchronization protocol incorporated into S-MAC. Even more, this overhead is required for the transmission of only one data frame. The authors of S-MAC discuss the possibility of performing resynchronization less frequently; however, this increases idle-listening due to earlier wake-up times.

7.3.2 Routing-Enhanced MAC (RMAC)

In [7], Du et al. have presented *RMAC* (*Routing-enhanced MAC*), a duty cycling protocol for wireless sensor networks that readdresses the throughput and end-to-end delay problems of S-MAC. Compared to S-MAC, RMAC achieves substantial improvements, by performing multiple and multi-hop signaling during the data phase.

Time structuring of RMAC is similar to S-MAC, with active, sleep, sync, and data phases (see Fig. 7.11). Unlike S-MAC, synchronization is not covered by RMAC. Instead, the authors propose to use an existing synchronization protocol, to be executed during the sync phase.

During the data phase, nodes passively contend for the medium. Nodes winning contention transmit a PION (PIONeer) frame, containing the addresses of sender, next hop on the route, previous hop (if applicable), and destination. Furthermore, PION frames contain a hop count n_{hops} and the transmission duration d_{DATA} of the subsequent DATA frame.

When a node on a route receives a PION frame, it forwards an updated PION frame. First, this serves as a confirmation to the sender of the previous PION frame. Second, it is received by nodes in range, which helps to solve the hidden station

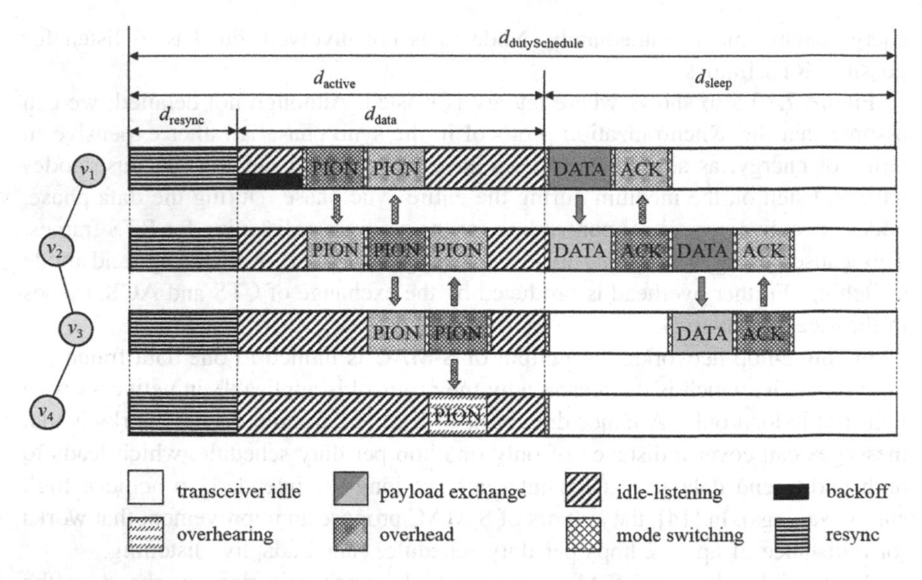

Fig. 7.11 Operation of RMAC

problem. Third, it extends the route by one hop, if the destination node has not yet been reached. This way, PION frames combine the functionality of RTS and CTS frames.

By sending a PION frame with hop count n_{hops}, a node reserves a data transfer interval of length $d_{dti} = 2 \cdot d_{SIFS} + d_{DATA} + d_{ACK}$, where d_{SIFS} and d_{ACK} denote the durations of short inter-frame space (SIFS) and ACK transfer duration, respectively. This data transfer interval starts at $t_{sleepPhase} + n_{hops} \cdot d_{dti}$, with $t_{sleepPhase}$ denoting the point in time where the next sleep phase starts.[4] When a node receives a PION frame with itself being the denoted receiver, it reserves a data reception interval starting at $t_{sleepPhase} + n_{hops} \cdot d_{dti}$. All other nodes determine network allocation vector (NAV) intervals to avoid interference in the confirmation PION segment, the DATA segment, and the ACK segment.

Figure 7.11 shows a scenario with a line topology and a frame to be routed from node v_1 to node v_3. Therefore, a sequence of PION frames is exchanged, followed by DATA and ACK frames in the reserved intervals of the sleep phase. Energy waste due to overhead occurs in the sync phase, and by the exchange of PION and ACK frames. Idle-listening occurs during the active phase, and due to synchronization inaccuracies, as receiving nodes have to switch to active mode earlier in order not to miss the DATA frame (not shown in Fig. 7.11).

Simulation experiments reported in [7] clearly show that RMAC outperforms S-MAC. This is due to the possibility to signal more than one message transfer per data phase, which increases throughput and reduces end-to-end delay. The price for

[4]For a simplified presentation, we abstract from clock offsets.

this is a longer data interval and therefore more energy consumption due to idle-listening.

A drawback of RMAC is that it builds on network status information such that nodes can determine next hops to destination nodes. To establish this kind of information, broadcast messages would have to be exchanged. However, as RMAC supports only unicast transfers, it is unclear how this information is to be established.

Another drawback of RMAC is the high synchronization overhead by the proposed synchronization protocol Reference Broadcast Synchronization (RBS) [8], which has a communication complexity of up to $O(n^2)$, where n is the number of nodes. In fact, RBS is not even suitable, as it only determines clock offsets among nodes belonging to the same time zone, but does not establish global reference points in time, which are needed for duty cycling. However, this problem could be remedied by selecting a suitable and more efficient synchronization protocol.

7.4 Conclusions

In this chapter, we have explained concepts, measurements, and requirements of duty cycling, have presented our duty cycling approach devised for ProNet 4.0 [1] and partially implemented in MacZ, and have surveyed related work. Duty cycling in MacZ is highly efficient due to the flexible time structure, which permits to place active virtual time regions where needed, and to switch the transceiver to energy-saving mode in-between. Given the type of virtual time region, further potential for energy saving can be exploited, with exclusive regions performing best. The high flexibility of the global time structure can be used to configure duty schedules and virtual time regions such that objectives regarding throughput and end-to-end latency are achieved. Furthermore, by building on Black Burst Synchronization (BBS, see Chap. 2), deterministic tick and time synchronization with low and bounded offsets and convergence delay are available.

Literature

Chair for Networked Systems

1. Gotzhein R (2014) ProNet 4.0—A wireless real-time communication system for industry 4.0. White Paper, Networked Systems Group, Department of Computer Science, University of Kaiserslautern. http://vs.informatik.uni-kl.de/publications/2014/Go14/whitePaperEN-ProNet4.0.pdf. Last accessed 27 Aug 2019
2. Christmann D, Gotzhein R, Krämer M, Winkler M (2013) Flexible and energy-efficient duty cycling in wireless networks with MacZ. In: Proceedings 10th annual international conference

on new technologies of distributed systems (NOTERE 2010), Tozeur, Tunisia, May 31–June 2, pp 121–128; J Concurr Comput Pract Exp 25(2):218–233
3. Fliege I, Geraldy A, Gotzhein R, Jaitner T, Kuhn T, Webel C (2006) An ambient intelligence system to assist team training and competition in cycling. In: Moritz EF, Haake S (eds) Developments in sports. The engineering of sports 6, vol 1. Springer Science and Business Media, New York, pp 103–108
4. Sefati H, Gotzhein R, Kramer C, Schloesser S, Weiss M (2018) Dynamic overlay line topology establishment and repair in wireless networks. In: IEEE wireless communications and networking conference (WCNC 2018), Barcelona, Spain, 15–18 Apr 2018

Further References

5. Bluetooth SIG (2010) Bluetooth core specification version 4.0. Specification of the Bluetooth System
6. Chipcon AS, CC2420 ZigBee-ready RF transceiver. http://www-inst.eecs.berkeley.edu/~cs150/Documents/CC2420.pdf. Last accessed 27 Aug 2019
7. Du S, Saha AK, Johnson DB (2007) RMAC: a routing-enhanced duty cycle MAC protocol for wireless sensor networks. In: 26th IEEE international conference on computer communications (INFOCOM 2007), Anchorage, Alaska, 6–12 May 2007, pp 1478–1486
8. Elson J, Girod L, Estrin D (2002) Fine-grained network time synchronization using reference broadcasts. In: Proceedings of the fifth symposium on operating systems design and implementation (OSDI 2002), Boston, MA, Dec 2002
9. Ferrari F, Zimmerling M, Mottola L, Thiele L (2012) Low-power wireless bus. In: 10th ACM conference on embedded network sensor systems (SenSys'12), Toronto, Canada, 6–9 Nov 2012, pp 1–14
10. Crossbow Technology Inc., MICAz wireless measurement system. Document Part Nr 6020-0060-04 Rev A. http://www.openautomation.net/uploadsproductos/micaz_datasheet.pdf. Last accessed 27 Aug 2019
11. Polastre J, Hill J, Culler D (2004) Versatile low power media access for wireless sensor networks. In: Proceedings of the 2nd ACM conference on embedded networked sensor systems (SenSys 2004), Baltimore, 3–5 Nov 2004, pp 95–107
12. PermaSense (2018) University of Zürich (UZH) and Eidgenössische Technische Hochschule Zürich (ETH). http://www.permasense.ch/en.html. Last accessed 27 Aug 2019
13. Ye W, Heidemann JS, Estrin D (2002) An energy-efficient MAC protocol for wireless sensor networks. In: 21st IEEE conference on computer communications (INFOCOM 2002), New York, 23–27 June 2002, pp 1567–1576
14. Ye W, Heidemann J, Estrin D (2004) Medium access control with coordinated adaptive sleeping for wireless sensor networks. IEEE/ACM Trans Netw 12(3):493–506
15. ZigBee™ Alliance (2005) ZigBee specification, version 1.0. www.zigbee.org. Last accessed 27 Aug 2019, p 378

Chapter 8
Quality of Service Multicast Routing with Mobility Support

In multi-hop networks, routing is required to provide end-to-end communication. In this chapter, we explain foundations of routing in wireless ad hoc networks, present our protocol *quality of service (QoS) multicast routing (QMR)* for partially mobile wireless time-division multiple access (TDMA) networks [3] implemented in ProRoute of ProNet 4.0 [1], survey and compare related work, and draw conclusions.

8.1 Foundations

In this chapter, we provide the context of QoS routing in wireless ad hoc networks, explain concepts and requirements, and address areas of operation.

8.1.1 Context

The objective of *routing* is to discover and operate routes between sets of nodes in order to provide end-to-end communication in multi-hop networks. Routes are requested by distributed applications, by specifying information to identify and localize nodes. Localization can be based on topology information about nodes and links, position information determined by geographical coordinates, or content information about stored data.

Depending on communication requirements, different kinds of routes can be distinguished. Unicast routes are established between pairs of nodes. Broadcast routes distribute messages to all nodes; whereas, multicast routes support message transfer to particular groups of nodes. More advanced schemes are *n*-hop cast, where distribution

---

The original version of this chapter was revised: "$\mathrm{FSF}(p) =_{\mathrm{df}} \{n_{\mathrm{Free}}(v, v') > 0 \wedge$" has been corrected to "$\mathrm{FSF}(p) =_{\mathrm{df}} \{(v, v') \in \mathrm{p} \mid (v, v') \in \mathrm{NoDec}(p) \wedge n_{\mathrm{Free}}(v, v') > 0 \wedge$". The correction to this chapter is available at https://doi.org/10.1007/978-3-030-33319-5_14

R. Gotzhein, *Real-time Communication Protocols for Multi-hop Ad-hoc Networks*, Computer Communications and Networks,
https://doi.org/10.1007/978-3-030-33319-5_8

of messages is to all nodes in n-hop neighborhood, and concast, where a group of senders delivers messages to a single receiver, e.g., a sink node of a sensor network.

In distributed real-time computer systems, routes may have to satisfy specific quality of service (QoS) requirements regarding performance, reliability, guarantee, and synchronicity. For instance, routes may have to guarantee minimal end-to-end transmission rates or maximal transmission delays. This requires that routes are operated deterministically, which is a particular challenge in multi-hop wireless ad hoc networks.

Another challenge is node mobility, which may be required, e.g., for the operation of autonomous mobile robots in a production environment. To establish stable routes with predictable communication behavior, stationary nodes are preferable. If nodes can move, routes may easily break, impairing or even preventing deterministic guarantees.

8.1.2 Concepts and Requirements

Routing in computer networks is a broad topic. In this chapter, we set the focus on QoS routing in wireless time-division multiple access (TDMA) networks, with node localization based on topology information.

We model a network topology as a directed graph $G = (V, E)$, consisting of a set V of nodes and a set $E \subseteq V \times V$ of links (see Chap. 4). We distinguish the following types of topologies:

- A *communication topology* is modeled as a graph $G_C = (V, E_C)$, where links $e = (v_i, v_j)$ express that v_j is in communication range of v_i, i.e., v_j can successfully receive messages from v_i.
- An *interference topology* is modeled as a graph $G_I = (V, E_I)$, where links $e = (v_i, v_j)$ express that v_j is in interference range of v_i, i.e., v_i can interfere with a reception at v_j.

We assume that a node v_j in communication range of v_i is also in interference range of v_i. The graphs G_C and G_I can be merged into a combined network topology $G = (V, L, E)$, where $L = \{c, i\}$ is a set of labels distinguishing between different types of links. We require that the complete communication subgraph[1] $G_C = (V, \{c\}, E_C)$ is connected, i.e., for all pairs $(v, v') \in V \times V$ of distinct nodes, there is a directed path $p = (v_1, \ldots, v_{|p|+1}) \in V^+$ of communication links such that for all i, $1 \leq i \leq |p|$, $(v_i, v_{i+1}) \in E_C$, $v = v_1$, and $v' = v_{|p|+1}$, with $|p|$ denoting the number of edges of p.

Figure 8.1 shows a symmetrical topology. Solid lines represent communication links, which are interference links too. To simplify presentation, only communication links are shown in these cases. Additional interference links are depicted as dashed lines.

Based on the network model $G = (V, L, E)$, communication routes between sets of nodes can be identified. A unicast route is a directed loop-free communication path $p \in V^+$ between a pair of nodes. A multicast route is a rooted communication

[1] Obtained by reducing G to the set of all nodes and all communication links.

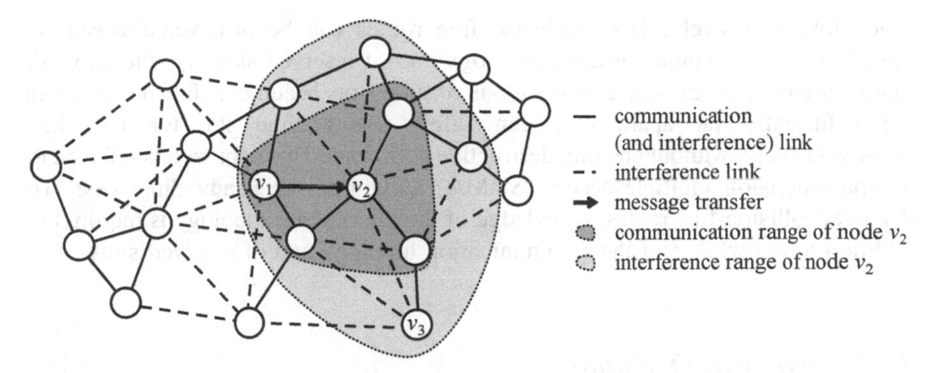

Fig. 8.1 Network topology and interference problem

tree $t \subseteq G$, i.e., a subgraph G' of G rooted at the source node forming a directed loop-free tree of communication links. A concast route can be characterized as an inverse rooted communication tree of G, consisting of links that lead from leaf nodes to the root node. Given these definitions, unicast routes are special cases of multicast routes.

Route discovery is based on network status information. Routing algorithms can be classified according to how detailed this information is, whether it is updated at runtime, and how these updates are triggered. A node has global network status information if it knows the entire network topology. For QoS routing, additional knowledge about link attributes, e.g., available bandwidth, average transmission delay, or statistical link reliability may be required. A node has aggregated network status information if only a summary of the network topology is available, e.g., hop distance or accumulated delays to other nodes.

In case of variable link quality or node mobility, it is necessary that network status information is updated while operating the network. A routing algorithm with this capability is called dynamic and static otherwise. Dynamic routing schemes are proactive if updates are performed independent of whether routes are being requested, e.g., periodically, and reactive otherwise.

In this book, we focus on core functionalities of wireless real-time networking with TDMA. The availability of network-wide TDMA, the possibility to reserve time slots, and information about communication and interference topologies form the basis for the discovery of routes that can be operated without the interference problem.

Definition 8.1 (*Interference problem*) Let $G = (V, L, E)$ be a network topology, v_1, $v_2 \in V$, and $(v_1, v_2) \in E_C$. Then, the transfer of a message from v_1 to v_2 can be disturbed by an overlapping transmission of another node v_3 in interference range of v_2, i.e., $(v_3, v_2) \in E_I$ (see Fig. 8.1).

The interference problem is caused by transmissions that overlap in time, frequency, and space, called *collisions*. To avoid collisions, it is sufficient to reserve time slots exclusively. Here, collision-free routes can be discovered based on knowledge of the communication topology and of reserved slots. As the network diameter increases, exclusive reservations of time slots become inefficient. If nodes are sufficiently far apart, they can safely use a time slot together, i.e., non-exclusively, without creating destructive collisions. This approach is referred to as space-division multiple access (SDMA) and improves bandwidth usage. To discover collision-free routes, knowledge of the interference topology is required in addition to knowledge of the communication topology and of reserved slots.

8.1.3 Areas of Operation

Routing is a core functionality of all kinds of multi-hop networks, including wireless ad hoc networks. In the Internet, exterior gateway protocols, e.g., Border Gateway Protocol (BGP) [4], determine routes between autonomous systems of providers; interior gateway protocols, e.g., Open Shortest Path First (OSPF) [8], are used inside autonomous systems. Wireless technologies such as ZigBee [12] incorporate routing protocols for the dynamic formation of mesh networks.

In wireless networked control systems, routing protocols developed for the Internet or for ZigBee are not suitable, due their best-effort characteristics. Instead, QoS routing protocols are required. WirelessHART [11] incorporates a proactive QoS routing protocol for the dynamic formation of mesh networks. Here, route discovery is performed by a network manager and based on global network status information, yielding routes with exclusively reserved time slots[2] and deterministic end-to-end delay.

In the task Adaptive Services in Networked Production Environments of the project Software Innovations for the Digital Enterprise (SINNODIUM) of the Software-Cluster [10], the demonstrator vertical integration of production processes (VI-P) has been developed and deployed (for details, see Sect. 12.3). Part of VI-P is a wireless mesh network of stationary sensor nodes monitoring the status of factory devices, room temperatures, and gas concentrations, and a gateway node collecting and aggregating sensor readings, and forwarding them to the enterprise resource planning (ERP) system. Furthermore, a mobile autonomous robot supporting remote factory monitoring and maintenance, which is part of the wireless mesh network, is operated. To provide real-time end-to-end group communication in this partially mobile wireless network, we have devised a QoS multicast routing algorithm based on TDMA [3].

[2]WirelessHART also applies frequency-division multiple access (FDMA) techniques.

8.2 QoS Multicast Routing (QMR)

In this chapter, we present *QoS multicast routing* (*QMR*) for partially mobile wireless TDMA networks [3]. QMR has been inspired by the communication requirements of industrial networked control systems consisting of stationary and mobile nodes. The protocol module ProRoute of the ProNet 4.0 protocol stack [1] implements QMR.

For the operation of QMR, we assume that the stationary nodes form a network, i.e., there is a path of communication links between all pairs of stationary nodes. Furthermore, we assume that each mobile node is in communication range of at least one stationary node at all times. In a closed industrial environment, both assumptions can be satisfied by placing stationary nodes until connectivity and spatial coverage are achieved.

We now present QMR in several steps, starting with the iterative construction of QoS routing trees in networks consisting of stationary nodes only, and then extending this algorithm to support mobile nodes too. We motivate and explain our design decisions, formalize constraints for collision-free operation, introduce heuristics for route selection and slot scheduling, and present the results of simulation experiments. As the problem of finding optimal QoS routing trees and slot schedules is NP-complete, QMR relies on heuristics either to achieve small end-to-end delay or to favor route feasibility and slot reuse.

8.2.1 Context and Design Decisions for Stationary Nodes

QMR is specifically devised for wireless networked control systems consisting of nodes exchanging real-time control information. Conceptually, sensor and actuator nodes provide periodical and event-triggered application services, e.g., temperature readings, alarm notifications, and control value settings. Controller nodes use these services to collect sensor values, and to determine and apply control values, thereby building feedback loops. In a usage scenario, sensor and actuator nodes register their services in a service registry (see Chap. 10). Controller nodes look up, subscribe to, and unsubscribe from services dynamically.

For each subscription to a service, a QoS route between the service provider and the subscriber, i.e., the service user, is to be established. If several controllers subscribe to the same sensor service, a multicast route from the service provider to all subscribers can decrease network load compared to one unicast route per subscription. Furthermore, if subscriptions and/or unsubscriptions occur at different points in time, existing multicast routes have to be extended and/or reduced during operation, to avoid service interruptions.

In a network consisting of a set V_s of stationary nodes only, it is sufficient to build a QoS routing tree consisting of a single branch when the first subscription to a service is made. For subsequent subscriptions and unsubscriptions, the tree is extended and reduced, respectively. Furthermore, to satisfy QoS requirements, suitable slot reservations along the branches are to be handled.

In wireless networked control systems, routes have to satisfy QoS requirements regarding timeliness, bandwidth, and reliability. In TDMA networks, this can be achieved by scheduling suitable time slots along the branches of a routing tree. To guarantee timeliness, time slots of each branch are to be scheduled such that the time span from the start of the slot on the first hop until the end of the slot on the last hop matches the specified end-to-end delay. To provide required bandwidth, slot reservations on each branch are to be repeated with sufficient periodicity. To achieve reliability, slot reservations have to be sufficiently exclusive such that the interference problem (see Definition 8.1) is solved.

For the discovery of QoS routing trees satisfying these requirements, precise information about the network status is needed. First, this concerns the knowledge of the communication topology $G_C = (V_s, E_C)$, to determine trees consisting of communication links $(v, v') \in E_C$. Second, to select time slots $s \in S$, information about their current reservation status $TX^s \subseteq E_C$ recording whether s is reserved for transmission on links $(v, v') \in E_C$ has to be available. Third, to solve the interference problem, the interference topology $G_I = (V_s, E_I)$ must be known.

At this point, design decisions about the provision of network status information and the location of routing decisions are due. In Chap. 4, we have presented Automatic Topology Discovery Protocol (ATDP) for the automatic detection of communication, interference, and sensing topologies. ATDP is executed before regular network operation is started and terminates once a stable topology, where all links are in states *stable* or *fluctuating*, has been determined and disseminated. Since we are dealing with a network of stationary nodes in a closed industrial environment, this solves the problem of acquiring knowledge of communication and interference topology, which is available to all nodes.

Discovery of QoS multicast routes can be performed by a centralized routing manager or in a decentralized way by distributing route discovery and slot scheduling among network nodes. With QMR, we have decided for a centralized solution, for a number of reasons:

- First, a centralized solution avoids the problem of mutual blocking of resources due to concurrent QoS route searches. When exploring route alternatives, distributed routing protocols have to pre-reserve resources, e.g., time slots, until a final route decision has been reached. If nodes have to contend for the same resources, route searches may fail despite the existence of feasible QoS routes. With a centralized routing manager, route searches can easily be serialized, which rules out mutual blocking of resources.

  Using Arbitration and Cooperative Transfer Protocol (ACTP, see Chap. 6), it is possible to serialize distributed route searches in a decentralized routing protocol. Before triggering a route search, nodes perform network-wide deterministic arbitration, with only the winner starting the search. In [2], we have devised Black Burst-based QoS routing (BBQR), a decentralized QoS routing protocol for wireless TDMA networks applying this idea. However, given its overhead and complexity, we have reached the conclusion that a centralized approach is preferable.

- Second, a centralized approach reduces management traffic substantially. For QoS routing, network status information has to be collected and updated. In addition to topology information G_C and G_I provided by ATDP, this concerns the current reservation status TX^s. In a centralized approach, the routing manager keeps track of this status, without sharing it with other nodes. In the decentralized case, reservations have to be negotiated with other nodes, and reservation decisions have to be exchanged, which produces an enormous amount of management traffic. In particular, status information has to be exchanged with interference neighbors, which is very difficult to achieve, as these nodes may not be in communication range. Furthermore, nodes repeatedly exchange route request and reply messages when searching for routes. With a centralized routing manager, this management traffic is obsolete.
- Third, unless using exclusively reserved time slots, management traffic is prone to collisions and loss. Route searches may fail because route request messages are corrupted. Furthermore, route searches may fail or yield unfeasible routes because the reservation status of nodes is inconsistent, due to loss or delay of status messages. Again, this problem is avoided by a centralized approach.

A drawback of centralized routing is its lack of scaling, as at some point, the routing manager may become a bottleneck. Furthermore, it is a single point of failure. However, given the advantages and considering the fact that we are dealing with relatively small wireless ad hoc networks consisting of up to 100 nodes, we have decided to adopt the centralized approach to QoS routing.

8.2.2 Slot Inhibited Policies

To solve the interference problem, information about communication and interference topologies $G_C = (V_s, E_C)$ and $G_I = (V_s, E_I)$ and about the current reservation status $TX^s \subseteq E_C$ of slots $s \in S$ is required. Here, V_s denotes the set of stationary nodes, and $(v, v') \in TX^s$ expresses that slot s is reserved for transmission from v to v'.

We now introduce a reservation criterion $F^s_{TX} \subseteq E_C$, with $(v, v') \in F^s_{TX}$ expressing that slot s is free for transmission from v to v', without causing any interference problems with already existing reservations. In [9], the constraints of F^s_{TX} are called *slot inhibited policies*.

Definition 8.2 (*Communication and interference distance*) Let G_C and G_I be communication and interference topologies of a network G. Then, the *communication distance* $d_{G_c}(v, v')$ between nodes v, $v' \in V_s$ is defined as

$$d_{G_c}(v,v') =_{\text{df}} \min_{p \in P_{G_c}(v,v')} |p|$$

where $P_{G_c}(v,v')$ is the set of all cycle-free communication paths starting in v and ending in v'. *Interference distance* $d_{G_I}(v,v')$ is defined analogously.

Definition 8.3 (*Communication and interference neighborhood*) Let G_C and G_I be communication and interference topologies of a network G, $v \in V_s$, and $i \geq 0$ be an integer value. The *i-hop communication neighborhood* $CN_{\leq i}(v)$ and *i-hop interference neighborhood* $IN_{\leq i}(v)$ of v are defined as follows:

$$CN_{\leq i}(v) =_{\text{df}} \{v' \in V_s | d_{G_c}(v,v') \leq i\}$$
$$IN_{\leq i}(v) =_{\text{df}} \{v' \in V_s | d_{G_I}(v,v') \leq i\}$$

Since we assume that nodes in communication range are also in interference range, we have $CN_{\leq i}(v) \subseteq IN_{\leq i}(v)$.

Definition 8.4 (*Reservation status*) Let G_C be communication topology of a network G, and $s \in S$ be an exclusive slot. The *reservation status* $TX^s \subseteq E_C$ of slot s defines for all links $(v,v') \in E_C$, whether s is reserved for transmission from v to v', provided $v' \in CN_{\leq 1}(v) \backslash \{v\}$. The following relations are derived from TX^s as follows:

$TX^s(v) =_{\text{df}} \exists v' \in V_s.TX^s(v,v')$	s is reserved by v for transmission
$RX^s(v,v') =_{\text{df}} TX^s(v,v')$	s is reserved by v for reception from v'
$RX^s(v) =_{\text{df}} \exists v' \in V_s.RX^s(v,v')$	s is reserved by v for reception

We call the subset $TX^s_v = \{TX^s(v_1,v_2) | v = v_1\}$ reservation status of v for s.

Definition 8.5 (*Reservation criterion*) Let G_I be the interference topology of a network G, $v, v' \in V_s$, $s \in S$ be an exclusive slot, and TX^s be the reservation status of s. The *reservation criterion* F^s_{TX} defines whether s is free for transmission from v to v' as follows:

$$F^s_{\text{TX}}(v,v') =_{\text{df}} \forall v_1 \in IN_{\leq 1}(v).\neg RX^s(v_1) \wedge \forall v_2 \in IN_{\leq 1}(v').\neg TX^s(v_2)$$

If the reservation criterion is satisfied for a slot $s \in S$ and nodes v, $v' \in V_s$, slot s can be reserved without causing an interference problem.

Figure 8.2 illustrates the reservation criterion, showing the interference ranges of nodes v_1 and v_2. To check the reservation criterion, the reservation status TX^s_v of all nodes $v \in IN_{\leq 1}(v_1) \cup IN_{\leq 1}(v_2)$ and of nodes v_3 and v_4 is required. Thus, despite a network diameter of seven communication hops, the reservation status of all but four nodes is needed for an accurate reservation decision. In a decentralized routing approach, this status information has to be exchanged before a reservation decision for link (v_1, v_2) can be made, causing substantial management traffic and bearing the risk of inconsistent decisions.

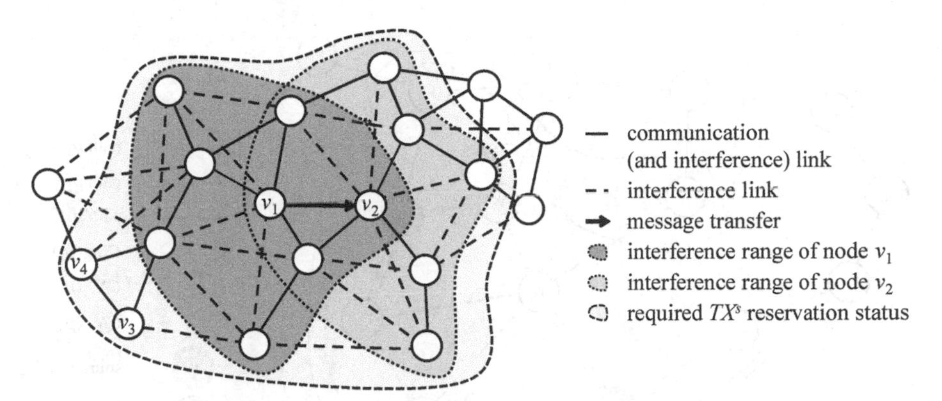

Fig. 8.2 Illustration of the reservation criterion

8.2.3 *Routing Tree Discovery for Stationary Nodes*

In QMR, routing tree discovery is an iterative process, starting with the search of a path from source node to the first destination node and extending this initial routing tree. This design decision is in line with the application context, with subscriptions and/or unsubscriptions taking place at different points in time. To determine the initial routing tree, QMR computes the set of shortest paths measured in communication hops. If more than one shortest path is found, a heuristics is applied to select a path running through denser parts of the network. To formalize this criterion, we extend the definition of communication neighborhood (see Definition 8.3) to routing trees, which includes paths as special case.

Definition 8.6 (*Communication neighborhood*) Let G_C be the communication topology of a network G, $t \subseteq G_C$ be a routing tree, and $i \geq 0$ be an integer value. The *i-hop communication neighborhood* $CN_{\leq i}(t)$ of t is defined as follows:

$$CN_{\leq i}(t) =_{\text{df}} \bigcup_{v \in t} CN_{\leq i}(v)$$

To construct an initial routing tree consisting of a single path, QMR selects a shortest path p with the largest single-hop communication neighborhood $CN_{\leq 1}(p)$. The rationale behind this heuristics is that when adding further destination nodes, it can be expected that the required extensions of the routing tree are smaller, which reduces bandwidth consumption, as less additional time slots are needed. A drawback of this heuristics is that in denser parts of the network, there are less free time slots, which could render a selected path unfeasible. So, a better heuristics could be to select a path with the smallest $CN_{\leq 1}(p)$. We leave this for further study.

Figure 8.3 illustrates the selection of the initial routing tree for source node v_{11} and destination node v_8. There are three shortest path candidates p_1, p_2, and p_3 of

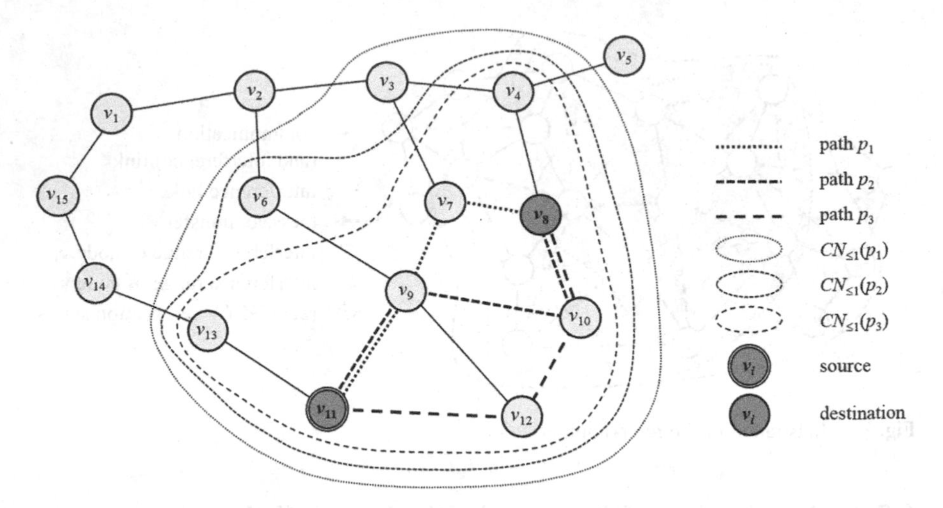

Fig. 8.3 Illustration of initial routing tree selection

length 3. Here, path p_1 has the largest single-hop communication neighborhood $CN_{\leq 1}(p_1)$, as it contains the neighborhoods of p_2 and p_3, and is therefore selected.

While the selection of a shortest path in hops has its merits in wired networks, it is in general not the first choice in wireless networks, as it gives preference to longer and therefore less reliable hops. However, QMR builds on a communication topology determined by ATDP, which consists of stable and therefore reliable communication links only. Therefore, in this particular case, selection of a shortest path is a good choice, especially as it optimizes bandwidth consumption measured as the number of required time slots.

To extend an existing routing tree t with source node v by a route to a new destination node v'', QMR starts by determining, for each node $v' \in t$, a shortest path $p_{v', v''}$ from v' to v''. If there are several shortest paths, a path with the largest single-hop communication neighborhood is chosen. Then, for each preselected path $p_{v',v''}$, the length of the path $p_{v'} = p_{v,v'} \bullet p_{v',v''}$ from source node v to new destination node v'' via v' is determined. Here, $p_{v,v'}$ is a branch of the existing routing tree, $\bullet$ denotes concatenation of paths. From the candidate set of shortest paths $p_{v'}$, a path that adds the smallest number of links is selected, because this optimizes the additional bandwidth consumption. If there still are several candidate paths, a path $p_{v'}$ adding the largest single-hop communication neighborhood to t is chosen, with $p_{v',v''}$ appended to the existing routing tree. As before, the rationale of the second selection criterion is that further extensions of the routing tree can be kept small.

Figure 8.4 illustrates the extension of an existing routing tree t rooted at node v_{11} by a branch to node v_1. The shortest paths $p_{4,1}$, $p_{7,1}$, $p_{9,1}$, and $p_{13,1}$ starting at nodes already belonging to t all have length 3, while shortest paths $p_{5,1}$, $p_{8,1}$, and $p_{11,1}$ (not highlighted in the figure) have length 4. Among the concatenated paths p_4, p_7, p_9, and p_{13}, p_9 and p_{13} are the shortest paths. Since p_{13} yields the largest extension of

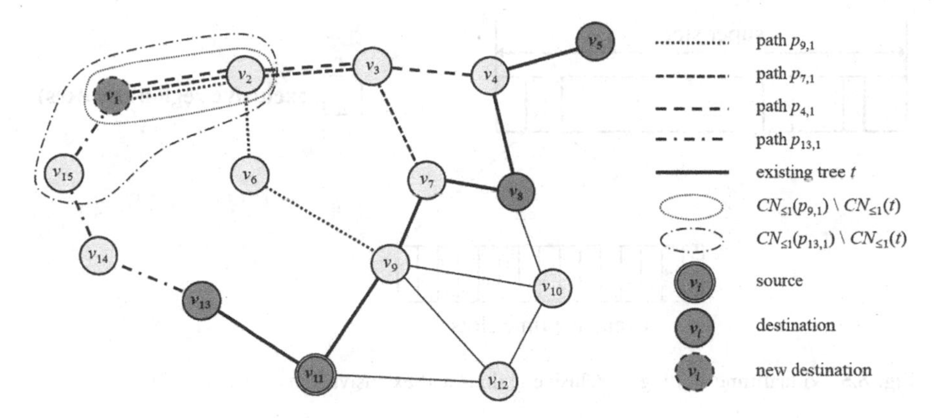

Fig. 8.4 Illustration of the extension of an existing routing tree

the single-hop neighborhood of t of the remaining candidates, it is selected, with $p_{13,1}$ added to the existing routing tree.

Up to this point, routing trees are formed without considering the availability of time slots. This implies that a selected routing tree may be unfeasible due to lack of suitable free slots, while another routing tree might be feasible. There are two possible strategies to overcome this problem. First, we could check the availability of time slots already when searching for a routing tree. Second, if a search fails, we can repeat the search by considering different trees. Both strategies increase the complexity of route search and may detect larger routing trees that consume more bandwidth. In case of QMR, we have decided against both strategies, as they are only required when network load is already rather high, which reduces the chances of finding another feasible path anyways. Instead, we optimize slot usage by keeping routing trees small.

8.2.4 Slot Scheduling for Stationary Nodes

We now extend the iterative construction of routing trees by an algorithm for slot scheduling, i.e., selection and reservation of time slots along the branches of a routing tree. We assume that there are exclusive regions decomposed into exclusive slots of the same size d_{slot} (see Fig. 8.5 and Sect. 5.2.1).

When a message is sent from node v_1 to node v_2 in communication range of v_1 in time slot s, we assume that node v_2 returns an acknowledgment in the same time slot. First, this requires symmetrical communication links, which are detected by ATDP (see Chap. 4). Second, this implies that nodes v_1 and v_2 are both senders and receivers in time slot s. Thus, to avoid interference, we have to strengthen the reservation criterion in Definition 8.5 as follows:

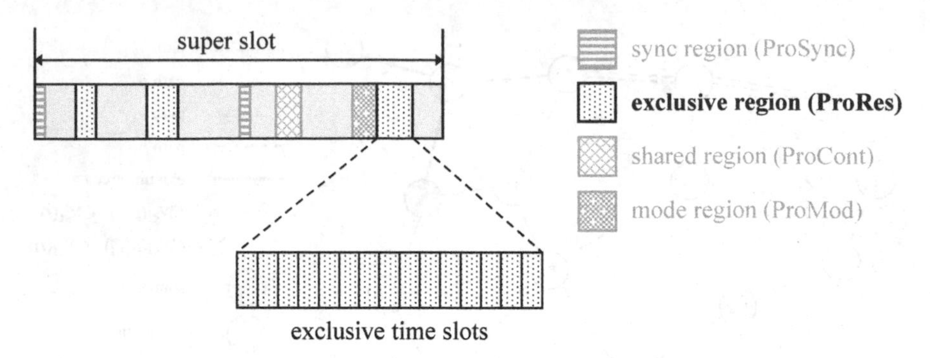

Fig. 8.5 Virtual time slotting—exclusive regions and exclusive slots

Definition 8.7 (*Strengthened reservation criterion*) Let G_I be interference topology of a network G, v, $v' \in V_s$, $s \in S$ be an exclusive slot, and TX^s be the reservation status of s. The *strengthened reservation criterion* $\overline{F}^s_{\mathrm{TX}}$ defines whether s is free for transmission from v to v' and vice versa as follows:

$$\begin{aligned}\overline{F}^s_{\mathrm{TX}}(v,v') &=_{\mathrm{df}} F^s_{\mathrm{TX}}(v,v') \wedge F^s_{\mathrm{TX}}(v',v)\\ &= \forall v_1 \in IN_{\leq 1}(v) \cup IN_{\leq 1}(v').\neg TX^s(v_1) \wedge \neg RX^s(v_1)\end{aligned}$$

QMR constructs routing trees iteratively, by determining an initial path and appending paths as required. In each iteration step, time slots along the path to be added are scheduled, i.e., selected and reserved. In the following, we assume that the bandwidth requirements of a route are satisfied by one time slot per hop and super slot. If this is not sufficient, several time slots can be reserved. Furthermore, we assume that there are no constraints regarding the placement of reserved slots of a routing tree in a super slot.

QMR offers two slot scheduling strategies: achieving small end-to-end delay and high route feasibility. Furthermore, QMR exploits local multicasting to reduce bandwidth consumption.

Small end-to-end delay—shortDelay

For a routing tree t, the theoretical lower bound for end-to-end delay is given as the product of routing tree depth and exclusive slot duration, i.e., $depth(t) \cdot d_{\mathrm{slot}}$. This lower bound is only reached if consecutive exclusive slots of the same exclusive region are reserved for all branches of t, in the order defined by t. In practice, to achieve small end-to-end delay, gaps between reserved slots belonging to the same or different exclusive regions should be kept small.

To achieve low end-to-end delays, QMR uses the slot scheduling strategy *shortDelay*, which applies the slot decision policy nearest slot first (NSF) to links in the order of the path. Informally, given the slot selected on the previous hop of a branch of a routing tree t, NSF selects the next slot that is free for transmission on the next hop.

Definition 8.8 (*Nearest Slot First* (*NSF*)) Let $s_0,\ldots,s_{n-1}$ be the sequence of exclusive slots located in the exclusive region(s) of a super slot. Let s_i be the starting point for searching a slot that is free for transmission from node v to node v'.

- When forming an initial routing tree, the starting point s_i for this search can be any slot, e.g., s_0 (First Fit) or slot s_k reached after the last search (Rotating First Fit).
- When extending an existing routing tree, starting point s_i is the slot $s_{(k+1) \bmod n}$ following the reserved slot s_k on the previous hop.

The *slot decision policy NSF* performs a rotating forward search starting at s_i and selecting the first slot $s_k = s_{(i+j) \bmod n}$, with $0 \leq j \leq n{-}1$, that is free for transmission from v to v' if such a slot exists. Formally,

$$
\begin{aligned}
\mathrm{NSF}(s_i, v, v') =_{\mathrm{df}} \quad & \textbf{if } \exists j.(0 \leq j \leq n-1 \wedge \overline{F}_{\mathrm{TX}}^{s_j}(v, v')) \\
& \textbf{then } s_{(i+j) \bmod n} \textbf{ such that} \\
& \qquad j = \min\{k \mid 0 \leq k \leq n-1 \wedge \overline{F}_{\mathrm{TX}}^{s_{(i+k) \bmod n}}(v, v')\} \\
& \textbf{else} \text{ undefined}
\end{aligned}
$$

If NSF returns valid time slots for all hops of the path to be appended to the routing tree, a feasible extended routing tree has been determined. Otherwise, the extension is unfeasible.

Table 8.1 shows slot reservations and blockings for a scenario based on the topology in Fig. 8.4, with five exclusive slots. To keep the scenario small, we assume that all nodes in interference range are also in communication range. In a first route search, the QoS route $p = (v_4, v_3, v_2, v_1)$ has been established, with the reservations shown in the table (regular font). In addition, the table shows for which nodes a slot is blocked due to reservation of a slot in interference range (regular font). If the scheduled slots belong to the same exclusive region, end-to-end delay for p is $3 \cdot d_{\mathrm{slot}}$, i.e., minimal.

Next, a slot schedule for the path $p' = (v_{11}, v_9, v_7, v_8)$ with small end-to-end delay is to be determined. Slot search starts on link (v_{11}, v_9) at s_3 (Rotating First Fit). Applying the strengthened reservation criterion, it turns out that slot s_3 is free for transmission from node v_{11} to node v_9, which is therefore scheduled (bold entries in the table). Furthermore, v_6, v_7, v_{10}, v_{12}, and v_{13} in interference range of v_9 or v_{11} are now blocked in s_3. On the next hop from v_9 to v_7, slot search starts in s_4, which is successfully scheduled as shown in the table.

On the last hop from v_7 to v_8, slot search starts in $s_{(4+1) \bmod 5}$, which is s_0. However, s_0 is already blocked for both nodes, i.e., the strengthened reservation criterion is not satisfied. Since the next slot s_1 is blocked for v_7, slot search continues with s_2, where communication between v_7 and v_8 does not interfere with v_1 and v_2. Therefore, s_2 can be scheduled again, increasing bandwidth utilization due to SDMA.

Table 8.1 Slot reservations and blockings "Small end-to-end delay"—QoS routes $p = (v_4, v_3, v_2, v_1)$ and $p' = (v_{11}, v_9, v_7, v_8)$ (for the topology, see Fig. 8.4)

Mode	Slot				
	s_0	s_1	s_2	s_3	s_4
v_1		x	$RX^{s2}(v_1, v_2)$		
v_2	x	$RX^{s1}(v_2, v_3)$	$TX^{s2}(v_2, v_1)$		
v_3	$RX^{s0}(v_3, v_4)$	$TX^{s1}(v_3, v_2)$	x		x
v_4	$TX^{s0}(v_4, v_3)$	x	x		
v_5	x				
v_6		x	x	x	x
v_7	x	x	**$TX^{s2}(v_7, v_8)$**	**x**	**$RX^{s4}(v_7, v_9)$**
v_8	x		**$RX^{s2}(v_8, v_7)$**		**x**
v_9			**x**	**$RX^{s3}(v_9, v_{11})$**	**$TX^{s4}(v_9, v_7)$**
v_{10}			**x**	**x**	**x**
v_{11}				**$TX^{s3}(v_{11}, v_9)$**	**x**
v_{12}				**x**	**x**
v_{13}				**x**	
v_{14}					
v_{15}			x		

$TX^s(v_i, v_j)$: slot s reserved for transmission from v_i to v_j
$RX^s(v_i, v_j)$: slot s reserved for reception by v_i from v_j
x: blocked due to reservation in interference range

If the scheduled slots belong to the same exclusive region, end-to-end delay for p' is $n \cdot d_{\text{slot}}$. The reason is that after slot s_4, slot s_2 is scheduled, which can only be used in the following super slot. In the given scenario and with the search heuristics, this is the best possible result. Starting slot search at s_0 (First Fit) would have produced a better result, with an end-to-end delay of only $4 \cdot d_{\text{slot}}$. However, First Fit leads to a concentration of scheduled slots at the beginning of a super slot, which is not always desirable. Furthermore, with only five exclusive slots per super slot, the options in this simple scenario are of course very much restricted.

High route feasibility—highReuse

To achieve high route feasibility and to enhance slot reuse, QMR uses the slot scheduling strategy *highReuse*, which applies slot decision policies Fewest Slots First (FSF), Least Conflicts First (LCF), and Most Reuse First (MRF), in that order. Similar slot decision policies have been introduced in [9], in a decentralized approach. In the following, we define these policies and illustrate their application.

Table 8.2 shows slot reservations and blockings for a scenario based on the topology in Fig. 8.4, with five exclusive slots. To keep the scenario small, we assume that all nodes in interference range are also in communication range. In preceding route searches, the QoS routes $p = (v_2, v_3, v_4, v_5)$ and $p' = (v_{12}, v_{11})$ have been established, with the reservations and blockings shown in the table (regular font).

Table 8.2 Slot reservations and blockings "high route feasibility"—QoS routes $p = (v_2, v_3, v_4, v_5)$, $p' = (v_{12}, v_{11})$, and $p'' = (v_{11}, v_9, v_7, v_8)$ (for the topology, see Fig. 8.4)

Mode	Slot				
	s_0	s_1	s_2	s_3	s_4
v_1	x				
v_2	$TX^{s0}(v_2, v_3)$	x			
v_3	$RX^{s0}(v_3, v_2)$	$TX^{s1}(v_3, v_4)$	x	x	
v_4	x	$RX^{s1}(v_4, v_3)$	x	$TX^{s3}(v_4, v_5)$	
v_5		x		$RX^{s3}(v_5, v_4)$	
v_6	x			x	
v_7	x	x	$TX^{s2}(v_7, v_8)$	$RX^{s3}(v_7, v_9)$	
v_8		x	$RX^{s2}(v_8, v_7)$	x	
v_9	$RX^{s0}(v_9, v_{11})$		x	$TX^{s3}(v_9, v_7)$	
v_{10}	x		x	x	
v_{11}	$TX^{s0}(v_{11}, v_9)$		$RX^{s2}(v_{11}, v_{12})$	x	
v_{12}	x		$TX^{s2}(v_{12}, v_{11})$	x	
v_{13}	x		x		
v_{14}					
v_{15}					

$TX^s(v_i, v_j)$: slot s reserved for transmission from v_i to v_j
$RX^s(v_i, v_j)$: slot s reserved for reception by v_i from v_j
x: blocked due to reservation in interference range

Next, a slot schedule for path $p'' = (v_{11}, v_9, v_7, v_8)$ with high route feasibility is to be determined. To select a link of p'' where we start slot selection, we apply the first slot decision policy FSF:

Definition 8.9 (*Fewest Slots First* (*FSF*)) Let S be the set of exclusive slots of a super slot, p be a path for which slots are to be selected, and $\text{NoDec}(p) \subseteq p$ be the non-empty set of links for which no slot decision has been made so far. The *slot decision policy* FSF determines the link to be considered next for slot selection as some link with the fewest number of slot choices:

$$\begin{aligned} \text{Free}(v, v') &=_{\text{df}} \{s \in S | \bar{F}^s_{\text{TX}}(v, v')\} \\ n_{\text{Free}}(v, v') &=_{\text{df}} |\text{Free}(v, v')| \\ \text{FSF}(p) &=_{\text{df}} \{(v, v') \in \text{p} \mid (v, v') \in \text{NoDec}(p) \wedge n_{\text{Free}}(v, v') > 0 \wedge \\ &\quad \forall (v_1, v_2) \in \text{NoDec}(p).\, n_{\text{Free}}(v, v') \leq n_{\text{Free}}(v_1, v_2)\} \end{aligned}$$

FSF captures the heuristics that to improve the chances of finding a feasible route, slot search should start on a link with the smallest number of slot choices. If there are no choices, i.e., $\text{FSF}(p) = \{\}$ although $\text{NoDec}(p) \neq \{\}$, the route search terminates without success. If there are one or more choices, one link is selected, and further slot decision policies are applied.

When starting with an empty tree t and $p'' = (v_{11}, v_9, v_7, v_8)$ (see Table 8.2 and Fig. 8.4), we get $n_{\mathrm{Free}}(v_{11}, v_9) = 4$ (free slots s_0, s_1, s_3, s_4), $n_{\mathrm{Free}}(v_9, v_7) = 2$ (free slots s_3, s_4), and $n_{\mathrm{Free}}(v_7, v_8) = 2$ (free slots s_2, s_4). FSF(p'') returns the links with the least number of choices, i.e., $\{(v_9, v_7), (v_7, v_8)\}$. We continue with one of these links, e.g., (v_9, v_7). To select a slot for (v_9, v_7), we apply the second slot decision policy LCF.

Definition 8.10 (*Least Conflicts First* (*LCF*)) Let p be a path for which slots are to be selected, NoDec(p) $\subseteq p$ be the non-empty set of links for which no slot decision has been reached so far, and $(v, v') \in$ FSF(p). Then, the *slot decision policy* LCF reduces the set of free slots Free(v, v') to those with the least conflicts with free slots of the previous two and next two hops of p, restricted to links for which no decision has been reached yet.

LCF captures the heuristics that to improve the chances of finding a feasible route, the options of links that may be affected by a slot selection should be reduced as little as possible. In the example, the choice of s_4 on link (v_9, v_7) reduces the options of links (v_{11}, v_9) and (v_7, v_8). However, selecting s_3 affects link (v_{11}, v_9) only, which is therefore the better choice.

Since there are further links for which no slot decision has been reached so far, we continue by applying FSF to the remaining links, yielding $n_{\mathrm{Free}}(v_{11}, v_9) = 3$ (free slots s_0, s_1, s_4) and $n_{\mathrm{Free}}(v_7, v_8) = 2$ (free slots s_2, s_4). Applying LCF to (v_7, v_8) leads to the selection of s_2 on link (v_7, v_8). After that, only link $(v_{11}, v_9) \in p''$ remains without slot decision, with free slots s_0, s_1, and s_4 left for selection after application of FSF and LCF. To select a slot for (v_{11}, v_9), we apply the slot decision policy MRF.

Definition 8.11 (*Most Reuse First* (*MRF*)) Let $S' \subseteq S$ be the set of free slots on link $(v, v') \in$ FSF(p) after applying LCF. The *slot decision policy MRF* reduces S' to the set of slots with the highest utilization so far as follows:

$$n_{\mathrm{Blocked}}(s, v, v') =_{\mathrm{df}} |\{v_1 \in IN_{\leq 1}(v) \cup IN_{\leq 1}(v') \mid \neg\exists\, v_2 \in CN_{\leq 1}(v_1).\bar{F}^{s}_{\mathrm{TX}}(v_1, v_2)\}|$$
$$\mathrm{MRF}(p, v, v', S') =_{\mathrm{df}} \{s \in S' \mid (v, v') \in \mathrm{NoDec}(p) \wedge$$
$$\forall s' \in S'.\, n_{\mathrm{Blocked}}(s, v, v') \geq n_{\mathrm{Blocked}}(s', v, v')\}$$

MRF captures the heuristics that high reuse increases the degree of SDMA and the chances of finding feasible routes in subsequent route searches. In the example, $n_{\mathrm{Blocked}}(s_0, v_{11}, v_9) = 2$, $n_{\mathrm{Blocked}}(s_1, v_{11}, v_9) = 1$, and $n_{\mathrm{Blocked}}(s_4, v_{11}, v_9) = 0$, yielding *MRF*($p''$, v_{11}, v_9, $\{s_0, s_1, s_4\}$) = $\{s_0\}$. Therefore, slot s_0 is selected for link (v_{11}, v_9). The slot schedule for path $p'' = (v_{11}, v_9, v_7, v_8)$ and the additional blockings are shown in Table 8.2 (bold font).

In the scenario in Table 8.2, seven transmissions are scheduled in four slots, which shows that the slot decision policies FSF, LCF, and MRF indeed foster SDMA and high slot usage. Up to now, slot s_4 is not reserved at all, and some of the other slots are still free for reservation on one or more links. For instance, slot s_1 can be reserved on links (v_{12}, v_{11}), (v_{11}, v_{13}), (v_{13}, v_{14}), and (v_{14}, v_{15}) in both directions. Furthermore, although already reserved twice, slots s_2 and s_3 are still free on link (v_1, v_{15}), in both directions.

We formally capture slot utilization by the utilization rate r_{util} defined as the number of time slots used for transmission or receptions divided by the number of time slots blocked for transmissions or receptions due to reservations in single-hop interference neighborhood (provided there is at least one slot usage):

$$r_{\text{util}} =_{\text{df}} \frac{|\{(s,v) \in S \times V \mid TX^s(v) \vee RX^s(v)\}|}{|\{(s,v) \in S \times V \mid \exists v' \in IN_{\leq 1}.(TX^s(v') \vee RX^s(v')\}|}$$

A high value of r_{util} increases the potential for SDMA, because it increases the options of reserving slots multiple times. For the scenarios in Tables 8.1 and 8.2, we obtain $r_{\text{util}} = \frac{6+6}{6+6+25} \approx 0.32$ and $r_{\text{util}} = \frac{6+2+6}{6+2+6+22} \approx 0.39$. To reach the theoretical optimum of 1, no slots should be blocked unless used for transmission or reception. In practice, this optimum cannot be reached in multi-hop networks.

We conclude by pointing out that the slot decision policies FSF, LCF, and MRF need a more rigorous and generalized treatment. To keep this chapter concise, we have presented the underlying ideas in a semi-formal way.

Local multicasting

When a path is added to an existing tree, suitable slots along the new branch have to be scheduled. At the node where the branch starts, QMR examines the possibility of applying local multicasting, to reduce bandwidth consumption. For instance, if path (v_9, v_6, v_2, v_1) is to be attached to the routing tree in Fig. 8.4, there is already a time slot s selected on link (v_9, v_7). If s can also be used on link (v_9, v_6), i.e., if $\forall v \in IN_{\leq 1}(v_6) \backslash \{v_9\}. \neg TX^s(v) \wedge \neg RX^s(v)$ holds, v_6 can receive the transmission of v_9 in the same time slot.

When using local multicasting, several acknowledgments are to be exchanged in the same time slot. As QMR supports up to three receivers per local multicast, exclusive time slots are sized such that up to three acknowledgments fit. The order of acknowledgments is determined by the routing master and forwarded to nodes on the route.

8.2.5 Design Decisions for Partially Mobile Networks

We now generalize QMR to support QoS multicast routing in wireless networked control systems consisting of sets of stationary and mobile nodes V_s and V_m. Here, mobile nodes, e.g., autonomous robots, may act as service providers, by hosting sensors and actuators, and as service users executing control functionalities.

A straightforward approach is to include mobile nodes in QoS routing trees, without making a distinction between stationary and mobile nodes. This, however, yields routes that may break quickly due to node movements. Furthermore, it renders SDMA for slot reuse unfeasible, as mobile nodes may appear anywhere in the network topology. Therefore, QoS multicast routing in partially mobile networks requires a more sophisticated approach.

For the operation of QMR, we assume that the stationary nodes form a stable communication topology, which is essential for achieving stable routes. Furthermore, we stipulate that each mobile node be always in communication range of at least one stationary node. For better performance, a minimal subset $V_a \subseteq V_s$ of stationary nodes called *access nodes* providing spatial coverage can be determined for this purpose. Thus, mobile nodes have single-hop access to the network of stationary nodes, a property that makes the establishment of stable QoS routing trees with mobile nodes as source or destination nodes feasible. In addition, it opens potential for efficient slot schedules.

8.2.6 Routing Tree Discovery in Partially Mobile Networks

To incorporate mobile nodes, we distinguish between outgoing routes, with a mobile node as destination, and incoming routes, with a mobile node as source node.

Outgoing routes

To establish an *outgoing route* from a stationary source node v_s to a mobile destination v_m, QMR first builds a multicast tree t of stationary nodes from v_s to all access nodes $v_a \in V_a$. Then, each branch of t is extended by an access hop (v_a, v_m) from each access node to the mobile node. As v_m is in range of at least one access node at all times, this provides single-hop access from t to v_m. During operation, management frames are exchanged in pre-reserved management slots to identify the access node in charge and to ensure that only one access hop is used. The special case of adding a mobile node to an already existing routing tree t is equivalent to adding all access nodes and access hops to t.

When operating outgoing routes, messages sent by a source node are forwarded to all access nodes. The access node currently in charge is responsible to deliver a received message to the mobile node via its access hop.

Figure 8.6 illustrates the discovery of outgoing routes. In the example, a multicast tree is iteratively built from source node v_{11} to access nodes v_2, v_8, and v_9. Route search starts with an access node in minimum hop distance to the source node, yielding path $p_1 = (v_{11}, v_9)$ as initial multicast tree. QMR extends this tree by adding access nodes with increasing hop distance to the source node, as explained in Sect. 8.2.3, yielding paths p_2 and p_3. Paths p_1, p_2, and p_3 are then extended by access hops to v_m. In the runtime topology in Fig. 8.6, v_m is currently in range of access node v_2; therefore, the access hop (v_2, v_m) is used.

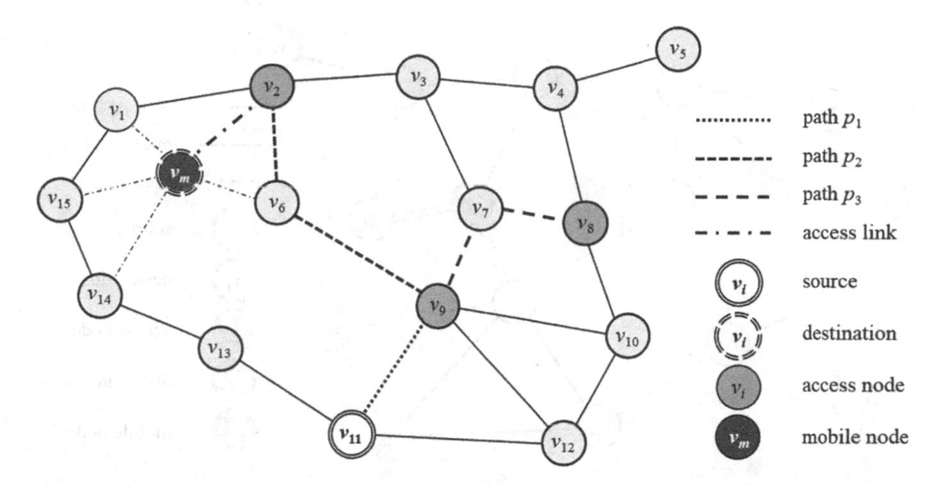

Fig. 8.6 Routing tree from source v_{11} to mobile node v_m via access nodes v_2, v_8, and v_9 (see [3], Fig. 4)

Incoming routes

To build an *incoming route* from a mobile source node v_m to a stationary destination v_d, QMR first chooses a stationary distributor node $v_{dn} \in V_s$, which is in charge of receiving data from v_m and forwarding them to v_d. Then, a concast tree t_c from all access nodes $v_a \in V_a$ to v_{dn} is built and extended by access hops (v_m, v_a). To build t_c, QMR determines a multicast tree from v_{dn} to all $v_a \in V_a$ as outlined before, which is then inverted. This is feasible, as QMR considers bidirectional links only. Next, QMR creates a multicast tree t_m from v_{dn} to v_d, with the option to extend this tree if further destination nodes are added later. Finally, t_c and t_m are concatenated at v_{dn} to form an incoming route.

There are several strategies to choose the distributor node $v_{dn} \in V_s$. A straightforward strategy is to select $v_{dn} = v_d$, which implies that initially, no multicast tree from v_{dn} to v_d is required. This choice works best if the mobile node is always close to v_d, and no further destination nodes are added later. Another strategy is to choose a distributor node in the center of the topology, yielding concast and multicast trees of low depth, i.e., short routes. QMR adopts this strategy and chooses a node with the minimum average distance to all other nodes as distributor node.

When operating incoming routes, messages sent by a mobile source node are received by the access node currently in charge and forwarded to all destinations via the distributor node. We observe that to reach the distributor node, only one branch of the concast tree is used.

Figure 8.7 illustrates the discovery of an incoming route from mobile source v_m to destination v_5, extended by destination v_{12} in the second step. In the example, v_7 is chosen as distributor node. Then, a concast tree t_c from access nodes v_2, v_8, and v_9 to distributor node v_7 is formed and extended by access hops (v_m, v_2), (v_m, v_8), and

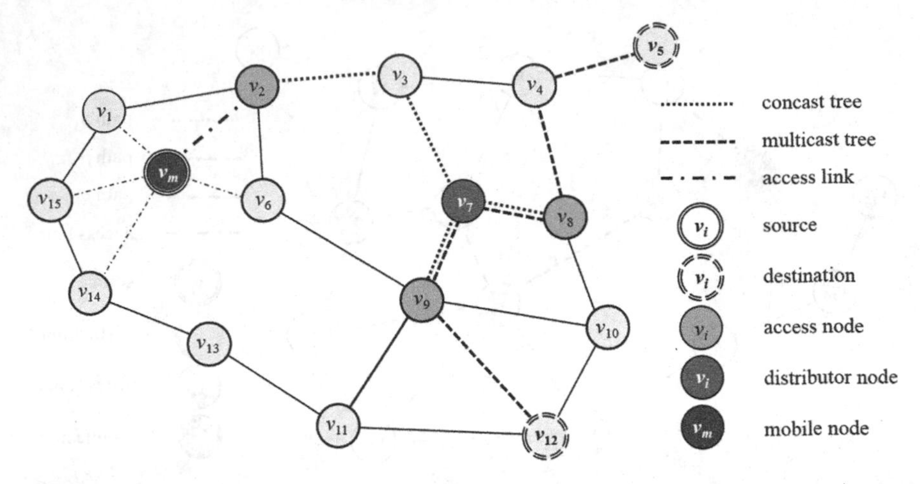

Fig. 8.7 Routing trees from mobile source v_m to destinations v_5 and v_{12} via distributor node v_7 (see [3], Fig. 5)

(v_m, v_9). Next, a multicast tree t_m from v_7 to v_5 is established and concatenated with t_c and the access hops. Later, destination v_{12} is added. In the runtime topology in Fig. 8.6, v_m is currently in range of access node v_2; therefore, the access hop (v_m, v_2) is used.

In the example, some links are part of the concast tree t_c and the multicast tree t_m, and therefore used in both directions, e.g., (v_7, v_8) and (v_8, v_7). While this might look inefficient, it is a consequence of our design decisions and systematical treatment, yielding functional multicast routes involving mobile source nodes. Also, only one branch of the concast tree is used during operation. In the example, this is the branch (v_m, v_2, v_3, v_7), which is disjoint with t_m; therefore, no link is used in both directions. If v_m uses access node v_8, then the link between v_7 and v_8 is used twice. However, if local multicasting is feasible, there is no loss of efficiency.

So far, we have considered outgoing and incoming routes with only one mobile node. This can be generalized to having several mobile nodes as destinations and/or sources, by establishing, for each mobile node, outgoing and/or incoming routes, and by composing them.

8.2.7 Slot Scheduling in Partially Mobile Networks

We now extend the iterative construction of routing trees in partially mobile networks by an algorithm for slot scheduling, i.e., selection and reservation of time slots along the branches of a routing tree. As before, we assume that super slots contain exclusive regions decomposed into exclusive slots.

In case of an outgoing route from a stationary source node v_s to a mobile destination v_m, there is a multicast tree consisting of stationary nodes only. Slot reservations for this tree are covered by the slot scheduling strategies in Sect. 8.2.4, to achieve small end-to-end delays or high route feasibility and slot reuse. For the last hop from access nodes to the mobile node, it is sufficient to reserve one additional time slot, as at any point in time, only one access node is in charge. However, since the location of the mobile node is not determined, the slot has to be free network-wide and is reserved exclusively, i.e., reuse through SDMA cannot be exploited.

In case of an incoming route from a mobile destination v_m to a group of stationary nodes, different kinds of slot reservations are needed. First, there are the access hops from v_m to access nodes. As the mobile node sends to exactly one access node at a time, it is sufficient to reserve one time slot. As in case of outgoing routes, this reservation must be exclusive, as location and interference range of v_m are not known. Second, slot reservations on the concast tree from access nodes to distributor node are required. Since only one access node is in charge at any time, only the corresponding branch of the concast tree is used for transmissions. It follows that transmissions on different branches of the concast tree are not in conflict. Therefore, the same scheduling strategies as in Sect. 8.2.4 can be applied, with the difference that blocking entries are not considered if they are caused by another branch of the concast tree only. Finally, reservations for the multicast tree from the distributor node to destinations are due (see Sect. 8.2.4).

8.2.8 Experimental Assessment

To assess and compare the slot scheduling strategies of QMR to achieve small end-to-end delays (*shortDelay*) or high route feasibility and slot reuse (*highReuse*), we have conducted a series of simulation experiments. Furthermore, we have investigated the benefits of using access nodes in partially mobile networks. We have decided against comparisons with other routing protocols, as they fundamentally differ from QMR in functionality and/or QoS support (see Sect. 8.3).

For the experimental assessment, we have used several topologies of stationary nodes. In the following, we provide details for the topologies shown in Fig. 8.8. Topology 1 represents dense networks with uniform placement of nodes and links. Topology 2 is a representative of a stretched network as found, for instance, in production lines. We assume that the 1-hop interference neighborhood of a node is a subset of its 2-hop communication neighborhood. For conservative slot scheduling decisions, we extend the interference neighborhood to be maximal, i.e., $\forall v \in V.\ IN_{\leq 1}(v) = CN_{\leq 2}(v)$.

To evaluate the slot scheduling strategies *shortDelay* and *highReuse*, we have randomly generated, for each topology, 50 route request scenarios with three sets of nodes consisting of one source node and three destination nodes each. Furthermore, we have varied the number of exclusive slots per super slot from 10 to 15. The

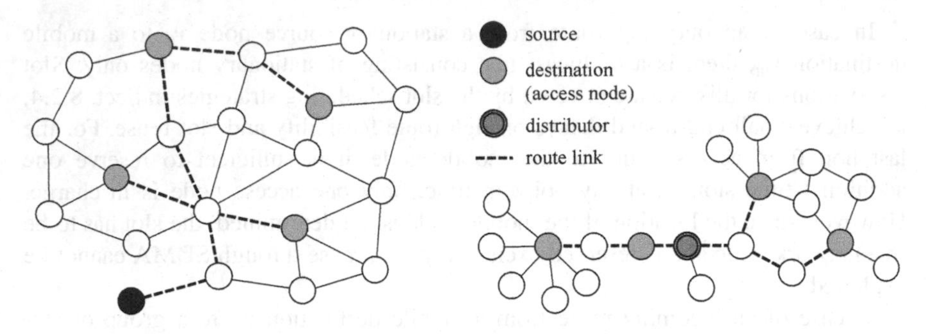

Fig. 8.8 Topology 1 (left), with multicast tree. Topology 2 (right), with concast tree (see [3], Figs. 6 and 7)

Table 8.3 Experimental assessment for topology 1 (see [3], Table III)

Strategy	Slots	Tree depth	Utilization [%]	Delay [slots]	Success rate [%]
shortDelay	10	2.52 ± 1.21	0.18 ± 0.02	3.16 ± 2.07	40
highReuse	10	2.59 ± 1.22	0.19 ± 0.02	4.66 ± 4.10	54
shortDelay	12	2.74 ± 1.24	0.17 ± 0.01	3.61 ± 2.48	94
highReuse	12	2.73 ± 1.23	0.18 ± 0.02	5.35 ± 5.08	96
shortDelay	15	2.75 ± 1.23	0.17 ± 0.01	3.56 ± 2.18	100
highReuse	15	2.74 ± 1.24	0.18 ± 0.02	6.06 ± 6.39	100

Table 8.4 Experimental assessment for topology 2 (see [3], Table IV)

Strategy	Slots	Tree depth	Utilization [%]	Delay [slots]	Success rate [%]
shortDelay	10	3.33 ± 2.12	0.23 ± 0.01	4.47 ± 3.63	10
highReuse	10	3.54 ± 2.30	0.24 ± 0.01	7.33 ± 7.28	16
shortDelay	12	3.63 ± 2.06	0.23 ± 0.01	5.36 ± 5.12	52
highReuse	12	3.96 ± 2.22	0.24 ± 0.01	9.08 ± 8.34	90
shortDelay	15	3.99 ± 2.21	0.23 ± 0.01	5.70 ± 4.34	100
highReuse	15	3.99 ± 2.21	0.24 ± 0.01	11.22 ± 11.15	100

results of the experiments for the topologies in Fig. 8.8 are shown in Tables 8.3 and 8.4.

First, we observe that the average depth of multicast trees is almost independent of the slot scheduling strategy. The reason for this is that multicast trees are built before slot scheduling occurs. Small variations can be explained by different success rates, which have some impact on calculated averages. Another finding is that both scheduling strategies achieve about the same average utilization rates. A possible reason is that both topologies are relatively dense, reducing the potential for slot reuse.

As expected, *shortDelay* establishes routes with substantially smaller end-to-end delays compared to *highReuse*. This can be explained by the slot decision policy of *shortDelay*, which handles links in the order of the path and performs a rotating forward search for free slots; whereas, *highReuse* applies other criteria to link handling and slot search. The strength of *highReuse* is the success rates for the discovery of feasible routes. For instance, in topology 2 with 12 exclusive slots (Table 8.4), *highReuse* is successful in 90% of all route requests, while *shortDelay* achieves only 52%. This can be explained by the slot decision policies of *highReuse*, which improve the chances of finding a feasible route.

Based on these findings, we conclude that the slot scheduling strategy should be chosen such that the application requirements are reflected. For a route request, both strategies return a deterministic delay in slots, which has to be augmented by gaps between exclusive regions to which these slots belong. An approach could be to use *highReuse* to find a feasible route and to apply *shortDelay* as backup strategy in case this does not match the specified maximum end-to-end delay.

To assess the advantages of having dedicated access nodes in order to incorporate mobile nodes into QoS multicast routes, simulation experiments with outgoing and incoming routes have been conducted. Topology 1 in Fig. 8.8 shows a multicast tree for an outgoing route from a source node to four access nodes. When applying *shortDelay*, we obtain a delay of up to six slots, depending on the current location of the mobile node, with 69 blockings. In comparison, a multicast tree to all stationary nodes, also applying *shortDelay*, yields a maximum delay of 11 slots and 128 blockings.

Topology 2 in Fig. 8.8 shows a concast tree that is part of an incoming route. Here, QMR optimizes slot usage by exploiting the fact that only one branch of the tree is used at any time. When using *shortDelay*, the maximum delay is three slots with the optimization, and seven slots without.

8.3 Related Work

In this chapter, we outline QoS routing protocols for wireless TDMA networks and compare them to QoS multicast routing (QMR). To keep the survey concise, we have selected one representative protocol for centralized and one for decentralized route discovery.

8.3.1 Distributed Slots Reservation Protocol (DSRP)

In [9], Shih et al. present *Distributed Slots Reservation Protocol* (*DSRP*), a representative for decentralized discovery of QoS unicast routes in TDMA networks. DSRP is a reactive routing protocol, which enforces slot inhibited policies and applies slot decision policies.

Route discovery of DSRP consists of two phases. During the route request phase, DSRP finds a set of feasible QoS routes and preselects time slots. In the route reply and reservation phase, it selects and finalizes a QoS route and reserves time slots. If time slots selected during the route request phase are no longer available for reservation, DSRP applies a slot adjustment protocol with a slot adjustment algorithm to negotiate slot exchanges with neighbor nodes.

The decision whether a time slot s is free for transmission on a link (v_1, v_2) is based on the transmit and receive reservation status TX^s and RX^s of nodes v_1 and v_2. This information is recorded and updated by each node and requires the exchange of reservation decisions with nodes in single-hop interference neighborhood. To make this feasible, DSRP assumes that nodes in interference range are also in communication range, i.e., $IN_{\leq 1}(v) \subseteq CN_{\leq 1}(v)$. Nevertheless, there still is the problem of local uncertainty about the reservation status of neighbor nodes due to possible collisions and delays of management traffic.

Slot s is free for transmission on a link (v_1, v_2) if it satisfies slot inhibited policies $SIP_{1.3}$ as follows:

$$\begin{aligned}
SIP_1 &: \neg TX^s(v_1) \wedge \neg RX^s(v_1) \wedge \neg TX^s(v_2) \wedge \neg RX^s(v_2) \\
SIP_2 &: \forall v \in CN_1(v_1). \neg RX^s(v) \\
SIP_3 &: \forall v \in CN_1(v_2). \neg TX^s(v)
\end{aligned}$$

Here, $CN_1(v) = CN_{\leq 1}(v) \setminus \{v\}$. Apart from the assumption $IN_{\leq 1}(v) \subseteq CN_{\leq 1}(v)$, this corresponds to the reservation criterion in Definition 8.5. For a given link, SIP_1 and SIP_2 are checked by v_1; whereas, v_2 verifies SIP_3. Thus, knowledge about the reservation status in single-hop communication neighborhood is sufficient.

The selection of slots that are free for transmission is guided by three slot decision policies. During route discovery, route request packets record the sets of free slots of the previous three hops (or less, if the distance of the sending node to the source node is smaller). Three-hop Backward Decision Policy (3BDP) determines slots for the last but two link such that unfeasible slot assignments due to the hidden station problem are avoided. On top of 3BDP, Least Conflicts First Policy (LCFP) assigns slots having the least conflict with free slots of the following two hops. LCFP corresponds to LCF in Definition 8.10, which, however, considers the following hops too. Finally, Most Reuse First Policy (MRFP), which corresponds to MRF in Definition 8.11, prefers slots with high utilization to increase the effects of SDMA.

The source node initiates the route request phase by broadcasting a RREQ packet, which contains the ids of source, destination, and transmitting node, route request id, bandwidth requirement, and the partial route with selected slots, free slots, and slot usage. Nodes receiving a route request for the first time try to extend the partial route. If applicable, they select slots for the last but two link, applying slot decision policies 3BDP, LCFP, and MRFP (in this order). If the bandwidth requirements can still be satisfied, they broadcast an updated RREQ packet to continue the route search.

On reception of a RREQ packet with a feasible route, the destination node initiates the route reply and reservation phase by returning a RREP packet on the unicast route recorded in the received packet. On each hop, time slots selected during the route request phase are reserved, if they are still free for transmission. However, if several concurrent route requests are processed, slots may have been selected several times and may therefore not be available anymore. In this case, DSRP starts a slot adjustment protocol with a slot adjustment algorithm to negotiate slot exchanges with neighbor nodes. However, slot adjustments may also be performed concurrently and may therefore lead to inconsistent results.

One strength of DSRP is the definition and application of slot inhibited and slot decision policies. Simulations show good success rates compared to other distributed QoS routing protocols. Major drawbacks are high communication overhead, problems caused by loss and delay of routing management messages, and the possibility of concurrent route request processing, which makes route discovery unreliable and unpredictable. Message loss and delay may also cause inconsistent reservation status information, resulting in the discovery of unfeasible routes. The authors claim that DSRP has good success rates even with mobile nodes. However, node mobility may lead to route breaking and rules out SDMA. Finally, DSRP only supports QoS unicast routing.

Despite its strengths, DSRP reveals that decentralization of QoS route discovery suffers severe problems that are difficult to solve. With Black Burst-based QoS routing (BBQR) [2], we have devised a decentralized QoS routing protocol that avoids some of these problems, but at the price of high overhead and complexity. Therefore, we argue that a centralized approach is preferable. In larger networks, several routing managers can share the task, based on suitable network clustering.

8.3.2 QoS Routing in WirelessHART

WirelessHART [11] is a protocol stack developed for wireless networking in industrial environments based on TDMA, supporting reservations for exclusive medium access and multi-hop networks. It extends the wired Highway Addressable Remote Transducer (HART) protocol developed by the company Rosemount in the 1980s. In WirelessHART, route discovery and slot scheduling are centralized.

WirelessHART operates mesh networks consisting of nodes with different roles and capabilities. The node taking the role of network manager is responsible for time synchronization, channel management, route discovery, slot scheduling, and slot reservation. The gateway (possibly the same node) couples wireless field devices to the wired plant automation network. It acts as sink for packets from and as source for packets to devices.

WirelessHART distinguishes the following main types of routing:

- *Source routing*: In this context, a source route is a complete path from a source node to a destination node. Source routes are computed by the network manager and distributed to the source node via the gateway. To send data to the destination node, a source node adds the complete path description to the packet. Intermediate nodes forward incoming packets according to this description.
- *Graph routing*: In this context, a graph is a loop-free directed subgraph of the communication topology with a single source or destination node. Graphs are computed by the network manager and distributed to nodes contained in the graph via the gateway. Furthermore, each graph has a unique identifier, which is added to packets sent via the corresponding graph. Three kinds of graphs are distinguished. A *broadcast graph* connects the gateway node to all other nodes of the wireless mesh network. An *uplink graph* connects a set of nodes to the gateway. A *downlink graph* connects the gateway to a single node. When using graph routing, intermediate nodes are supposed to forward incoming packets on the first available outgoing link of the corresponding graph.

In addition to the route information provided by a path or graph, a node has to observe the slot schedule, which determines the assigned time slots that may be used to forward incoming packets on a given link.

The specification of WirelessHART defines source and graph routing, as well as frame formats and a protocol for the distribution of source and graph routes to network nodes. However, algorithms to generate paths, graphs, and slot schedules are not part of the standard, and therefore, have to be added by the system developer.

WirelessHART favors centralized route and slot schedule computation, performed by the network manager [5]. For this, it is assumed that the network manager knows the network topology. It is, however, not defined in the standard how this knowledge is established and maintained. Furthermore, given the industrial application context, it is assumed that nodes are stationary. Chen et al. [5] consider the use of quarantine or start-up phase, where routes and slot schedules are computed and distributed. This would imply that communication requirements remain static during system operation.

In [7], Nobre et al. provide an extensive survey of WirelessHART routing and slot scheduling schemes. Among the surveyed *routing algorithms* is the algorithm by Han et al. [6], which computes a broadcast graph, an uplink graph for all nodes, and downlink graphs for individual nodes. For a directed graph $G = (V, E)$, a node $v \in V$ is (n, m)-reliable if and only if it has at least n incoming and m outgoing edges. This criterion is then applied as follows:

- A broadcast graph is reliable if all nodes (except gateway and access points) are (2, 0)-reliable. This increases the chances that broadcast messages from the gateway reach all wireless devices.
- An uplink graph is reliable if all nodes (except gateway and access points) are (0, 2)-reliable. Again, this adds redundancy in case individual links fail.

- A downlink graph from gateway v_g to node $v \in V$ is reliable if v_g and all intermediate nodes are (0, 2)-reliable.

Han's algorithm then constructs minimal reliable subgraphs of G, by reducing edges and, in case of downlink graphs, nodes. If there are nodes not satisfying the required (n, m)-reliability in the original graph G, the algorithm tries to attach these nodes via a single link to the resulting graph.

Among the *slot scheduling algorithms* surveyed in [7] is the algorithm by Han et al. [6], which uses the Fastest Sampling Rate First (FSRF) policy. Starting points are broadcast, uplink, and downlink graphs. Free slots are assigned from source to destination using depth first search, exploiting the concurrent use of different channels for different branches of the graphs.

In summary, the QoS routing algorithms that are devised for WirelessHART separate route discovery and slot scheduling. First, paths and/or graphs to support general communication scenarios are determined. Then, time slots for QoS end-to-end communication along paths and branches of graphs are exclusively assigned, based on sampling rates. This may not yield optimal QoS routes; however, given the complexity of the problem, it is a very good compromise.

Flexibility of QoS routing in WirelessHART is reduced in several ways. First, for different sessions between the same nodes, the same path and/or graph is used. This may create slot shortages that could be avoided by having separate routes. Second, WirelessHART mainly considers unicast and broadcast communication between the gateway and the complete set of devices. In a networked control system, there may be several controllers acting as sink nodes for a subset of devices, which would require more flexible communication structures. Furthermore, if several controllers share sensors, multicast routes help reducing slot usage. Third, WirelessHART assumes that communication requirements are static. Thus, slot assignment has to be static, too, leading to a waste of resources in case of temporary sessions. Finally, mobile nodes are not supported.

8.4 Conclusions

In this chapter, we have explained the concepts and requirements of routing in wireless ad hoc networks, have presented QoS multicast routing (QMR) for partially mobile wireless TDMA networks, and have surveyed related work. QoS routing is a challenging task with solutions that are often customized to a specific problem domain. QMR has been inspired by the communication requirements of industrial networked control systems, as found in a setting for remote maintenance in a production plant.

For the operation of QMR, we have assumed that there is a set of stationary nodes forming a wireless multi-hop TDMA network, and that additional mobile nodes are in communication range of at least one stationary node at all times. Furthermore, we have assumed that communication sessions may be of dynamic nature, with dynamically varying groups of nodes and bandwidth requirements. Finally, we have considered flexible scenarios with several controllers that may share sensors and actuators.

QMR copes with these requirements by incrementally constructing and pruning routing trees and assigning and releasing exclusively reserved slots at runtime. Compared to offline computation of routes and slot schedules, this approach is highly flexible. In particular, routes and slot schedules can be adapted during operation especially in contexts with varying communication requirements.

In QMR, routing and slot scheduling decisions are taken by a centralized routing manager, which has important advantages compared to decentralized solutions: It prevents mutual blocking of resources due to concurrent route searches, substantially reduces management traffic, and avoids the problem of inconsistent reservation status due to lost or delayed management messages.

Literature

Chair for Networked Systems

1. Gotzhein R (2014) ProNet 4.0—A wireless real-time communication system for industry 4.0. White Paper, Networked Systems Group, Department of Computer Science, University of Kaiserslautern. http://vs.informatik.uni-kl.de/publications/2014/Go14/whitePaperEN-ProNet4.0.pdf. Accessed on 27 Aug 2019
2. Becker P, Birtel M, Christmann D, Gotzhein R (2011) Black burst-based quality-of-service routing (BBQR) for wireless Ad-Hoc networks. In proceedings of the 11th international conference on new technologies in distributed systems (NOTERE 2011), Paris, France, pp 1–8
3. Gebhardt J, Gotzhein R, Igel A, Kramer C (2015) QoS multicast routing in partially mobile wireless TDMA networks. In proceedings of IEEE global communications conference (Globecom 2015), San Diego, USA, Dec 6–10, 2015

Further References

4. Rekhter R, Hares S, Li T (2006) A border gateway protocol 4 (BGP-4), IETF RFC 4271, Jan 2006
5. Chen D, Nixon M, Mok A (2010) WirelessHART™—real-time mesh network for industrial automation, Springer
6. Han S, Zhu X, Mok A, Chen D, Nixon M (2011) Reliable and real-time communication in industrial wireless mesh networks. In 17th IEEE real-time and embedded technology and applications symposium, Chicago, Illinois, USA, April 11-14, pp 3–12
7. Nobre M, Silva I, Guedes LA (2015) Routing and scheduling algorithms for WirelessHART networks: a survey. Sens—Open Access J 15(5):9703–9740
8. Moy J (1998) Open shortest path first (OSPF) version 2. IETF RFC 2328, April 1998

9. Shih KP, Chang CY, Chen YD, Chuang TH (2006) Dynamic bandwidth allocation for QoS routing in TDMA-based mobile ad-hoc networks. Comput Commun 29:1316–1329
10. http://software-cluster.org/
11. International Electrotechnical Commission (IEC) (2010) Industrial communication networks—wireless communication network and communication profiles—WirelessHART (IEC 62591 ed 1.0). Geneva, Switzerland, April 2010
12. ZigBee™ Alliance (2005) ZigBee specification, version 1.0, June 2005 p 378. www.zigbee.org. Accessed on 27 Aug 2019

Chapter 9
Network Clustering

Network clustering is the activity of dividing a network into groups of nodes. In this chapter, we explain foundations of clustering, present *heterogeneous network clustering (HNC)* [3] implemented in ProMid of ProNet 4.0 [1], survey and compare related work, and draw conclusions.

9.1 Foundations

In this chapter, we provide the context of clustering in wireless ad hoc networks, explain concepts, and address areas of operation.

9.1.1 Context

Clustering is a basic functionality of wired and wireless computer networks to reduce the complexity of communication topologies by creating overlay structures. This can be used, for instance, for hierarchical routing to decrease management traffic and the size of routing tables, thereby achieving scalability. Overlay structures also exist in control systems, where controllers collect, aggregate, and exchange data of sensor nodes located in their neighborhood. A well-known functionality based on clustering is Domain Name System (DNS), a hierarchical distributed naming service used in the Internet. Other clustering applications are energy balancing to extend lifetime of networks consisting of battery-driven nodes and redundancy measures to improve fault tolerance.

The criteria for cluster formation depend, i.e., on clustering objectives, node capabilities, node deployment, and network dynamics:

R. Gotzhein, *Real-time Communication Protocols for Multi-hop Ad-hoc Networks*, Computer Communications and Networks,
https://doi.org/10.1007/978-3-030-33319-5_9

- *Clustering objectives*: To support intra- and inter-cluster routing, connectivity is a primary concern. For load balancing, equally sized clusters are sought. To keep the number of resource-rich and therefore expensive cluster manager nodes small, the objective is a minimal cluster count. For maximal network lifetime, energy consumption is to be minimized and/or evenly distributed.
- *Node capabilities*: This concerns, for instance, computing power, communication speed and range, energy supply, and software functionalities. In homogeneous networks, nodes have similar capabilities, while in heterogeneous networks, there may be resource-rich nodes selected as cluster managers.
- *Node deployment*: In industrial networks, node deployment is typically controlled, to achieve well-balanced topologies and/or spatial coverage. In other scenarios, nodes may be deployed randomly, yielding non-uniform node distributions that require self-organizing approaches.
- *Network dynamics*: Networks may consist of stationary and/or mobile nodes. Node mobility is a challenge for clustering, as it requires dynamic cluster reformation, which is time-consuming. Network dynamics also concerns activities, which may be event-triggered or time-triggered.

Clustering algorithms have been proposed for various combinations of these and further criteria and are, therefore, targeting different application contexts.

9.1.2 Concepts

Clustering divides a network into subsets of nodes called *clusters*, with the purpose of achieving defined clustering objectives. The nodes of a cluster are called *cluster members*. Cluster membership may be fixed or variable. Clustering may be performed offline or during network operation by executing a clustering algorithm, possibly repeatedly to cope with node mobility. Depending on clustering objectives, clusters may be overlapping or disjoint.

Cluster members are classified as cluster heads or followers. A *cluster head* is a distinguished node of a cluster, which is pre-assigned or elected by other nodes. *Followers* are cluster members other than the cluster head. Followers may act as *gateways* to establish inter-cluster connectivity. Figure 9.1 shows an example.

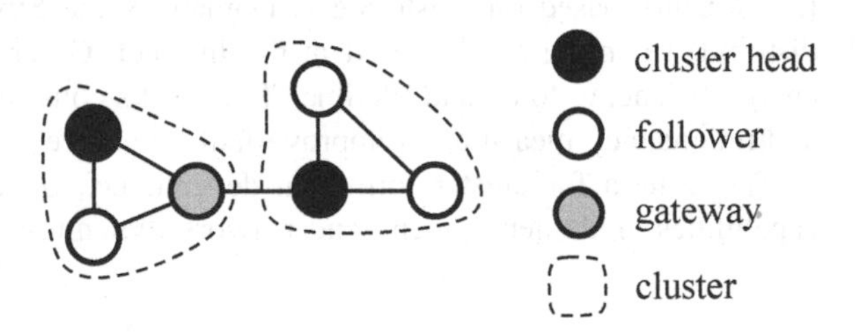

Fig. 9.1 Network clustering —node types and (disjoint) clusters

We now introduce basic clustering concepts:

Definition 9.1 A *graph* G is a tuple (V, E) consisting of a set V of vertices and a relation $E \subseteq V \times V$ of edges. It is *undirected* (or *symmetrical*) if and only if $\forall v, v' \in V. ((v, v') \in E \Rightarrow (v', v) \in E)$. Otherwise, it is *directed.*

Definition 9.2 Let $G = (V, E)$, $G' = (V', E')$ be graphs. G' is a *subgraph* of G if and only if $V' \subseteq V$ and $\forall v, v' \in V'. ((v, v') \in E' \Rightarrow (v, v') \in E)$. G' is a *complete subgraph* of G if and only if in addition, $\forall v, v' \in V'. ((v, v') \in E \Rightarrow (v, v') \in E')$ holds.

Definition 9.3 A *path* p of $G = (V, E)$ is a finite, non-empty sequence $(v_1, \ldots, v_{|p|+1}) \in V^+$ such that for all i, $1 \leq i \leq |p|$, $(v_i, v_{i+1}) \in E$, with the *length* $|p| \geq 0$ denoting p's number of edges. A path p is *cycle-free* if vertices occur at most once in p.

Definition 9.4 A graph $G = (V, E)$ is *(strongly) connected* if and only if for each pair (v, v') of vertices, there exists a path p with v and v' being the first and the last vertices, respectively. Otherwise, G is *unconnected.*

Definition 9.5 Let $G = (V, E)$ be a graph, $G' = (V', E')$ be a subgraph of G. G' is a *maximal connected subgraph* of G if and only if G' is connected and $\forall v \in V'$. $\forall v' \in V \setminus V'$ $((v, v') \notin E \wedge (v', v) \notin E)$. If G is connected, then $V' = V$ holds.

Definition 9.6 A *communication topology* is modeled as a connected graph $G = (V, E)$, where vertices are called *nodes*, and edges are called *links*.

Definition 9.7 The *hop distance* $d_G(v, v')$ between nodes $v, v' \in V$ is defined as

$$d_G(v, v') =_{\text{df}} \min_{p \in P_G(v,v')} |p|$$

where $P_G(v, v')$ is the set of all cycle-free paths starting in v and ending in v'. The *i-hop communication neighborhood* $CN_{\leq i}(v)$ of $v \in V$, $i \geq 0$, is defined as

$$CN_{\leq i}(v) =_{\text{df}} \{v' \in V \mid d_G(v, v') \leq i\}$$

Definition 9.8 A *d-hop dominating set* is a set of nodes $V' \subseteq V$ such that every node $v \in V$ is in d-hop communication neighborhood $CN_{\leq d}(v')$ of some node $v' \in V'$. Elements of V' are called *dominating nodes*. A minimal d-hop dominating set[1] is a d-hop dominating set consisting of a minimal number of nodes.

Definition 9.9 A *k-hop connected d-hop dominating set* is a d-hop dominating set $V' \subseteq V$ where every dominating node is connected to at least one other dominating node by a path of length k or less, and where all dominating nodes are connected via sequences of such paths.

[1]The dominating set problem of determining a minimal dominating set is considered to be NP-complete.

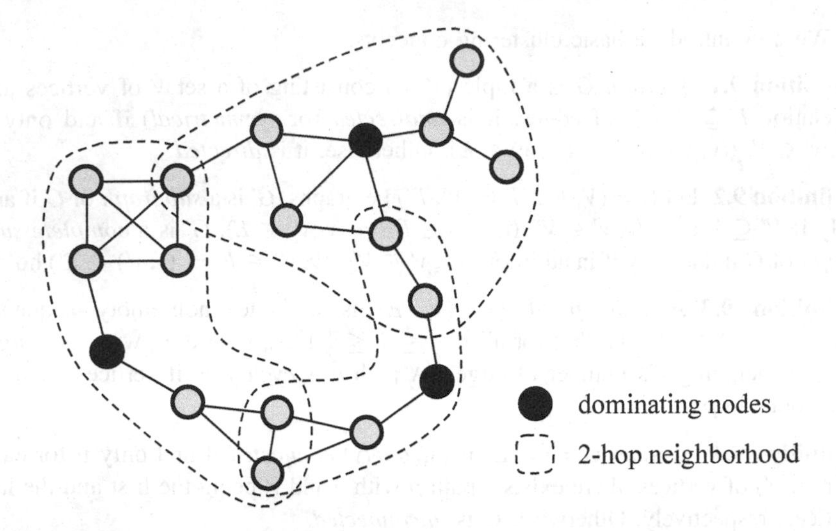

Fig. 9.2 Minimal 4-hop connected 2-hop dominating set

Figure 9.2 shows an example of a 4-hop connected 2-hop dominating set of size 3. It is minimal, because there is no 2-hop dominating set with less than three nodes.

9.1.3 Areas of Operation

In wireless ad hoc networks, clustering is applied to tackle a variety of problems. A core problem is limited lifetime of networks with battery-driven nodes. Battery replacement increases the cost of maintenance and should, therefore, occur in rare intervals only. In harsh environments, it may not even be possible to recharge or replace batteries. Measures based on proper clustering can extend network lifetime substantially. In wireless sensor networks, cluster heads may act as data collectors for their followers. This reduces the tasks and therefore energy consumption of followers in two ways. First, computation is restricted to pure sensing. Second, communication is with the cluster head only, i.e., highly localized. Depending on the kind of application, this may even open up potential for duty cycling, by switching hardware components from active mode to sleep or idle mode when not needed (see Chap. 7). In addition, cluster heads may aggregate data received from followers, which reduces frequency and amount of data to be forwarded to other parties and therefore limits energy consumption due to communication. To balance energy consumption, elected cluster heads may pass their role to other nodes after some period of activity. Here, a cluster head selection criterion could be the available energy of nodes.

In industrial control applications, wireless networks are typically heterogeneous in the sense that they consist of nodes with different capabilities and resources.

From an operational perspective, nodes with full functionality, for instance, controllers, and nodes with reduced functionality, such as sensors and actuators, may be distinguished. From an energy point of view, controllers may be (partially) attached to a power network, while sensors and actuators operate battery-driven. This makes controllers natural candidates for the role of permanent cluster heads, hosting computationally expensive control algorithms and service registries. Furthermore, the overlay structure established by clustering can be used for hierarchical routing schemes, with routes formed by cluster heads and a minimal set of followers acting as gateways.

In [2], we have presented a protocol for dynamic overlay line topology establishment and repair in wireless networks based on two levels of network clustering, with the objective to satisfy application requirements and extend network lifetime. Our work has been triggered by the requirements of the Asset Monitoring for Rail Applications (AMRA) project at Bosch BEG. The purpose of the project is to track cargo wagons and to monitor their status in order to detect malfunctions and to plan maintenance activities. On each wagon, several sensor nodes are placed to detect, for instance, shocks, flats, cargo status, and wagon location. Furthermore, a dedicated master node collects and aggregates sensor data and exchanges data with master nodes of neighboring wagons. Here, a first level of network clustering is given by the nodes of each cargo wagon, with the master node as cluster head. On the second level, master nodes are clustered, and subline topologies are established within each cluster, with the cluster head acting as data collector for a set of cargo wagons.

9.2 Heterogeneous Network Clustering (HNC)

In this chapter, we present *heterogeneous network clustering (HNC)*, a clustering algorithm for wireless TDMA networks consisting of stationary nodes [3]. HNC has been inspired by the requirements of industrial networked production and control environments, in particular the requirement to establish distributed service registries in heterogeneous wireless networks. We have implemented HNC as part of the production middleware ProMid of the ProNet 4.0 protocol stack [1].

9.2.1 Context and Design Decisions

In a research project with industrial partners, we have explored methods to provide high-level services for distributed control applications. Conceptually, sensor and actuator nodes provide periodical and event-triggered services such as temperature readings and control value settings. Controller nodes use these services to build up feedback loops in order to achieve well-defined control objectives.

To manage services, we have decided to build up a service registry, where sensors and actuators register their services. Controllers can then look up services and contact the service provider in order to subscribe to particular services. If properly supported by a middleware, the distributed nature of a networked control system remains transparent for control applications. For scalability, the service registry can be distributed.

HNC is a clustering algorithm specifically devised for heterogeneous wireless networks, as, for instance, in wireless networked control systems. In the considered industrial setting, nodes have different capabilities and resources. Accordingly, HNC distinguishes between nodes that must, may, or must not become cluster heads or gateways. To apply HNC in homogeneous networks, all nodes are classified as may-nodes.

To perform network clustering, up-to-date information about the communication topology is required. Usually, clustering protocols incorporate functionality to collect and maintain this information. In case of HNC, this information is already provided by ATDP, our protocol for the automatic detection of communication, interference, and sensing topologies (see Chap. 4). Even more, ATDP terminates once a stable topology has been detected and disseminated to all network nodes. For this reason, we have made the design decision to devise a centralized algorithm. Since all nodes have consistent information about the communication topology and because HNC is conceived to yield deterministic results, it is sufficient to execute HNC locally, without sharing the outcome via message exchange.

9.2.2 Overview

HNC clusters a network consisting of nodes that must, may, or must not become cluster heads or gateways into a 3-hop connected 1-hop dominating set, if such a clustering is feasible. Clustering objectives are a small number of clusters and efficient connectivity, achieved in six steps. In the first step, HNC selects all must-nodes as clusters heads and determines their sets of followers. In the second step, may-nodes are added as cluster heads until a 1-hop dominating set is established. In the third and fourth step, HNC selects may-nodes as gateway nodes until 3-hop connectivity is given. In the remaining steps, connectivity is optimized.

Starting point of HNC is a communication topology $G = (V, E)$ consisting of a set V of nodes and a set E of links. Topology information is provided by ATDP (see Chap. 4) and reduced to bidirectional communication links. Nodes are classified as follows:

- V_{mand}: Set of mandatory nodes, i.e., nodes that must be selected as cluster heads.
- V_{opt}: Set of optional nodes, i.e., nodes that may be selected as cluster heads or gateways.
- V_{excl}: Set of excluded nodes, i.e., nodes that must not become cluster heads or gateways.

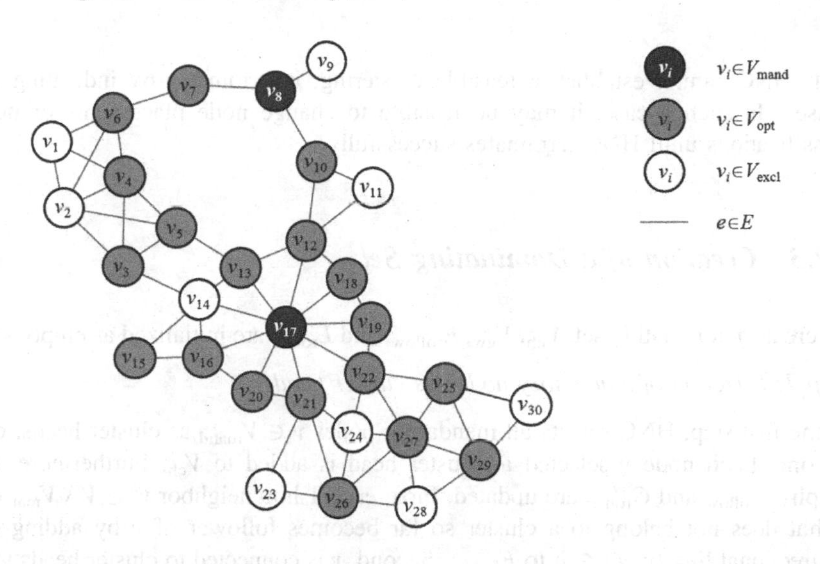

Fig. 9.3 Sample topology with mandatory, optional, and excluded nodes

The sets V_{mand}, V_{opt}, and V_{excl} are mutually disjoint, with their union yielding V. Furthermore, we require that the complete subgraph $G' = (V_{\text{mand}} \cup V_{\text{opt}}, E')$ of $G = (V, E)$ is connected, to ensure 3-hop connectivity if a 1-hop dominating set can be established. An example satisfying these constraints is shown in Fig. 9.3.

If HNC is able to establish a feasible clustering, the result is a 5-tuple $(V_{\text{ch}}, V_{\text{gw}}, E_{\text{follow}}, G_{\text{router}}, G_{\text{out}})$ with:

- V_{ch}: Nodes selected as cluster heads, forming a 1-hop dominating set of the initial topology G, with $V_{\text{mand}} \subseteq V_{\text{ch}}$ and $V_{\text{ch}} \subseteq V_{\text{mand}} \cup V_{\text{opt}}$.
- V_{gw}: Nodes selected as gateways, with $V_{\text{gw}} \subseteq V_{\text{opt}}$.
- $G_{\text{follow}} = (V, E_{\text{follow}})$: Cluster topology of the 1-hop dominating set:
  - E_{follow}: Set of bidirectional communication links connecting each follower with its cluster head, with $E_{\text{follow}} \subseteq E$.
- $G_{\text{router}} = (V_{\text{router}}, E_{\text{router}})$: Routing topology of the optimized 3-hop connected 1-hop dominating set:
  - $V_{\text{router}} = V_{\text{ch}} \cup V_{\text{gw}}$: Nodes acting as routers, i.e., nodes selected as cluster heads or gateways.
  - $E_{\text{router}} = \{(v_1, v_2) \in E \mid v_1, v_2 \in V_{\text{router}}\}$: Set of bidirectional communication links between router nodes.
- $G_{\text{out}} = (V, E_{\text{out}})$: Overlay intra- and inter-cluster communication topology of the initial topology $G = (V, E)$, with $E_{\text{out}} = E_{\text{router}} \cup E_{\text{follow}}$.

If HNC cannot establish a feasible clustering, it terminates by indicating the causes. In such a case, it may be feasible to change node placements or node classifications until HNC terminates successfully.

9.2.3 Creation of a Dominating Set

To create a dominating set, V_{ch}, V_{gw}, E_{follow}, and E_{router} are initialized as empty sets.

Step 1: Selection of mandatory nodes as cluster heads

In the first step, HNC selects all mandatory nodes $v \in V_{mand}$ as cluster heads, one by one. Each node v selected as cluster head is added to V_{ch}. Furthermore, the graphs G_{follow} and G_{router} are updated. First, each 1-hop neighbor $v' \in V \setminus V_{mand}$ of v that does not belong to a cluster so far becomes follower of v by adding the bidirectional link $(v, v') \in E$ to E_{follow}. Second, v is connected to cluster heads v' in 1-hop distance by adding $(v, v') \in E$ to E_{router}. Thus, clusters formed so far are maximal connected subgraphs (see Definition 9.5) of G_{follow}.

During execution, HNC forms *partitions* $G' = (V', E')$, which are maximal connected subgraphs of G_{out}. With each partition, the smallest index number i of all cluster heads $v_i \in V_{ch}$ belonging to this partition is associated. If $V = V'$, i.e., if the partition covers all nodes of G, a feasible clustering has been found.

If, after Step 1, there is a single partition covering all nodes of G, HNC continues with Step 5. Otherwise, if a 1-hop dominating set has already been established, the algorithm continues with Step 3. If not, Step 2 is next.

Figure 9.4 shows the sample topology of Fig. 9.3 after Step 1 has been performed. Mandatory nodes v_8 and v_{17} have been selected as cluster heads, nodes in their 1-hop neighborhood have become followers. Furthermore, partitions 8 and 17 are created, which coincide with the clusters of v_8 and v_{17}, respectively. At this point, V_{ch} is not a dominating set.

Step 2: Selection of optional nodes as additional cluster heads

In the second step, nodes $v \in V_{opt}$ are selected as cluster heads in order to establish a 1-hop dominating set, if needed. From $V_{opt} \setminus V_{ch}$, nodes that cover most previously uncovered nodes, i.e., nodes that neither belong to V_{ch} nor to the current set of followers, are selected first. The objective of this heuristic is to keep the number of clusters small. In case of multiple cluster head candidates with the same number of previously uncovered nodes, nodes already belonging to the current set of followers are preferred. This way, the new cluster is 1-hop connected to an existing cluster, which reduces the set of required gateway nodes. If there are still several candidate nodes v_i, the one with the smallest index number i is chosen, to obtain deterministic results.

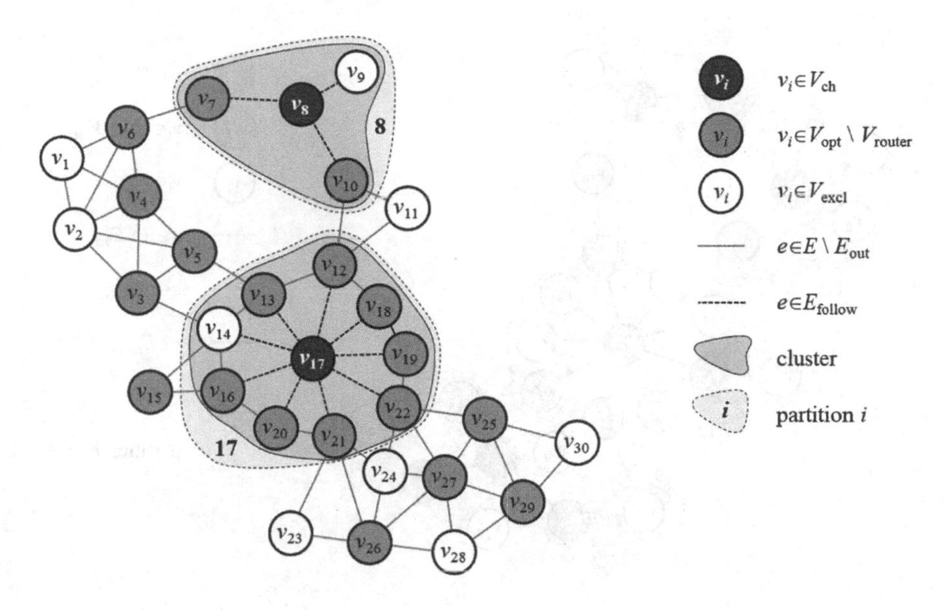

Fig. 9.4 Result after step 1

For each node $v \in V_{opt}$ selected as cluster head, data structures V_{ch}, G_{follow}, and G_{router} are updated as follows. First, v is added to V_{ch}. Second, 1-hop neighbors v' of v that do not belong to a cluster so far become followers of v by adding the bidirectional link $(v, v') \in E$ to E_{follow}. If v has been a follower to some cluster head v' before becoming cluster head itself, the link (v, v') is removed from E_{follow}. Third, v is connected to cluster heads v' in 1-hop distance by adding $(v, v') \in E$ to E_{router}. Finally, partitions are formed as defined in Step 1.

If, after Step 2, there is a single partition covering all nodes of G, HNC continues with Step 5. Otherwise, if a 1-hop dominating set has been established, the algorithm continues with Step 3. If not, HNC terminates unsuccessfully, and a report with incomplete clustering is generated.

Figure 9.5 shows the result after performing Step 2. Now, every node either is cluster head or in 1-hop neighborhood of a cluster head. This implies that V_{ch} is a 1-hop dominating set of $G = (V, E)$. Still, there are four partitions; therefore, the algorithm continues with Step 3.

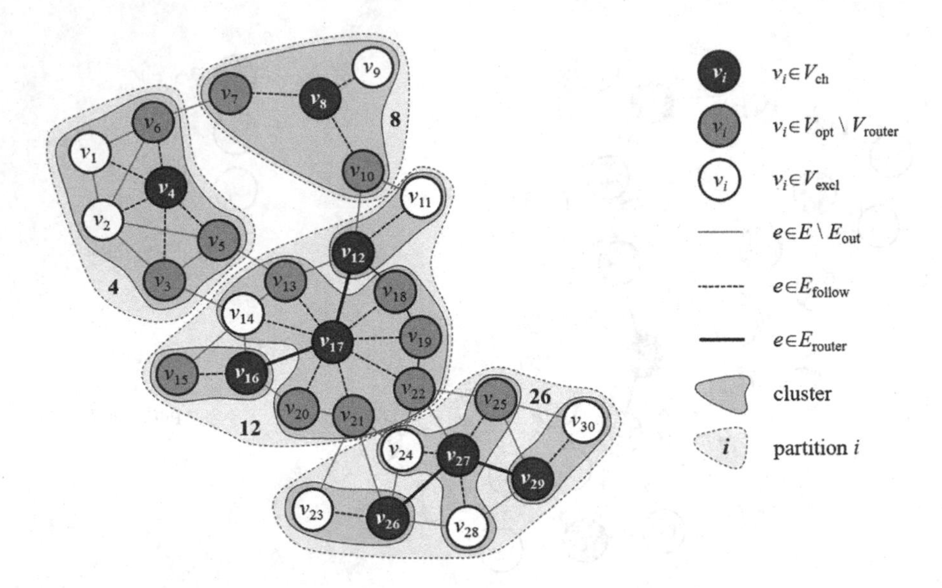

Fig. 9.5 Result after step 2

9.2.4 *Establishment and Optimization of Connectivity*

Step 3: Selection of optional nodes as gateways (1)

In the third step, nodes $v \in V_{opt} \setminus V_{router}$ are selected as gateways if they have at least one cluster head belonging to another partition in 1-hop neighborhood, thereby connecting two or more partitions. To keep the number of gateways small, nodes connecting most partitions are selected first. In case of several gateway candidates, the node v_i with the smallest index number i is chosen, to obtain deterministic results.

For each node $v \in V_{opt} \setminus V_{router}$ selected as gateway, data structures V_{gw} and G_{router} are updated as follows. First, v is added to V_{gw}. Second, 1-hop neighbors $v' \in V_{ch}$ of v are connected by adding the bidirectional links (v, v') to E_{router}. This includes the link to the cluster head of v, which now belongs to cluster and router topology. Finally, partitions are updated.

If, after Step 3, there is a single partition covering all nodes of G, HNC continues with Step 5. Otherwise, the algorithm continues with Step 4.

Figure 9.6 shows the result after Step 3. According to Fig. 9.5, nodes v_{10} and v_{22} are gateway candidates, connecting two partitions each. As v_{10} has the smaller index number, it is selected first, followed by v_{22}. This connects partitions 8, 12, and 26 to form the new partition 8. After Step 3, two partitions remain, therefore, the algorithm continues with Step 4.

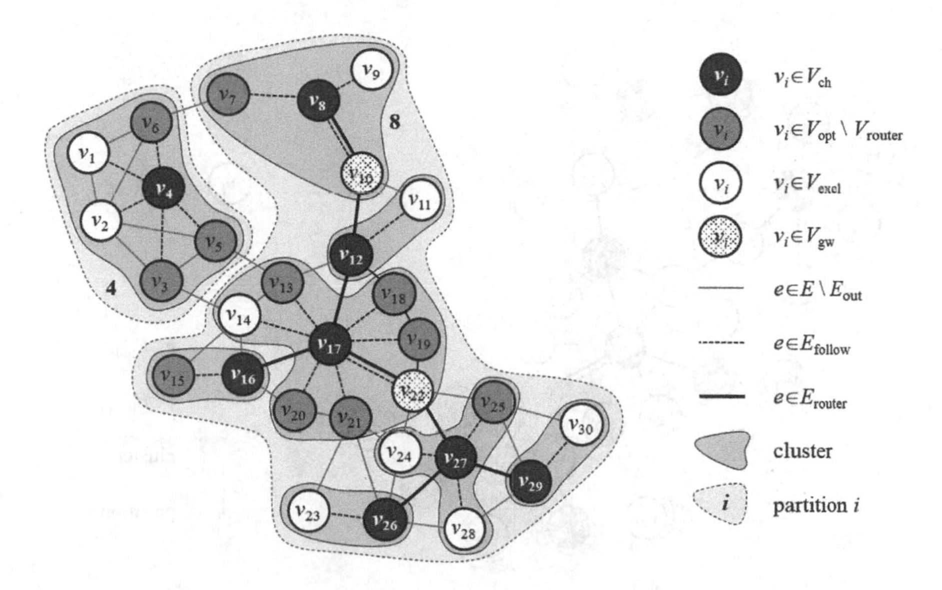

Fig. 9.6 Result after step 3

Step 4: Selection of optional nodes as gateways (2)

In the fourth step, pairs of nodes $v \in V_{\text{opt}} \setminus V_{\text{router}}$ are selected as gateways if they together can connect two or more partitions. If the complete subgraph $G' = (V_{\text{mand}} \cup V_{\text{opt}}, E')$ of $G = (V, E)$ is connected, and since a 1-hop dominating set of a connected graph is always 3-hop connected, this will finally yield a single partition covering all nodes of the communication topology $G = (V, E)$. In case of several gateway pair candidates, the pair containing the nodes v_i and v_j with the smallest index number i and j is chosen, to obtain deterministic results. Data structures V_{gw} and G_{router} are updated accordingly. Then, HNC continues with Step 5.

Figure 9.7 shows the result after Step 4. According to Fig. 9.6, (v_5, v_{13}) and (v_6, v_7) are gateway pair candidates to connect partitions 4 and 8. Nodes v_5 and v_{13} are selected, because v_5 has the smallest index number. After Step 4, only one partition remains, which implies that HNC has found a feasible clustering.

Steps 5 and 6: Optimization of connectivity

If Steps 1–4 have been successful, a 3-hop connected 1-hop dominating set has already been established. However, it may contain routes between cluster heads that are rather long and therefore inefficient. The objective of Steps 5 and 6 is to enhance routing efficiency by selecting further nodes $v \in V_{\text{opt}} \setminus V_{\text{router}}$ as gateways, while keeping the number of routers low. In Step 5, single nodes are selected; in Step 6, pairs of nodes are added as gateways. In both steps, data structures V_{gw} and G_{router} are updated accordingly.

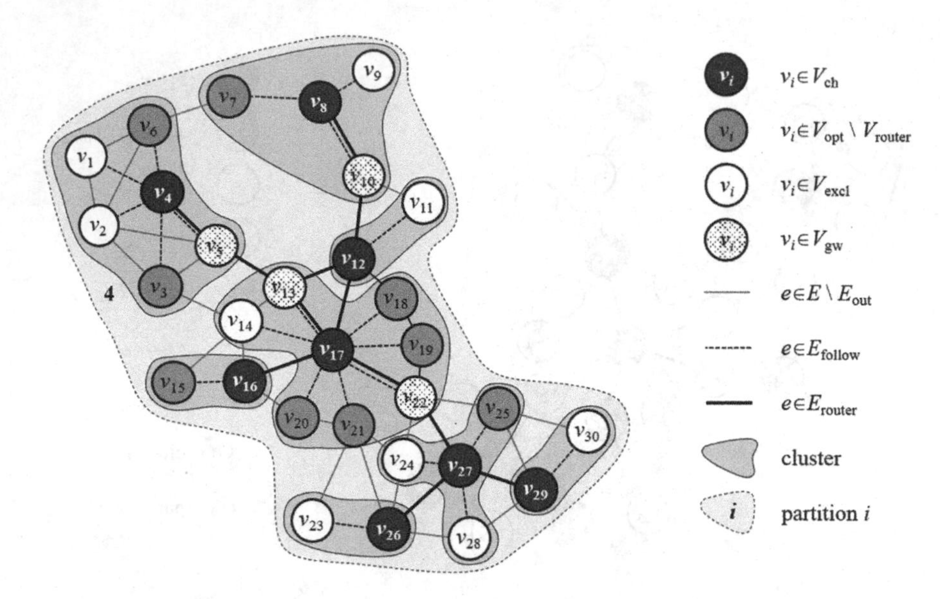

Fig. 9.7 Result after step 4

Figure 9.8 shows the result after Steps 5 and 6. In Step 5, node v_{21} is selected as additional gateway, shortening the route between v_{17} and v_{26} from three to two hops. In Step 6, nodes v_6 and v_7 become gateways, leading to a 3-hop route between v_4 and v_8.

9.2.5 *Evaluation*

In this chapter, we present the results of simulation experiments to assess the performance of HNC w.r.t. its clustering objectives, i.e., a small number of clusters and efficient connectivity [17]. In the experiments, we have generated random topologies and have applied HNC. Furthermore, we have applied HNC', which is HNC without optimization of connectivity, and adjusted versions of LCA [5] and MaxMinD [4] (see also Sect. 9.3) to cope with heterogeneous networks.

Linked clustering algorithm (LCA) is a deterministic algorithm for the formation of 3-hop connected 1-hop dominating sets in homogeneous wireless networks. To extend LCA to heterogeneous networks, mandatory nodes are selected in decreasing order of their node identifiers before other nodes are selected. Furthermore, we ensure that excluded nodes are not selected as cluster heads or gateways. We call this adjusted version LCA'.

MaxMinD is a deterministic algorithm for the formation of d-hop dominating sets that are $(2 \cdot d + 1)$-hop connected. For comparison with HNC, d is set to 1. To

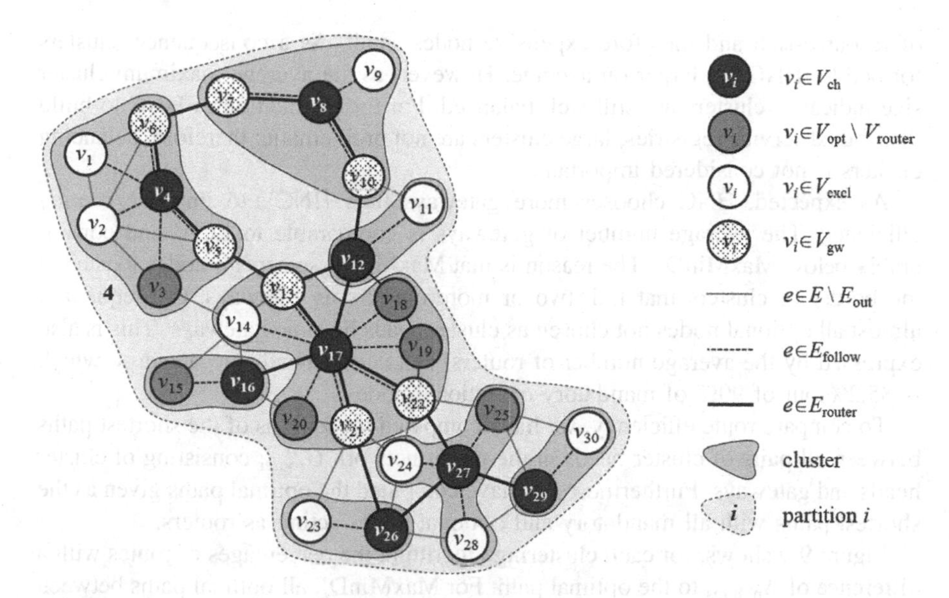

Fig. 9.8 Result after steps 5 and 6

adjust MaxMinD to heterogeneous networks, mandatory nodes always keep their own node identifier in the floodmax and floodmin phases to ensure they become cluster heads. Excluded nodes refrain from broadcasting their node identifier and are not considered in the gateway selection phase. We call this adjusted version MaxMinD'.

For the simulation experiments, we have generated random topologies. In each topology, 100 nodes are placed randomly in a 100 × 100 grid. Communication links between pairs of nodes are added if their distance is at most 14 grid units. Mandatory and excluded nodes are determined randomly, too, with a chance of 5% and 10%, respectively. Topologies are checked whether they are connected and dropped if not.

To compare the performance of HNC, HNC', LCA', and MaxMinD', they were applied to 1000 random topologies. Table 9.1 shows the average number of cluster heads, gateways, and routers, and the average maximum cluster size. Compared to LCA' and MaxMinD', HNC selects far less cluster heads, which keeps the number

Table 9.1 Comparison of performance w.r.t. to clustering objectives (see [3, Table 1])

Algorithm	*avg* $\lvert V_{ch}\rvert$	*avg* $\lvert V_{gw}\rvert$	*avg* $\lvert V_{router}\rvert$	*avg max* $\lvert V_{cluster}\rvert$
HNC	22.4	38.9	61.3	10.8
HNC'	22.4	18.1	40.5	10.8
LCA'	29.8	36.1	65.9	9.5
MaxMinD'	32.7	52.5	85.2	8.4

of resource-rich and therefore expensive nodes small. As a consequence, clusters formed by HNC are larger on average. However, as the average maximum cluster size indicates, clusters are still well-balanced. For the application of HNC to build distributed service registries, large clusters are not problematic; therefore, balancing clusters is not considered important.

As expected, HNC chooses more gateways than HNC', to improve routing efficiency. The average number of gateways is comparable to LCA' and considerably below MaxMinD'. The reason is that MaxMinD' selects all nodes located at the border of clusters that link two or more clusters as gateways. Consequently, almost all optional nodes not chosen as cluster heads become gateways. This is also expressed by the average number of routers, i.e., cluster heads or gateways, which is 85.2% out of 90% of mandatory or optional nodes.

To compare route efficiency, we have computed the lengths of the shortest paths between all pairs of cluster heads in the router network G_{router} consisting of cluster heads and gateways. Furthermore, we have computed the optimal paths given as the shortest paths with all mandatory and optional nodes acting as routers.

Figure 9.9 shows, for each clustering algorithm, the percentages of routes with a difference of Δ_{n_hops} to the optimal path. For MaxMinD', all optimal paths between pairs of cluster heads are contained in the router network, i.e., Δ_{n_hops} is always 0. This is not surprising, as MaxMinD' selects almost all optional nodes as gateways. Even though HNC selects considerably less gateways (see Table 9.1), 97.7% of all optimal paths are part of the router network, and no shortest route exceeds the optimal path by more than three hops. The results also show that the optimization of connectivity in Steps 5 and 6 of HNC has a significant effect: without these steps, only 49.8% of all optimal paths are part of the router network, and 2.6% of the paths are 10 or more hops longer. In summary, we observe that the objective of efficient connectivity while keeping the number of gateways low is achieved.

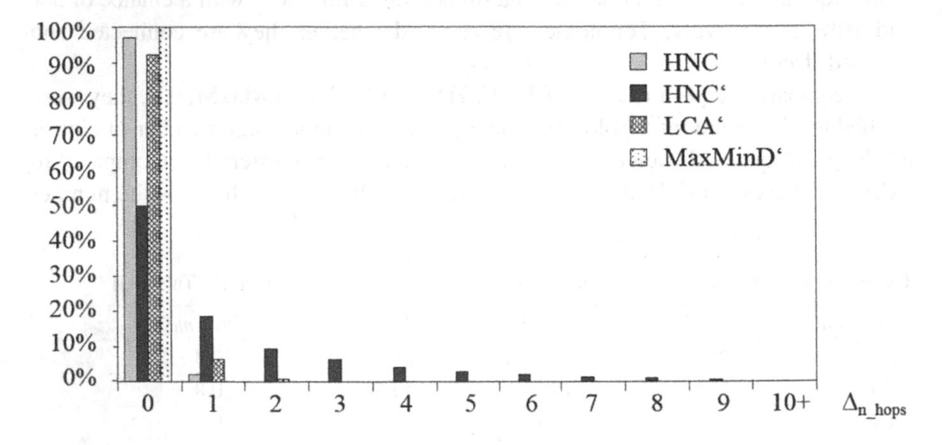

Fig. 9.9 Comparison of route efficiency (see [3, Fig. 13])

9.3 Related Work

In this chapter, we outline clustering protocols for wireless ad hoc networks and compare them to our heterogeneous network clustering (HNC) algorithm. To keep the survey concise, we have selected two representative protocols that establish intra- and inter-cluster connectivity in TDMA and non-TDMA networks. We have used adjusted versions of these protocols in the performance evaluation of HNC (see Sect. 9.2.5).

9.3.1 Linked Clustering Algorithm (LCA)

In [5], Baker and Ephremides present *linked clustering algorithm (LCA)*, a clustering protocol that has been tailored to the HF intra-task force communication network of the US Navy. Clustering objectives are intra- and inter-cluster connectivity, high availability, and fault tolerance. To achieve these objectives, clusters are formed on different radio channels. As HNC, LCA establishes a 3-hop connected 1-hop dominating set. Different from HNC, LCA assumes a homogeneous network, i.e., all nodes may become cluster heads.

The authors start by presenting a centralized version of LCA, which is later decentralized. A communication topology is modeled as a connected graph $G = (V, E)$, with nodes $v_i \in V$ having unique node identifiers $i \in \{1,\ldots, n\}$. The node with the highest node identifier becomes the first cluster head, with all 1-hop neighbors assigned as followers. As long as there are unclassified nodes, the algorithm continues, in decreasing order of node identifiers. If a node has one or more unclassified neighbors, it becomes cluster head, with its unclassified 1-hop neighbors assigned as followers. In addition, gateways may have to be selected to achieve inter-cluster connectivity. Figure 9.10 shows the final clustering for a small network.

The centralized version of LCA is only applicable if the network topology is already known. In case the topology is not available, there is a decentralized version of LCA that incorporates topology detection, based on TDMA. For this purpose,

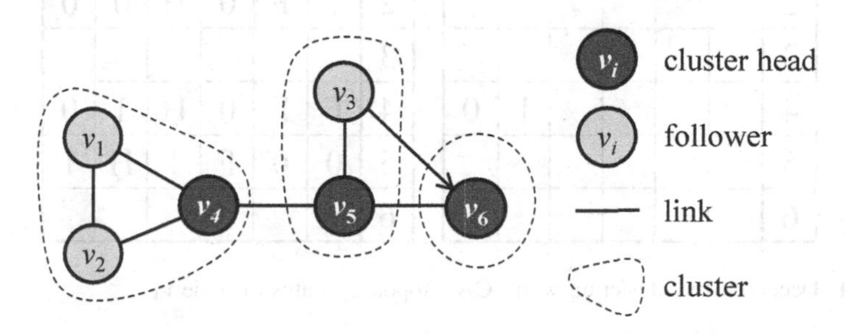

Fig. 9.10 Centralized clustering with LCA

time is decomposed into m epochs, two frames, and n slots, where m is the number of radio channels, and n is the highest node identifier. In each epoch k, LCA is performed on radio channel k. In each frame of an epoch, a phase of LCA is executed. Each phase consists of n slots assigned to nodes v_1 to v_n.

In the first frame of each epoch, each node v_i locally broadcasts its nodes heard set nhs_i, which is initially empty, in its assigned slot. In each slot $i \neq j$, nodes v_j update their nodes heard set nhs_j and their connectivity matrix cm_j based on whether they have received the message from v_i. At the end of the frame, each node v_i is aware of its 1-hop neighbors and of the existence or non-existence of bidirectional links for all v_j with $j > i$. For the topology in Fig. 9.10, Fig. 9.11a shows the nhs_4 and cm_4 at the end of the first frame. An entry 1 in cell $cm_4(4, 5)$ denotes a bidirectional link to node v_5, 0 in $cm_4(4, 6)$ indicates that there is no such link to node v_6.

In its assigned slot of the second frame, each node v_i decides whether it becomes a cluster head or a follower, and locally broadcasts its role ns_i and connectivity matrix row $cm_i(i)$. By then, a node has received all missing entries from its neighbors with lower node identifier, i.e., $cm_i(i)$ is complete, and the role ns_i can be determined. In addition, decisions about gateways are made. For the topology in Fig. 9.10, Fig. 9.11b shows ns_4 and cm_4 at the end of the second frame. In $cm_4(4)$, the local connectivity as well as the roles of 1-hop neighbors (H = cluster head, F = follower) are captured. Matrix rows $cm_4(1)$, $cm_4(2)$, and $cm_4(5)$ are received from these neighbors. This way, each node becomes aware of its 2-hop network topology. The final clustering is the same as in the centralized case (see Fig. 9.10).

Topology detection of the decentralized version of LCA is based on single observations. In its assigned slot of each frame, a node locally broadcasts one message. Topology decisions are then reached based on whether this message is received. This may lead to situations where bidirectional links are incorporated into the connectivity matrix, although they are actually too weak for reliable operation.

(a) $nhs_4 = \{ v_1, v_2, v_5 \}$

v_4	1	2	3	4	5	6
1						
2						
3						
4					1	0
5						
6						

(b) $ns_4 = \text{CH}$

v_4	1	2	3	4	5	6
1	F	1	0	H	0	0
2	1	F	0	H	0	0
3						
4	F	F	0	H	1	0
5	0	0	F	1	H	1
6						

Fig. 9.11 Decentralized clustering with LCA—topology status of node v_4

In HNC, topology detection is performed by ATDP (see Chap. 4), which collects extensive statistical information before deciding about link stability.

With its features—establishment of a 3-hop connected 1-hop dominating set, deterministic results—LCA is a potential candidate to replace HNC. With some adjustments (see LCA' in Sect. 9.2.5), it can even cope with heterogeneous networks consisting of nodes that must, may, or must not become cluster heads or gateways. Since the network topology is already available, the centralized version of LCA would be sufficient. Nevertheless, as we have shown in Sect. 9.2.5, HNC outperforms LCA' w.r.t. small number of clusters and efficient connectivity, and therefore is the better choice.

9.3.2 *MaxMinD*

In [4], Amis et al. introduce *MaxMinD*, a frequently quoted clustering protocol for wireless ad hoc networks. Clustering objectives are intra- and inter-cluster connectivity, and balanced cluster sizes. MaxMinD establishes a $(2 \cdot d + 1)$-hop connected d-hop dominating set, where d is a parameter to control the number of clusters and cluster sizes. With $d = 1$, HNC-like clusters can be configured.

MaxMinD works in four phases. In the floodmax phase, cluster heads are predetermined. During the floodmin phase, cluster sizes are balanced. In the cluster head selection phase, cluster heads and their followers are determined. Finally, inter-cluster routes are established in the gateway selection phase.

The floodmax phase is decomposed into d rounds. In each round r, nodes locally broadcast their node identifier and the identifier of the current winner, which is the $(r - 1)$-hop neighbor with the highest node identifier. Thus, after d rounds, each node has learned about the d-hop winner, which is a candidate for the role of becoming cluster head. Since this would lead to unbalanced cluster sizes, MaxMinD propagates the smaller node identifiers in each node's d-hop neighborhood remaining as winners in the subsequent floodmin phase. To reach local decisions about cluster heads and followers based on network status information collected in the previous phases, four rules are applied in the cluster head selection phase. Finally, based on local network status information, gateways are selected.

As in LCA, topology detection of MaxMinD is based on single observations only. Even worse, different from LCA, MaxMinD messages are not broadcasted in exclusive time slots, but contend with other traffic. This may lead to destructive collisions, which go undetected by the sending node. In addition, links that are actually too weak for sufficiently reliable message exchange may be considered for cluster formation, if there was a single successful message exchange during the execution of MaxMinD on such a link. Thus, in a real wireless ad hoc network, the outcome of MaxMinD is highly unreliable and unpredictable.

Another problem not addressed in the original paper is the need for synchronization. As MaxMinD works in phases and rounds, nodes have to agree on the state of execution. Furthermore, rounds have to be timed such that all nodes have a chance to successfully access the medium.

If the network topology is already available, a centralized version of MaxMinD with d set to 1 is a potential candidate to replace HNC. With some adjustments (see MaxMinD' in Sect. 9.2.5), it can even cope with heterogeneous networks. Nevertheless, as shown in Sect. 9.2.5, HNC outperforms MaxMinD' w.r.t. small number of clusters and gateways, while still providing connectivity of similar efficiency, and therefore is the better choice.

9.4 Conclusions

In this chapter, we have explained clustering concepts, have presented our heterogeneous network clustering (HNC) algorithm for wireless ad hoc networks, and have surveyed related work. HNC has been inspired by the communication requirements of industrial networked control systems.

HNC is a clustering algorithm specifically devised for the establishment of distributed service registries in heterogeneous wireless networks consisting of nodes with different capabilities and resources. Therefore, we have distinguished mandatory nodes that are to be selected as cluster heads, optional nodes that may be selected as cluster heads or gateways, and excluded nodes that can only act as followers.

HNC clusters a heterogeneous network into a 3-hop connected 1-hop dominating set, if such a clustering is feasible. Clustering objectives are a small number of clusters and efficient connectivity. By comparing HNC to adjusted versions of other clustering algorithms, we have shown that these objectives are achieved.

An extension of HNC could be the construction of (d + 1)-hop connected d-hop dominating sets. As the global communication topology is available, this extension seems straightforward. However, it would require that excluded nodes, which are not supposed to act as routers, be placed at the boundaries of clusters only. Therefore, this extension would not be beneficial in the context considered here.

Literature

Chair for Networked Systems

1. Gotzhein R (2014) ProNet 4.0—a wireless real-time communication system for industry 4.0, White Paper, Networked Systems Group, Department of Computer Science, University of Kaiserslautern. http://vs.informatik.uni-kl.de/publications/2014/Go14/whitePaperEN-ProNet4.0.pdf. Last accessed 27 Aug 2019

2. Sefati H, Gotzhein R, Kramer C, Schloesser S, Weiss M (2018) Dynamic overlay line topology establishment and repair in wireless networks. IEEE wireless communications and networking conference (WCNC 2018), Barcelona, Spain, April 15–18, 2018
3. Kramer C, Christmann S, Gotzhein R (2016) A clustering algorithm for distributed service registries in heterogeneous wireless networks. In: Proceedings of wireless days 2016, Toulouse, France, 23–25 Mar 2016

Further References

4. Amis A, Prakash R, Vuong T, Huynh D (2000) Max-min d-cluster formation in wireless ad hoc networks, INFOCOM 2000. In: 19th annual joint conference of the IEEE computer and communications societies, vol 1, Tel Aviv, Israel, pp 32–41
5. Baker D, Ephremides A (1981) The architectural organization of a mobile radio network via a distributed algorithm. IEEE Trans Commun 29(11):1694–1701

Chapter 10
Middleware for Networked Production and Control Systems

Middleware is a software layer providing abstractions for applications. In this chapter, we explain foundations of middleware, present *production middleware* (*ProMid*), a middleware for networked production and control systems that is part of the ProNet 4.0 protocol stack [1], survey and compare related work, and draw conclusions.

10.1 Foundations

In this chapter, we point out the importance of middleware, explain concepts, and address areas of operation.

10.1.1 Context

Middleware provides abstractions for applications. First, it hides platform-specific aspects such as heterogeneity and distribution of resources. Second, it establishes a high-level domain-specific context, by offering concepts and interfaces that are familiar to the application developer. Altogether, this fosters productivity of developers as well as quality and interoperability of application software.

Middleware is of high practical and commercial importance and therefore often developed and marketed by companies and consortia. Well-known middleware initiatives are Open Software Foundation (OSF) with Distributed Computing Environment (DCE), Object Management Group (OMG) with Common Object Request Broker Architecture (CORBA), World Wide Web Consortium (W3C) with Web Services, and Microsoft with Distributed Component Object Model (DCOM). For networked production and control systems, the Open Platform Communications (OPC) Foundation has released OPC UA, the OPC Unified Architecture [6]. Some wireless communication standards, for instance, ZigBee [12], comprise a middleware layer.

R. Gotzhein, *Real-time Communication Protocols for Multi-hop Ad-hoc Networks*, Computer Communications and Networks,
https://doi.org/10.1007/978-3-030-33319-5_10

In distributed real-time computer systems, applications have Quality of Service (QoS) requirements regarding performance, reliability, guarantees, and synchronicity. To satisfy such requirements, middleware has to build on real-time functionalities of the underlying operating and communication systems. In 2016, two OPC members have launched an OPC UA over time-sensitive networks (TSN) initiative for real-time extensions via Time-Sensitive Networks, which the OPC Foundation joined in 2018. Apart from that, real-time aspects are not well supported by commercial middleware.

10.1.2 Concepts

In distributed systems, middleware is often based on a *client–server model*. Here, servers—also called *service providers*—offer application services to clients— also called *service users*. Services are application-oriented and may be, for instance, database services, Web-based services, or domain-specific services. The task of a service-oriented middleware is to support service definition and composition (also called orchestration or choreography), service publication, and service usage.

Figure 10.1 shows two abstract service architectures, each consisting of a number of service users and service providers, and their interaction. The difference between these architectures is how services are made public. In Fig. 10.1a, service publication is either proactive, i.e., by announcement of the service provider, or reactive, i.e., by inquiry of the service user. By sending a call, the service user triggers the execution of a service, which may then lead to a reply containing service data.

In Fig. 10.1b, service publication is via a *service registry*, where service providers enter their services. Service users can then retrieve available services by performing lookups and receiving a list of available services in a result message. With this information, service users subscribe to and call-services and may get replies containing service data.

The middleware literature defines a number of Message Exchange Patterns (MEPs) for the interaction in message-based client–server architectures. An MEP is the generic definition of a form of interaction by message passing. MEPs exist in various contexts including Web Services, Software-Oriented Architecture (SOA), and Windows Communication Foundation (WCF). In a client–server architecture, we can distinguish the following MEPs:

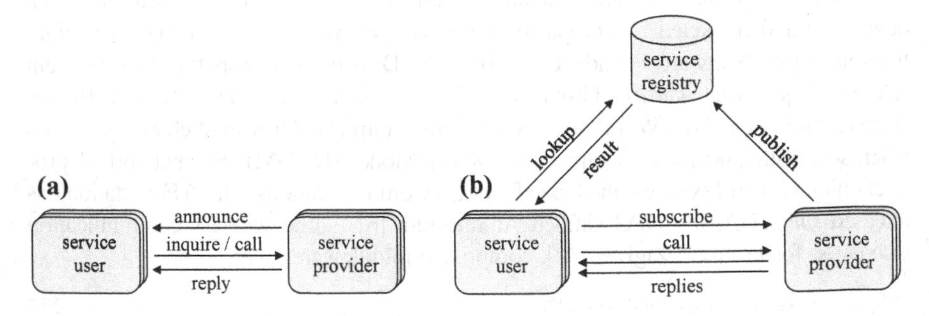

Fig. 10.1 Service architectures—interaction for service provision

- *Fire and Forget*: A service provider is called, without reply. Used, e.g., to set an actuator value.
- *Request–Response*: A service provider is called and returns a single reply. Used, e.g., to collect a sensor value.
- *Callback*: A service provider is called and then returns zero or more replies. Used, e.g., to collect a sequence of sensor values or event notifications.
- *Publish–Subscribe*: A service is made public by a service provider and can then be subscribed to by one or more service users. Until a service user unsubscribes, the service provider issues replies without further requests. Depending on the kind of service, the following reply patterns can be distinguished:
  - *Time-triggered*: The service provider issues replies at specified points in time, e.g. periodically. For instance, a temperature sensor sends a value in intervals of 10 s.
  - *Event-triggered*: The service provider issues replies when specified events occur. For instance, a photoelectric sensor sends a message when triggered.

Service definition and composition are a complex topic with a variety of approaches in different application domains. In fact, composition, orchestration, and choreography of services are an active field of research. To keep our treatment of concepts concise, we will not further address this topic.

10.1.3 Areas of Operation

Middleware is widely used for the development of distributed applications. Some programming languages offer a communication-oriented middleware to support the remote execution of code. Java, for instance, provides an abstract mechanism called Java Remote Method Invocation (RMI) [9]. When an RMI client calls a remote method, the call is transferred to the responsible RMI server on the target machine, executed, and a result is returned to the client. To realize Java RMI, client-stubs and server-skeletons, which exchange serialized parameters and results via messages on behalf of client and server, are created at runtime. For the application developer, remote method calls are syntactically similar to local method calls. However, there are semantical differences regarding, for instance, parameter usage and the possibility of server failure and message loss. Furthermore, the application developer has to program publication and lookup of remote methods in a service registry called Java Remote Object Registry (rmiregistry), remote interfaces, creation and export of remote objects, creation and configuration of a security manager, and, in some cases, routines for the serialization of objects.

The World Wide Web Consortium (W3C) [8] provides a middleware standard for the integration of Web-based applications called Web Services. Its architecture has similarities with the service architecture in Fig. 10.1b, with service providers offering services, service requesters acting as service users, and a service broker taking the role of the service registry. Web Services use the XML, SOAP, WSDL, and UDDI open standards to remain platform-independent. Simple Object Access Protocol (SOAP) is a protocol for exchanging XML data between service requester and service provider. Web Services Description Language (WSDL) is an XML-based interface description language used to characterize services. Universal Description, Discovery, and Integration—UDDI refers to a standardized registry service.

Middleware for the development of distributed applications such as Java RMI and Web Services is usually based on Internet protocols such as TCP or UDP and therefore abstracts from the specifics of communication technologies. For instance, it is not visible on application level whether wired and/or wireless technologies are used. Furthermore, as TCP and UDP provide best-effort communication only, the development of distributed real-time applications, for instance, in the production and control domain, is out of scope.

10.2 Production Middleware (ProMid)

In this chapter, we present *Production Middleware* (*ProMid*), our middleware tailored to networked production and control systems that is part of the ProNet 4.0 protocol stack [1, 4]. ProMid is an evolution of Wireless Networked Control Systems Communication Middleware (WNCS_CoM) [2, 3].

10.2.1 Concepts and Design Decisions

In ProMid, application-level services are characterized by *service offer specifications*. They are offered by *service providers* and subscribed to by *service users*. Service users identify a service by a *service requirement specification*. When a service user subscribes to a service, the middleware is in charge of resolving the service requirement specification, by determining corresponding service providers. Thus, on application level, service users are decoupled from service providers, which fosters abstraction and flexibility as follows. First, service users address services logically, without having to know where a service is physically located. Second, if service requirement specifications are coupled to service providers dynamically, the role of service provider can be passed between nodes at runtime, supporting redundancy measures, and load balancing.

A service is characterized by a service offer specification, which consists of a service type and a list of parameters further distinguishing the service. For instance, a temperature-sensing service may be associated with a geographic location such as a room number and has a minimum response interval in case of a periodical service. When subscribing to a service, the service user specifies the required service by its type, and by restricting the values of the associated parameters.

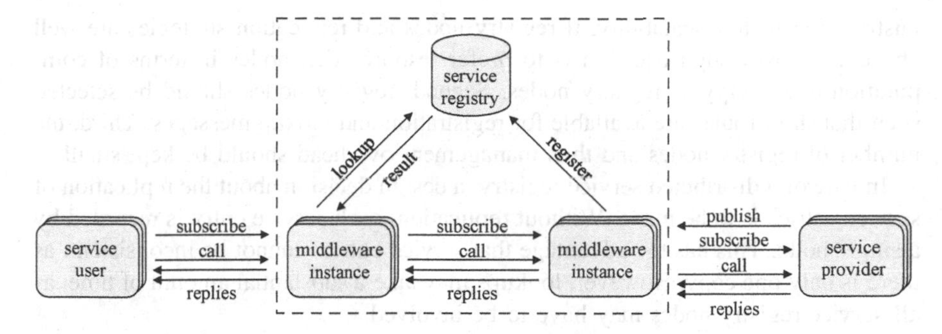

Fig. 10.2 Interaction for service provision—ProMid

Coupling of service offer specifications to service providers is by registration in a service registry. To register a service, a service provider publishes the service offer specification. The middleware is in charge of handling the registration process and of interacting with the (possibly distributed) service registry. Service subscriptions of service users are also handled by the middleware. First, the service requirement specification is to be resolved, which requires a lookup in the service registry. Second, a global service reference is created, which is used to identify the specific service. Third, the responsible service provider is informed.

Figure 10.2 illustrates the flow of control between service users and service providers via middleware instances, and their interaction with the service registry. Conceptually, service users subscribe to a service by providing a service requirement specification. It remains hidden to the service user how the local middleware instance resolves subscriptions. If the subscription is successful, the service user can call the service or gets replies.

10.2.2 Distribution and Replication of Service Registries

In ProMid, services are registered and looked up in a service registry. To deploy a service registry, design decisions regarding distribution and replication are to be made. A service registry may be centralized or distributed. In a distributed service registry, entries may be replicated.

If a service registry is centralized, all services are registered in the same location. This simplifies its operation, as service registration and lookup are executed by a single node. However, this solution does not scale well. First, the service registry node may become a bottleneck. Second, it produces substantial communication overhead, as registration and lookup messages are to be sent across the network. Therefore, a centralized service registry suits small-sized networks only.

If a service registry is distributed, services are registered in different locations; furthermore, replication of service entries in several locations becomes feasible.

Distribution fosters scalability, if registry nodes and replication strategies are well chosen. A first design decision is to prefer resource-rich nodes in terms of computation and energy as registry nodes. Second, registry nodes should be selected such that short routes are available for registration and lookup messages. Third, the number of registry nodes and thus management overhead should be kept small.

In case of a distributed service registry, a design decision about the replication of service entries is to be made. Without replication, each service entry is recorded by a single node. This has the advantage that service entries cannot be inconsistent, as there is only one copy. However, lookups may take a substantial amount of time, as all service registry nodes may have to be involved.

With replication, service entries are recorded by several nodes. This may lead to temporary inconsistencies, if service entries are modified or deleted. Furthermore, replication produces additional management traffic. However, in production and control systems, such changes are comparably rare. The advantage of replication is that lookups are more efficient, in terms of both delay and communication. Assuming that lookups occur more frequently than changes of service entries, replication is beneficial.

ProMid supports distribution of the service registry as well as replication of service entries. Distribution is determined by executing Heterogeneous Network Clustering (HNC; see Chap. 9 and Sect. 10.2.3). Replication is configured by setting a replication radius of n hops, which means that service entries are replicated on all registry nodes in n-hop neighborhood of the service registry node where the service has been registered. By setting n to 0, replication is disabled. If n equals the network diameter, all service entries are replicated on all registry nodes. If we assume a certain subscription locality, i.e., nodes interested in a particular service are in short hop distance to the service provider, replication should be kept local, too, by choosing small values for n.

10.2.3 Establishment of Distributed Service Registries

In ProMid, service providers publish service offer specifications in a distributed service registry. To establish this service registry, all nodes execute heterogeneous network clustering (HNC; see Chap. 9), a clustering algorithm for wireless TDMA networks, which is part of ProMid. HNC has been specifically devised for establishing distributed service registries in heterogeneous networks consisting of nodes with different capabilities and resources, as in production and control networks.

Production and control systems are composed of sensor, actuator, and controller nodes. For better connectivity, additional router nodes may be added. Typically, sensor and actuator nodes are devices with reduced functionality and resources, while controller and router nodes have full functionality and less resource constraints. Therefore, we assume that controller and router nodes are basically capable of acting as service registry and/or router nodes, whereas pure sensor and actuator nodes are not.

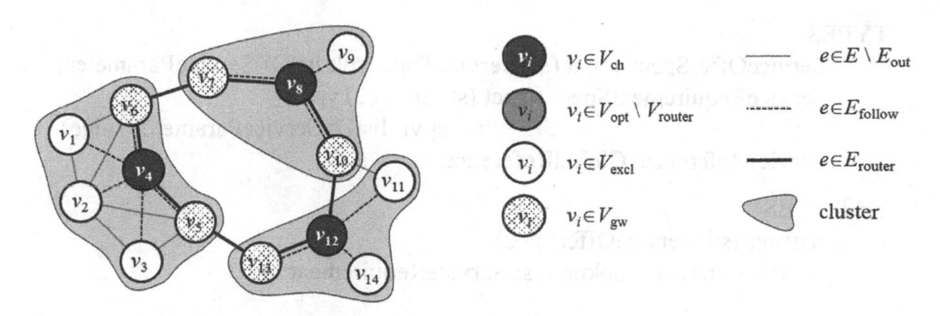

Fig. 10.3 Establishment of a distributed service registry with HNC

To apply HNC, we classify resource-rich controller and router nodes as mandatory or optional nodes, i.e., nodes that must or may become cluster heads, with the intention that all selected cluster heads act as registry nodes. Additionally, optional nodes not selected as cluster heads may become gateways between clusters. Furthermore, we classify all pure sensor and actuator nodes as excluded nodes, which makes them followers in terms of HNC.

HNC clusters a heterogeneous network into a 3-hop connected 1-hop dominating set, if such a clustering is feasible. Thus, all followers, in particular, all pure sensor and actuator nodes, are in single-hop distance of at least one cluster head. From this, it follows that nodes with reduced functionality and resources are not required to act as routers, which enables low duty cycles for minimal energy consumption (see Chap. 7).

Figure 10.3 shows an example of a communication topology after application of HNC, with $V_{mand} = \{v_4, v_8\}$, $V_{opt} = \{v_5, v_6, v_7, v_{10}, v_{12}, v_{13}\}$, and $V_{excl} = \{v_1, v_2, v_3, v_9, v_{11}, v_{14}\}$. To form a 1-hop dominating set, HNC selects all mandatory nodes and the optional node v_{12} as cluster heads, which act as service registry nodes. Furthermore, optional nodes are selected as gateways, as shown in the figure.

10.2.4 Operation of the Service Registry

The service registry offers confirmed bootstrap services to register and look up application services. To register a service, a service offer specification is provided, which indicates service type and a list of parameters further characterizing the service (see Listing 10.1). For instance, a sensor node may offer a temperature reading service, parameterized with sensor location and shortest reading interval.

To look up a service, a service requirement specification is provided, which indicates service type and a list of parameter values identifying the requested service. The service requirement specification must match the parameters of the service type and may restrict the set of service providers offering the service. For instance, a service user may be interested in a temperature reading service of a

```
TYPES
    ServiceOfferSpec: struct (st: ServiceType, spl: listOf ServiceParameter)
    ServiceRequirementSpec: struct (st: ServiceType,
                                    spvl: listOf ServiceParameterValue)
    ServiceReference: GlobalReference

SERVICES
    register (sd: ServiceOfferSpec)
    ServiceReference lookup (ss: ServiceRequirementSpec)
```

Listing 10.1 Bootstrap services of the service registry—register, lookup

specific location. The outcome of a successful lookup is a global service reference, which identifies the service provider network-wide.

Distribution of the service registry and replication of service entries are kept transparent to its users. Service registry instances run a protocol realizing the replication strategy and maintaining consistency of replicated service entries. Furthermore, they forward lookups in case these cannot be resolved locally. For message exchange among service registry instances, the overlay routing topology determined by HNC connecting all service registry nodes is used. For brevity of exposition, we omit further details about the operation of the protocol.

10.2.5 Service Publication and Subscription

To publish a service, a service provider uses the confirmed publish service offered by the middleware, parameterized with a service offer specification (see Listing 10.2). The local middleware instance then registers the service by sending a registration request to its cluster head, which hosts a service registry instance. For brevity of exposition, we omit confirmation parameters, which may contain information about the result of an operation or message exchange. Several service

```
TYPES
    ServiceOfferSpec: struct (st: ServiceType, spl: listOf ServiceParameter)
    ServiceRequirementSpec: struct (st: ServiceType,
                                    spvl: listOf ServiceParameterValue)
    ServiceReference: GlobalReference
SERVICES
    publish (sd: ServiceOfferSpec)
    ServiceReference subscribe (ss: ServiceRequirementSpec)
    unsubscribe (sr: ServiceReference)
```

Listing 10.2 Middleware services—publish, subscribe

providers may publish services of the same type, with possibly different service offer specifications. For instance, several sensor nodes may offer a temperature reading service in different locations.

To subscribe to a service, a service user applies the confirmed subscribe service offered by the middleware, parameterized with a service requirement specification. To resolve the service requirement specification, the local middleware instance issues a lookup to the service registry instance of its cluster and waits for the result, which is a global service reference in case of success. This reference is used by the service user to interact with the service provider and to unsubscribe from the service.

Once a service provider has been determined, the middleware requests routes between service user and service provider. Route requests are issued to ProRoute, the routing protocol module of the ProNet 4.0 protocol stack implementing QMR (QoS Multicast Routing, see Chap. 8). Depending on the service, routes in both directions may be required, with different QoS requirements. QMR creates or extends QoS routing trees, depending on whether a service already has other service users. Furthermore, time slots are exclusively reserved.

The message sequence chart in Fig. 10.4 illustrates application service publication and service subscription. Sensor and Controller are application instances, providing and using the periodic service of a sensor measuring the angle of an inverted pendulum (see Sect. 1.2), respectively. The middleware ProMid offers publish and subscribe services. On this level of abstraction, ProMid is represented as a single instance, to hide distribution.

Sensor publishes the service by issuing a publish request to ProMid, indicating service type Angle and minimal period 10 ms. After successful publication, ProMid returns a publish confirmation. In the second part of the scenario, Controller issues a subscribe request to ProMid, specifying service type, maximum period, and maximum delay. The subscribe confirmation informs Controller about the success of the subscribe request.

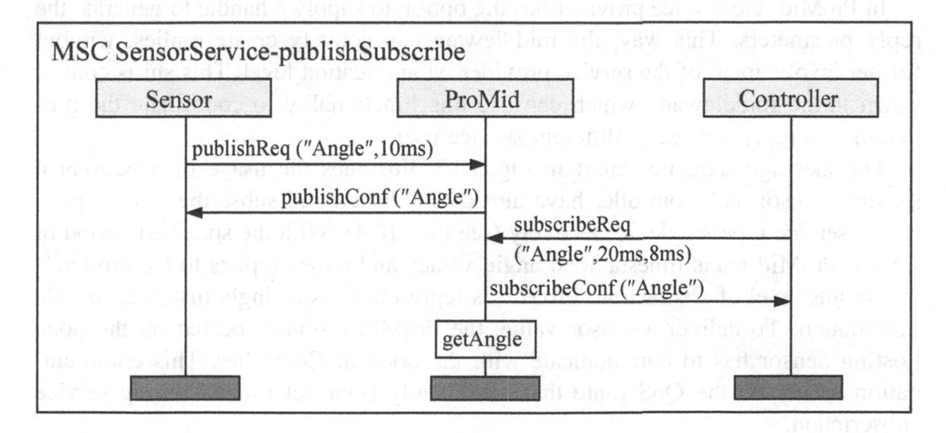

Fig. 10.4 Service publication and subscription

To provide publish and subscribe services, ProMid uses the bootstrap services of the distributed service registry. After the service provider Sensor has been located, a QoS route between the nodes hosting Sensor and Controller has to be established, satisfying the specified maximal delay of 8 ms.

10.2.6 Application Service Provision

ProMid supports the provision of different kinds of application services. We distinguish between call- and reply-services. A call-service is a service where, after subscription, the service user takes the initiative and contacts the service provider. Call-services are used, for instance, to apply steering values to actuators. A reply-service is a service where, after subscription, the service provider contacts the service user(s). Reply-services are used, for instance, to report sensor values.

Call- and reply-services can be scheduled at specified points in time, i.e., time-triggered, or by the occurrence of events, i.e., event-triggered. Time-triggered scheduling often follows a strictly or weakly periodic pattern. For instance, a temperature sensor reports a value in intervals of 10 s. Event-triggered scheduling is sporadic; however, it can be mapped to a periodic pattern supporting the minimum reaction time when assigning resources. For instance, a photoelectric sensor reports its status when triggered by movement, with a maximum reaction delay of 1 s.

When using a call-service, the service user is in charge of deciding when to contact the service provider. For this, ProMid accepts call-operations containing a valid service reference and a list of service parameter values. Based on subscription and established QoS route, ProMid may assert a maximum interaction delay.

If the service user has subscribed to a reply-service, the service provider is responsible to decide when to issue replies. Depending on the service, replies may be triggered by events or at specified points in time. The service user simply waits for incoming replies, i.e., no polling is needed.

In ProMid, the service provider has the option to supply a handle to generate the reply parameters. This way, the middleware can actively create replies, without further involvement of the service provider on application level. This shifts control down to the middleware, which can use this functionality to coordinate the provision of a reply-service to different service users.

The message sequence chart in Fig. 10.5 illustrates the usage of a periodical service. Sensor and Controller have already published and subscribed to the periodical service type Angle, respectively (see Fig. 10.4). With the specified period of 20 ms, ProMid reads timestamped angle values and issues replies to Controller.

On this level of abstraction, ProMid is represented as a single instance, to hide distribution. To deliver a sensor value, the ProMid instance located on the node hosting Sensor has to communicate with the node of Controller. This communication occurs on the QoS route that has already been determined during service subscription.

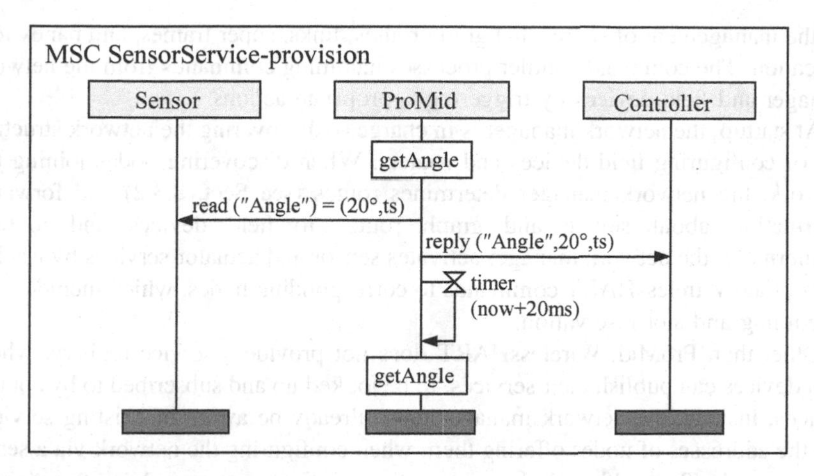

Fig. 10.5 Usage of a periodical service

10.3 Related Work

In this chapter, we briefly present and discuss protocol stacks for wireless ad hoc networks that comprise an application layer and compare them with our production middleware ProMid. To keep the survey concise, we have selected two protocol stacks that are state-of-the-practice.

10.3.1 WirelessHART

WirelessHART [10] is a protocol stack for wireless real-time networking in industrial environments, extending the wired Highway Addressable Remote Transducer (HART) protocol developed in the 1980s. The current version is WirelessHART 3.0 [11]. WirelessHART operates mesh networks consisting of nodes with different roles and capabilities. Field devices sample the state of the technical system and apply steering values. Routers provide network connectivity. The network manager is responsible for time synchronization, channel management, route discovery, exclusive slot scheduling, and slot reservation. The gateway couples wireless field devices to the wired plant automation network and acts as destination for packets from and as source for packets to field devices.

The architecture of the WirelessHART application layer consists of a command handler and supporting functionality. Commands are classified into a number of categories. HART commands comprise universal and common practice commands, to be supported by devices, for instance, what data is to be sent at what rate, and when to start and stop sending. The device manufacturer can define device-specific commands. WirelessHART commands comprise, in addition, network commands for the management of source and graph routes, links, super frames, and bandwidth allocation. The command handler processes incoming commands from the network manager and field devices by triggering appropriate actions.

At startup, the network manager is in charge of discovering the network structure and of configuring field devices and routers. When discovering nodes joining the network, the network manager determines routes (see Sect. 8.3.2) and forwards information about source and graph routes to field devices and routers. Furthermore, the network manager activates sensor and actuator services by issuing appropriate WirelessHART commands to corresponding nodes, which includes slot scheduling and slot reservation.

Other than ProMid, WirelessHART does not provide a service registry, where field devices can publish their services, to be looked up and subscribed to by control devices. Instead, the network manager must already be aware of existing services and the addresses of nodes offering them when configuring the network via a series of commands. Thus, although there is a notion of client–server applying to activated services, the level of abstraction is rather low. Furthermore, the network manager handles network topology discovery, determines routes and exclusive slot reservations, and transfers the resulting status updates to field devices and routers by issuing commands. Thus, network management is part of the application program, which offers much flexibility, but further reduces the level of abstraction. Finally, distribution is not hidden. Therefore, strictly speaking, the application layer of WirelessHART does not qualify as application middleware.

10.3.2 ZigBee

ZigBee [12] is a protocol stack for wireless networking in home and industrial automation, with low transmission rates and low energy consumption. It incorporates IEEE 802.15.4 [5] and adds routing, security, and application functionalities. ZigBee standards are developed by the ZigBee Alliance, the standard bearer of the open Internet of Things (IoT). The current version is ZigBee 3.0 [13]. ZigBee operates mesh networks consisting of nodes with different roles and capabilities. Full Function Devices (FFDs) are nodes supporting the complete set of ZigBee functionalities, other than Reduced Function Devices (RFDs). A coordinator is an FFD responsible for network formation. Routers are FFDs building up network connectivity. End devices are FFDs or RFDs hosting, for instance, sensors and actuators. The mesh topology of ZigBee networks can be characterized as a clustered tree rooted at the coordinator, with end devices as leaves. In clustering terminology, this is a 1-hop connected 1-hop dominating set (see Chap. 9).

The architecture of the ZigBee application layer consists of several components. Application objects implement end applications, using an application framework providing a set of functions for interaction with other components of the application layer. The APplication Support sublayer (APS) is in charge of connection management and message transport between application objects. The ZigBee device object (ZDO) implements role functionality, discovers device services and establishes connections.

For route discovery, ZigBee uses Ad hoc On-demand Distance Vector (AODV) [7], a reactive routing protocol for wireless networks. Routes discovered with AODV are best-effort routes, i.e., routes without guarantees regarding real-time and reliability. This implies that despite the possibility to configure guaranteed time slots, the ZigBee application layer does not support real-time services.

ZigBee employs a client–server model, where application objects act as service providers and/or service users. Elementary input and output services are combined into descriptors referring to an application profile and an application device and are associated with application objects. Other than ProMid, ZigBee does not provide a service registry, where service providers can publish their services, to be looked up and subscribed to by service users. Rather, service discovery is by inquiry, using identifiers of application profiles and elementary services. First, a node discovers a device, followed by a query to the application objects hosted on that device. To use services, clients send commands to service providers.

10.4 Conclusions

In this chapter, we have explained middleware concepts, have presented Production Middleware (ProMid) that is part of the ProNet 4.0 protocol stack, and have surveyed related work. ProMid is an application-specific middleware devised for networked production and control systems. It adopts a client–server architecture, with service providers and service users publishing and subscribing to services, respectively. Part of the middleware is a distributed service registry, where service entries can be replicated by setting a replication radius. However, the existence of a service registry and therefore distribution, replication, registration, and lookup remain transparent to service providers and service users on application level.

ProMid supports the provision of real-time call- and reply-services, which can be triggered at specified points in time or by the occurrence of events. The provision of these real-time services is mapped to communication functionalities of the ProNet 4.0 protocol stack, in particular, ProRes and ProRoute (see Chaps. 5 and 8). Again, the distributed system nature remains hidden to service providers and service users on application level.

Literature

Chair for Networked Systems

1. Gotzhein R (2014) ProNet 4.0—A Wireless Real-time Communication System for Industry 4.0. White Paper, Networked Systems Group, Department of Computer Science, University of Kaiserslautern. http://vs.informatik.uni-kl.de/publications/2014/Go14/whitePaperEN-ProNet4.0.pdf. Accessed on 27 Aug 2019
2. Chamaken A, Litz L, Krämer M, Gotzhein R (2009) Cross-layer design of wireless networked control systems with energy limitations. In: European Control Conference (ECC 2009), Budapest, Hungary
3. Haupt A, Gotzhein R et al. (2014) Control and communication co-design. In: Lunze J (ed) Control theory of digitally networked dynamic systems, Springer, pp 328–348
4. Berg M, Gotzhein R et al. (2015) Vertical integration and adaptive services in networked production environments. In: Proceedings of ERP future 2015, Munich, Germany, Springer LNBIP 245 Nov 16, pp 1–16

Further References

5. Institute of Electrical and Electronics Engineers IEEE (2011) Standard 802 Part 15.4: Low-Rate Wireless Personal Area Networks (LR-WPANs). IEEE Computer Society, New York, USA, Sep 2011
6. Mahnke W, Leitner SH, Damm M (2009) OPC unified architecture, Springer
7. Perkins C, Royer E (1999) Ad hoc On-demand distance vector routing. In: Proceedings of the 2nd IEEE Workshop on Mobile Computing Systems and Applications, New Orleans, USA, Feb 1999, pp 90–100
8. World Wide Web Consortium (2019). https://www.w3.org/. Accessed on 27 Aug 2019
9. Java Remote Method Invocation (2019). https://www.oracle.com/technetwork/java/javase/tech/index-jsp-136424.html. Accessed on 27 Aug 2019
10. International Electrotechnical Commission (IEC) (2010) Industrial Communication Networks—Wireless Communication Network and Communication Profiles—WirelessHART (IEC 62591 ed 1.0), Geneva, Switzerland, April 2010
11. Linear Technology (2019) SmartMesh WirelessHART User's Guide. https://www.analog.com/media/en/technical-documentation/user-guides/SmartMesh_WirelessHART_User_s_Guide.pdf. Accessed on 27 Aug 2019
12. ZigBee™ Alliance (2019) ZigBee Specification, Version 1.0, June 2005 p 378. URL: www.zigbee.org. Accessed on 27 Aug 2019
13. NXP Semiconductors (2018) ZigBee 3.0 Stack User Guide, JN-UG-3113, Revision 1.5, Sep 11, 2018

Chapter 11
Implementation Aspects of ProNet 4.0

In this chapter, we outline *Black burst integrated Protocol Stack (BiPS)* [2, 3], a real-time-capable implementation framework specifically devised for the protocols of ProNet 4.0 [1], and its implementation on the Imote2 [15] hardware platform.

11.1 BiPS—Black Burst Integrated Protocol Stack

Black burst integrated Protocol Stack (*BiPS*) [2, 3] comprises both operating system and communication functionalities. Instead of building on an existing operating system, BiPS provides real-time-capable operating system functionality tailored to the Imote2 target platform. In particular, BiPS comprises a real-time scheduler controlling the timely execution of processes and protocols. Furthermore, BiPS implements core protocol functionalities, in particular, tick and time synchronization, global time slotting, topology detection, and several medium access schemes. Altogether, BiPS classifies as a bare-metal solution.

11.1.1 Conceptual Design Decisions

In real-time systems, time-critical functionality is to be executed under stringent timing constraints, with predictable timing behavior. In a distributed real-time system, this applies to both real-time applications and real-time protocols. Therefore, BiPS runs directly on top of the Imote2 hardware platform, without any underlying operating system. In state-of-the-practice operating systems, MAC protocols are executed in the same conceptual domain as less time-critical functionalities. Even when using a real-time operating system, it is still difficult to realize schedules that satisfy the runtime requirements of real-time communication regarding, for instance, synchronization and hardware delays. To this end, BiPS

R. Gotzhein, *Real-time Communication Protocols for Multi-hop Ad-hoc Networks*, Computer Communications and Networks,
https://doi.org/10.1007/978-3-030-33319-5_11

provides a small set of operating system functionalities customized for real-time applications and real-time protocols.

The conceptual design of BiPS strictly distinguishes between real-time and non-real-time functionalities. For this, BiPS comprises two schedulers. The *BiPS real-time scheduler* (*BRTS*, see Sect. 11.1.4) is responsible for the timely execution of time-critical functionalities such as MAC protocols, core functionalities of BiPS, and real-time application tasks. The *BiPS application scheduler* (*BAS*, see Sect. 11.1.5) runs under control of BRTS and is concerned with the execution of non-real-time applications and non-time-critical higher-layer protocols. If for complex applications, the limited capabilities of BAS are insufficient, it can be replaced by schedulers of other operating systems such as FreeRTOS [9], to be incorporated into BiPS (see Sect. 11.1.6).

11.1.2 Overview of the BiPS Framework

BiPS is an implementation framework for wireless networked production and control systems customized for the Imote2 hardware platform [15]. This platform hosts an Intel XScale PXA271 processor with 256 KiB SRAM, 32 MiB SDRAM, and 32 MiB FLASH. It supports clock rates up to 416 MHz and integrates the widely used IEEE 802.15.4-compliant CC2420 transceiver [5].

Figure 11.1 shows the architecture of the BiPS framework. Shaded parts of the figure mark functionality covered by BiPS. Other parts are to be added before deployment. The overall structure of BiPS follows a layered approach, where higher layers abstract from lower layers by providing abstract interfaces.

- Layer 0 implements low-level functionality to interact with hardware components, such as hardware timers and general-purpose input/output (GPIO) pins.

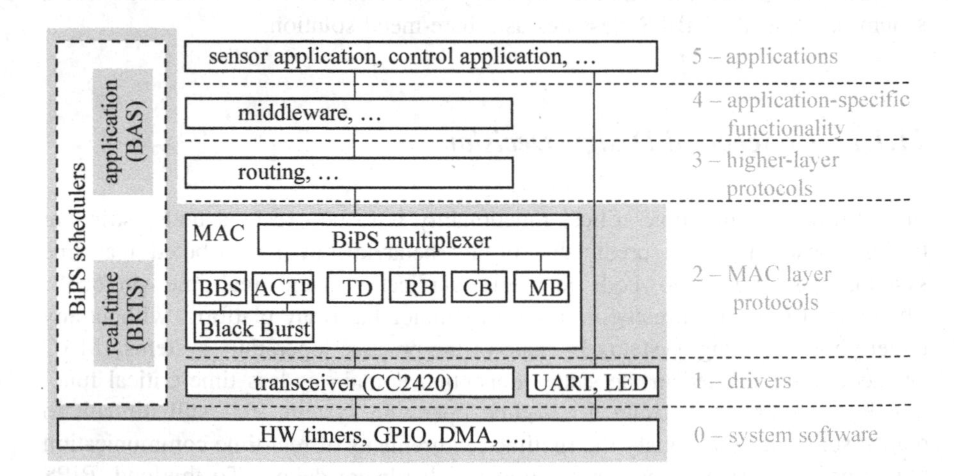

Fig. 11.1 Architecture of the BiPS framework (see [2, Fig. 1 and 3, Fig. 2])

- Layer 1 comprises hardware drivers for peripheral devices, in particular, an optimized driver for the CC2420 transceiver using direct memory access (DMA).
- Layers 2 consists of the synchronization protocol Black Burst Synchronization (BBS, see Chap. 2) providing network-wide synchronization in wireless multi-hop networks, virtual time region establishment (see Chap. 3), topology detection (TD, see Chap. 4), and four MAC protocols (RB, CB, MB, ACTP, see Chaps. 5 and 6). In addition, the BiPS multiplexer provides a unique interface to the upper layers.
- Layer 3 hosts higher-level protocols, e.g., routing protocols (see Chap. 8).
- On layer 4, application-specific functionality, for instance, a customized middleware for production and control systems providing time- and event-triggered real-time application services hosting a distributed service registry (see Chaps. 9 and 10), is located.
- Layer 5 comprises application instances, e.g., sensor operation and control algorithms.

Furthermore, the BiPS framework provides two schedulers:

- The BiPS real-time scheduler (BRTS) controls execution of MAC layer protocols and real-time applications in their assigned virtual time regions.
- The BiPS application scheduler (BAS) is responsible for the execution of non-real-time applications and higher-level functionality.

The BiPS framework is the basis for the implementation of ProNet 4.0 [1]. In fact, the conceptual architecture of ProNet 4.0 (see Fig. 1.3) is mapped to the architecture of the BiPS framework and extended by specific protocols for real-time multicast routing (see Chap. 8), clustering (see Chap. 9), and a middleware for networked production and control systems (see Chap. 10).

11.1.3 Access to MAC Protocols—BiPS Multiplexer

The protocol stack ProNet 4.0 supports several medium access schemes, located on MAC layer: exclusive access, shared access with priorities, mode access, and access for n-hop deterministic arbitration and value transfer. These access schemes are implemented by protocol modules ProRes (RB), ProCont (CB), ProMod (MB) (see Chap. 5), and ProArb (ACTP, see Chap. 6), which are activated during dedicated virtual time regions.

The lower part of Fig. 11.2 shows an example of virtual time slotting (see Chap. 3) for synchronization, medium access, and real-time applications for a single-hop network consisting of three nodes. Time slotting for communication is global, i.e., all nodes have the same perception of the time structure. In addition, there are real-time (RT) application regions, which are scheduled locally such that they do not overlap with virtual time regions for communication.

Access of real-time (RT) and non-real-time (NRT) applications and higher-layer protocols to MAC protocols in virtual time regions is provided by so-called *transmission opportunities* (*TOs*), located in the BiPS multiplexer (see Fig. 11.1).

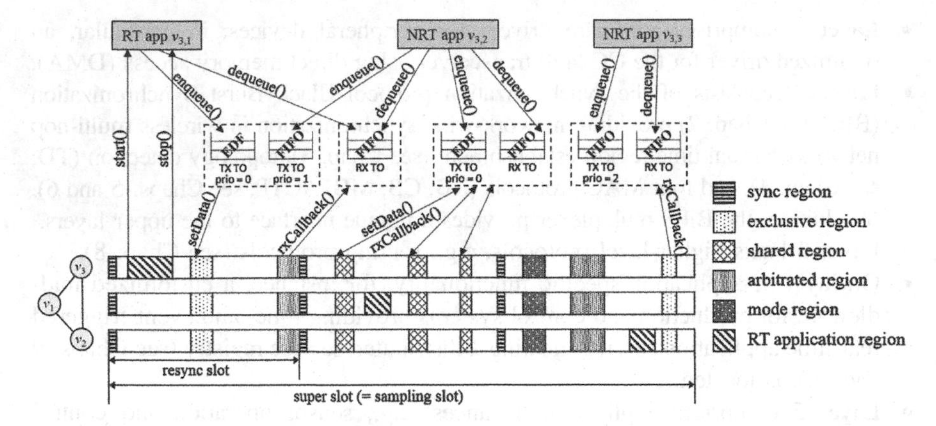

Fig. 11.2 Virtual time regions of nodes v_1, v_2, and v_3, and transmission opportunities of v_3 (see [2], Fig. 2 and [3], Fig. 6)

The multiplexer realizes common MAC functionalities, in particular, queuing of messages, thereby facilitating application development as well as the integration of further MAC protocols. Figure 11.2 shows 4 TOs associated with node v_3, containing one or more message queues, depicted as dashed rectangles. Each TO is associated with one or more virtual time regions of the same type, which determines the medium access scheme to be applied to data to be sent or received. The BiPS multiplexer provides a homogeneous interface to pass data between higher layers and MAC layer, where metadata such as addressing schemes may differ among MAC protocols. Thus, the multiplexer realizes temporal decoupling between real-time communication and higher-layer instances.

To send data, the BiPS multiplexer provides *TX TOs* (*Sending TOs*), which are shown as queues marked TX TO in Fig. 11.2. Higher-layer instances enqueue data into a TX TO queue via method call. When a MAC protocol is executed in a virtual time region and requests data to be sent, the message with highest priority is dequeued from a TX TO queue associated with this virtual time region and passed to this protocol. The multiplexer determines message priorities as follows. First, a TO can contain more than one TX TO queue, each with a unique priority. Among non-empty TX TO queues, the multiplexer selects the queue with highest priority. Next, the queue's scheduling strategy is considered. A first-in, first-out (FIFO) queue selects the oldest message, an earliest deadline first (EDF) queue returns the message with the closest deadline.

TX TOs directly interact with the MAC protocol. In particular, the MAC protocol returns a notification whether a transmission is successfully completed or not. In case of transmission failure, the message can be rescheduled up to a configurable maximum number of transmission attempts, and is reinserted into the TX TO queue. After completion of a message transfer—whether successful or not—the BiPS multiplexer notifies the higher-layer instance.

To receive data, the BiPS multiplexer provides *Receiving TOs* (*RX TOs*), which are shown as queues marked RX TO in Fig. 11.2. Higher-layer instances are notified about receptions and can dequeue and process frames according to the scheduling strategy associated with the RX TO queue.

11.1.4 BRTS—The BiPS Real-Time Scheduler

The BiPS real-time scheduler (BRTS) is in charge of realizing the schedule defined by virtual time slotting (see Fig. 11.2). In communication regions (exclusive regions, shared regions, mode regions, and arbitrated regions), BRTS activates and deactivates the associated MAC protocol at start and end of the region, respectively. In RT application regions, it transfers control to the associated application [see RT app $v_{3,1}$ in Fig. 11.2, methods start() and stop()]. In idle regions, it passes control to the BiPS application scheduler (see Sect. 11.1.5). All tasks of BRTS are non-interruptible and are executed with maximal priority.

Figure 11.3 shows the structure of a communication region and a message sequence chart (MSC) illustrating the actions of BRTS to schedule this region. Let $d_{\mathrm{communicationRegion}}$ be the configured duration of the virtual time region. For correct operation, it is essential that all nodes respect the boundaries of virtual time regions. In particular, this means that despite synchronization inaccuracies, all nodes can receive transmissions of other nodes in the same virtual time region. To satisfy this constraint, we adopt the following rules:

- At the beginning of a communication region, the transceiver is in receive mode[1], to be able to receive incoming frames of nodes with faster clocks.
- There is a guard interval $d_{\mathrm{start}} =_{\mathrm{df}} \max\{d_{\mathrm{maxTickOffset}}, d_{\mathrm{maxCfg}}\}$ at the beginning of a communication region where transmission is disabled. Here, $d_{\mathrm{maxTickOffset}}$ is the maximum synchronization inaccuracy between pairs of nodes; d_{maxCfg} is the maximum duration to finish configuration of the MAC protocol executed during this virtual time region. In particular, the guard interval ensures that transmission cannot overlap with the preceding virtual time region of nodes with slower clocks.
- There is a guard interval $d_{\mathrm{stop}} =_{\mathrm{df}} \max\{d_{\mathrm{maxTickOffset}}, d_{\mathrm{switch}}\}$ at the end of a communication region where transmission is disabled. Here, d_{switch} is the duration to switch the transceiver to receive mode, to ensure that it is in this mode when the subsequent communication region starts. Furthermore, the guard interval ensures that transmission cannot overlap with the subsequent virtual time region of nodes with faster clocks.

[1]Depending on the kind of communication region, the transceiver may be switched to sleep mode temporarily during this region.

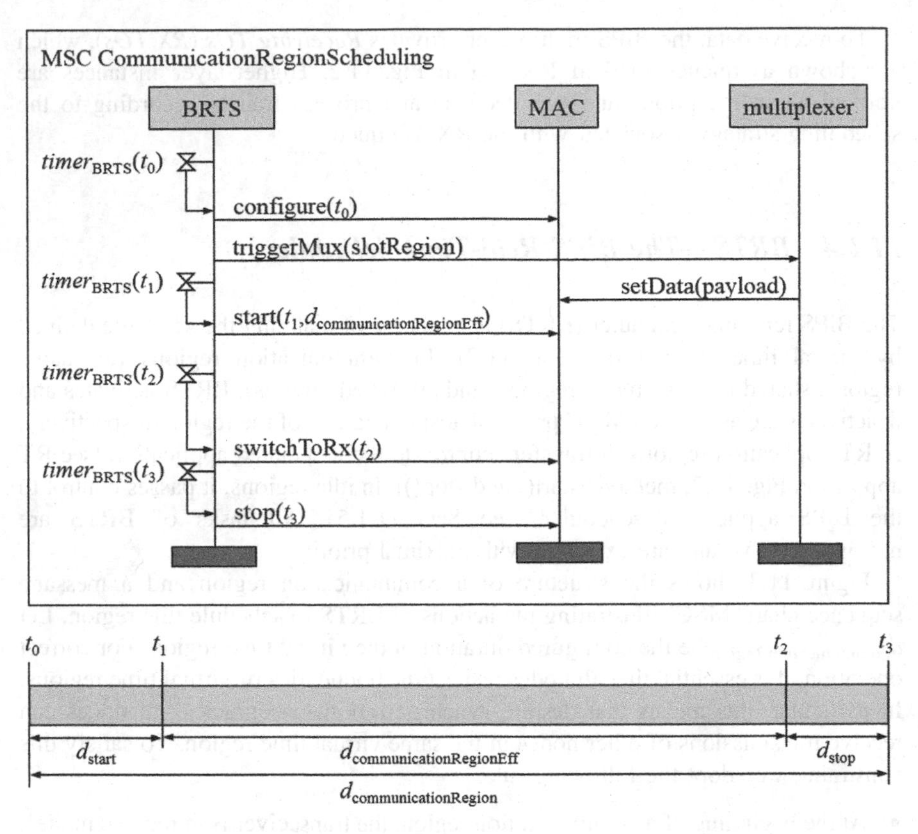

Fig. 11.3 Scheduling of communication regions (see [2], Fig. 3 and [3], Fig. 7)

From these considerations, it follows that the effective duration of a communication region that can be used for sending is given by $d_{\text{communicationRegionEff}} =_{\text{df}} d_{\text{communicationRegion}} - d_{\text{start}} - d_{\text{stop}}$ (see Fig. 11.3).

To activate MAC protocols, and to realize the proper timing behavior, BRTS is triggered by the expiry of the timer $\text{timer}_{\text{BRTS}}$. The MSC in Fig. 11.3 shows a scenario where $\textit{timer}_{\text{BRTS}}$ expires at time t_0, which is the beginning of a communication region. When executing the associated interrupt routine, BRTS triggers the configuration of the MAC protocol associated with the virtual time region and the multiplexer, which may already pass the message to be sent next in this virtual time region. When obtaining this message, the MAC protocol may already build a MAC frame and start to pass this frame to the transceiver's queue, such that sending can start as soon as the guard interval ends, i.e., at t_1. Furthermore, BRTS sets $\textit{timer}_{\text{BRTS}}$ to $t_1 = t_0 + d_{\text{start}}$.

At t_1, $\textit{timer}_{\text{BRTS}}$ expires and triggers BRTS, which starts the MAC protocol, indicating the effective duration $d_{\text{communicationRegionEff}}$ of the virtual time region for sending frames. Furthermore, BRTS sets $\text{timer}_{\text{BRTS}}$ to $t_2 = t_1 + d_{\text{communicationRegionEff}}$. Upon expiry of the timer at t_2, BRTS triggers the MAC protocol to switch the transceiver to receive mode, and sets the timer to $t_3 = t_2 + d_{stop}$. At t_3, BRTS deactivates the MAC protocol.

11.1.5 BAS—The BiPS Application Scheduler

The BiPS application scheduler (BAS) is in charge of scheduling non-real-time (NRT) applications, such as control applications and higher-layer protocols. It can be classified as cooperative, event-based, and non-preemptive. To be scheduled by BAS, applications register events, such as data receptions of the multiplexer or timer expiries. When activated by BRTS, BAS checks for the occurrence of registered events and passes control to the corresponding components. As BAS is non-preemptive, activated components may run to completion. However, they may be interrupted by BRTS, thereby providing high priorities to RT (Real-Time) applications and MAC protocols.

BAS supports interaction between NRT applications by events only. Therefore, synchronization primitives such as mutexes are not needed, as BAS tasks execute cooperatively and can only be interrupted by hardware interrupts. There are, however, critical sections where components use low-level functionalities for interaction. Examples are transmissions of messages via universal asynchronous receiver–transmitter (UART) or exchanges of data between application and BiPS multiplexer. Here, access to data structures has to be exclusive, to prevent data races. In state-of-the-practice operating systems, this is achieved by disabling interrupts temporarily. However, this would imply that BRTS would be delayed during exclusive access. We have adopted two measures to mitigate this problem:

- First, we distinguish between hardware interrupts that are used or not used by BRTS or RT applications. We then apply nested interrupts on two levels, which is supported by the PXA271 processor on hardware level. With this scheme, disabling of non-BRTS or non-RT interrupts does not affect real-time behavior. In particular, BRTS can still interrupt other interrupt handlers of lower priority.
- Second, we consider cases where the first measure cannot be applied. For instance, when data is exchanged with the BiPS multiplexer, BRTS has to be disabled temporarily, to prevent data races between different RT applications and the multiplexer. To meet deadlines, we measure the worst-case execution times $d_{maxCrit}$ of critical sections. Then, whenever a time-critical event is about to happen, BRTS takes over control $d_{maxCrit}$ before that event. BRTS then waits for the time-critical event to happen, handles the event, and returns control to BAS.

If BAS has no further tasks to schedule, it could switch the CPU to sleep mode in order to save energy (duty cycling). However, depending on the kind of sleep mode, the PXA271 processor has different activation delays. Furthermore, BAS is not aware of the real-time tasks to be scheduled next. Therefore, to maintain real-time operation, the decision about duty cycling is passed to BRTS.

11.1.6 Nesting of State-of-the-Practice Operating Systems

BiPS comprises BRTS, the BiPS real-time scheduler, and BAS, the BiPS application scheduler. For more complex applications, the limited capabilities of BAS—cooperative and event-based scheduling, no preemption, no task concept—are insufficient. For instance, in control applications performing complex calculations, it would be preferable to preempt processes after defined time slices, such that other tasks are treated fairly and do not suffer from starvation.

In BiPS, BAS can be replaced by incorporating another operating system (OS) together with its scheduling mechanisms, called *nested OS*. It is then possible to use all capabilities of the nested OS, such as preemptive priority scheduling, multi-tasking, and process interaction.

When porting an operating system to a target platform, the engineer has to adapt the generic code of the OS. In particular, this is required when nesting an OS into BiPS. In our work so far, we have nested FreeRTOS [9] and RIOT [11]. Both are state-of-the-practice real-time operating systems and support tasks, task priorities, synchronization primitives, and task preemption. We now address a number of porting steps and provide details about nesting FreeRTOS and RIOT:

- At startup, an operating system executes initialization routines to set up hardware and runtime environment. As initialization is controlled by BiPS, the code required by the nested OS is reduced to some function calls into the BiPS core. FreeRTOS needs an implementation for its xPortStartScheduler stub, which configures and starts a periodic timer and jumps into the first task. Our code for nesting FreeRTOS is similar to the code for a standalone porting, with the exception that we were able to exploit some BiPS functions, e.g., to set up the system timer. Similar considerations apply to the nesting of RIOT.
- To register and handle interrupts, and to disable them when executing critical sections, code for interrupt management is required. When porting FreeRTOS and RIOT, this was translated into calls of the BiPS core. This way, BiPS maintains some control about interrupt handling. In particular, a nested OS does not disable BiPS-related interrupts. Thus, BiPS can interrupt processes and core functions of a nested OS.
- Further code is required to manipulate a system timer for scheduling purposes of the nested OS. Since BiPS provides an interface for timer manipulation, this code simply calls BiPS functions. On the Imote2 hardware platform, nine hardware timers are available, one of which is reserved for the nested OS.
- Nesting of an OS requires code for context switching and task creation. For instance, if a task with a higher priority becomes runnable, or if the time slice of a running task expires, the CPU is rescheduled. The code implements saving and

restoring of CPU registers and flags, and of runtime-specific parts such as libc's _impure_ptr to support reentrancy of libc functions. Here, BiPS offers some helper functions, which have been sufficient to realize code for FreeRTOS and RIOT, without additional programming on assembler level.

- To access hardware components such as UART, an operating system incorporates device drivers. As BiPS already contains device drivers for the Imote2 hardware platform, nesting of an OS mostly consisted of providing wrappers that translate concepts of the nested OS to BiPS, and vice versa. For FreeRTOS, we have added mechanisms to transfer frames between transmission opportunities of BiPS and FreeRTOS queues.

Nesting of an OS into BiPS is similar to the realization of an OS on a virtual machine. Instead of interacting with the hardware platform directly, the nested OS interacts with BiPS. Since BiPS controls the execution of the virtual machine, real-time behavior of MAC protocols and RT applications is ensured. On the other hand, if the nested OS is a real-time OS—such as FreeRTOS—its real-time capabilities are reduced, as execution of tasks is delayed when BiPS takes over control.

11.1.7 Memory Management and Fault Handling

The core hardware of the Imote2 platform offers features that can be used to speed up execution and to facilitate debugging of applications. In this regard, the availability of instruction and data caches and an MMU (Memory Management Unit) is crucial. In BiPS, we use the MMU to detect and analyze common faults and provide tools to examine failure situations. In particular, we can detect overwriting of code and read-only data, accidental writes to system memory such as page tables, access of null pointers, and execution of invalid code provoked by illegal jumps.

If a failure occurs, the in-system fault handler of BiPS reports possible reasons. Furthermore, it prints a backtrace and offers a shell to examine memory content and to print current virtual memory layout, physical memory allocation, and information about heap and stacks. There is also a GDB (GNU Debugger) server included, so GDB can be invoked on a PC attached to the Imote2 via serial line. This can provide event more insight, since GDB can enhance pointer addresses by debug information.

By enabling the MMU, we can also enable the data cache of the Imote2, which speeds up average execution times significantly. On the downside, caches introduce jitter, which is conflicting with time-critical protocols and real-time applications. However, in BiPS, this jitter is largely reduced for time-critical functions by using cache locking, a feature preventing individual data and instructions from being evicted from the caches.

Using the MMU reduces development times, as BiPS and its applications work on a common virtual memory layout and are therefore independent of their physical memory placement. As result, an image with BiPS can be stored on non-volatile memory and can also be executed from SDRAM, which is much faster and saves write cycles of the flash memory.

11.1.8 Evaluation

In this chapter, we present results of experiments comparing BiPS with state-of-the-practice operating systems for wireless sensor networks. The objective of these experiments has been to evaluate BiPS as framework for real-time capable protocols and applications. In particular, we have evaluated memory usage, predictability of timer expiry processing delays, and delays to detect (start of frame delimiters (SFDs).

Since there are only few operating systems supporting the XScale PXA271 processor of the Imote2, quantitative comparisons are very limited. In fact, we only found direct support for this processor in TinyOS (used in Version 2.1.2). To compare BiPS with a classical real-time OS, we have ported RIOT [11] to the Imote2. However, since RIOT does not provide a driver for the CC2420 transceiver, comparisons are restricted to communication-independent aspects. In all experiments, the Imote2 was operated with 104 MHz.

11.1.8.1 Memory Usage

Memory usage can be divided into static and dynamic memory usage. Static memory is needed for the application image and stores code, read-only data, and pre-initialized read–write data. Its size is determined by the file size of the image after linking all object files. Dynamic memory is additionally needed during program execution. Its size is determined by memory usage for global variables, heap, and stacks. Since dynamic memory usage varies during execution, it is not straightforward to determine its size.

The Imote2 platform offers 32 MiB flash memory, which is the upper bound for static memory usage. Furthermore, it provides 256 KiB SRAM and 32 MiB SDRAM, which are the upper limits for dynamic memory usage. Since this is ample of memory, we had not much incentive to optimize memory usage. Features like virtual memory handling and a powerful fault handler consumed only about 59 KiB static memory and 26 KiB dynamic memory.

A BiPS application needs at least 140 KiB static and 26 KiB dynamic memory. An application using BiPS excessively consumes about 350 KiB static and 26 KiB dynamic memory, i.e., 1 and 0.1% of the available memory, respectively. By compiling for size instead of performance, static memory consumption can be reduced by 20%.

Table 11.1 Timer expiry processing delays and instruction counts (see [2], Table 1)

		Without load						With load					
		Delay [μs]			Instructions			Delay [μs]			Instructions		
		Min	Avg	Max	Min	Avg	Max	Min	Avg	Max	Min	Avg	Max
TinyOS	Alarm	13.5	13.6	13.7	112	112	112	13.5	16.8	18.4	112	112	112
	Timer	18.1	18.3	18.6	177	177	177	19.3	46.2	1.3	192	497	15,382
RIOT	hw timer	2.3	2.3	2.4	49	49	50	2.3	3.5	7.6	49	49.3	50
	Virtual timer	22	22	22	663	663	663	26.7	49.9	63.4	933	933	1,254
BiPS	hw timer	1.5	1.5	1.6	13	13	13	1.5	1.7	2.5	13	13.3	14
	Event timer	2.3	3.3	7.9	49	49.3	50	5.3	8.8	594	53	86.9	13,260

11.1.8.2 Timer Expiry Processing Delays

In real-time computer systems, reaction delays must be low and predictable. In this chapter, we analyze timer expiry processing delays. In the experiments, we have used one Imote2 node, on which TinyOS, RIOT, and BiPS were installed consecutively. Two timer variants were evaluated. In the first variant, timer expiries were processed in interrupt mode; this variant is called *alarm* in TinyOS, and *hw timer* in RIOT and BiPS. In the second variant, timer expiries were processed in non-interrupt mode; this variant is called *timer* in TinyOS, *virtual timer* in RIOT, and *event timer* in BiPS. Timer expiry processing delays were measured by performance counters of the Imote2. For each operating system and timer variant, experiments were performed without and with background load and repeated 10,000 times.

Table 11.1 shows the results of the experiments. Without load, there is no concurrency. In general, timer expiry processing delays are shorter and less variable in interrupt mode compared to non-interrupt mode. Due to its costly runtime model and its abstraction layers, hardware timer processing delays with TinyOS are significantly higher than with RIOT and BiPS. Here, RIOT and BiPS achieve low and stable delays, as they benefit from optimized implementations. A surprising result is the high delay of virtual timer expiry processing of RIOT, which is due to the need to preempt an idle task and to switch context, summing up to 22 μs.

When additional random background load is generated by a second timer,[2] evictions of data from the Imote2's instruction and data caches are provoked. This increases the delays for timer expiry processing, in particular, for timers in non-interrupt mode. Here, RIOT has the advantage of supporting priority-based preemptive scheduling, while TinyOS and BiPS suffer from cooperative scheduling. Yet, BiPS outperforms TinyOS due to better use of hardware features such as

[2]This timer expiry is processed in non-interrupt mode and with lowest priority in case priorities are supported by the operating system.

data caches. For timer expiries processed in interrupt mode, the increase of timer expiry processing delays with load is substantially smaller compared to the results without load. Here, BiPS performs best due to cache locking. In summary, reaction delays to timer expiries are low and highly predictable for BiPS hardware timers, which are therefore first choice for the execution of MAC protocols.

11.1.8.3 Detection Delays of Start of Frame Delimiters

According to the datasheet of the CC2420 transceiver, detection of a start of frame delimiter (SFD) is signaled with a delay of 3 μs. However, this does not include the delay until the SFD is recognized by software running on the CPU. To signal detection of an SFD, the CC2420 provides an output pin. On the Imote2 platform, this pin is connected to a general-purpose input/output (GPIO) pin of the micro-controller, thereby enabling the triggering of hardware interrupts when an SFD is signaled.

In the experiments, three Imote2 nodes were placed in close proximity, one node v_s acting as sender and two nodes v_{r1} and v_{r2} as receivers. Each receiver v_{ri} was connected by two GPIO pins sfd_{ri} and $slot_{ri}$ with a logic analyzer running with a sampling rate of 100 MHz. After detecting a rising edge at the SFD pin of the transceiver, receiver v_{ri} raised the GPIO pin sfd_{ri}. The second GPIO pin $slot_{ri}$ was raised 10 ms after the detection of the SFD to emulate virtual medium slotting and to evaluate the local accuracy of slot bounds. Therefore, a timer was activated, with the expiration time set 10 ms after detection of the SFD. Apart from control of the GPIO pins, no further application was executed. In total, 1,500 observations with BiPS and with TinyOS were made.[3]

Figure 11.4 shows the accuracy of SFD detection in software. For BiPS and TinyOS, the median of differences is almost 0 μs, which is due to the usage of identical receivers. However, the observed variation with TinyOS is much larger than with BiPS, with empirical standard deviations of 1.5 μs versus 0.06 μs. This indicates that detection accuracy with TinyOS suffers from diverging execution delays, while it is almost constant with BiPS.

Figure 11.5 illustrates the differences in the perception of slot starts 10 ms after the SFD event. Here, the differences with TinyOS are substantial and spread over a range of [−36.23, 30.58 μs], whereas differences with BiPS are almost negligible. One reason for this is that BiPS uses the μs hardware timers of the Imote2 platform, which is not the case for TinyOS.

[3]Experiments reported in this chapter were conducted with TinyOS and BiPS only, as RIOT does not provide a CC2420 driver.

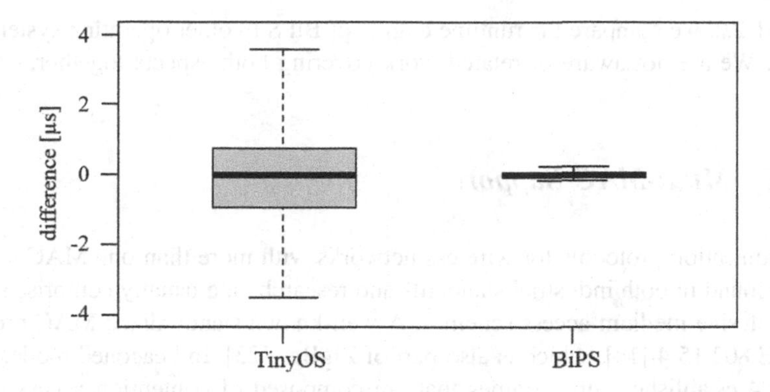

Fig. 11.4 Accuracy of SFD detection (see [2], Fig. 5)

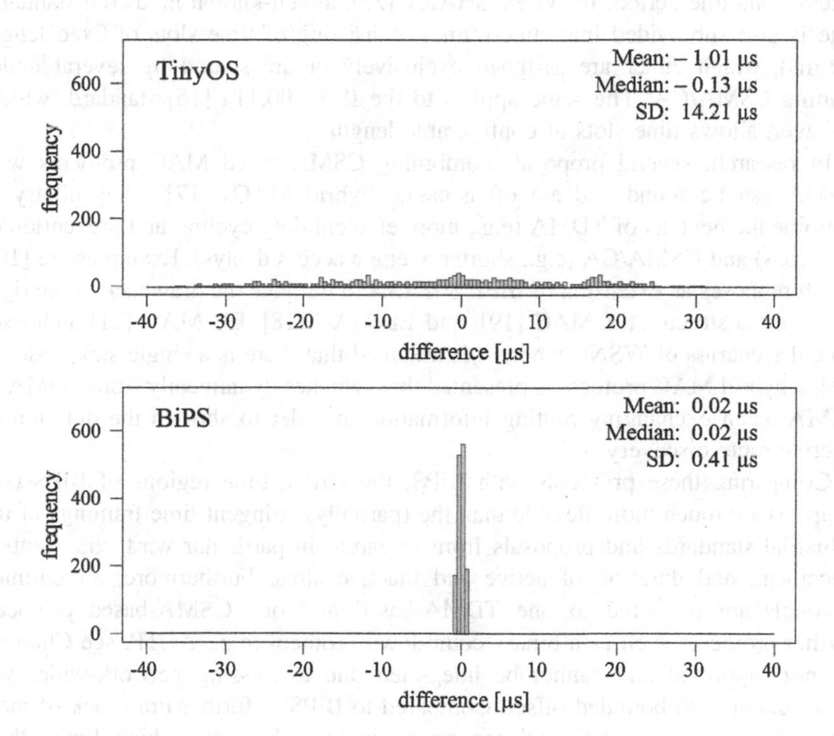

Fig. 11.5 Accuracy of slot start detection (see [2], Fig. 6)

11.2 Related Work

In this chapter, we discuss related work addressing one of the two aspects covered by BiPS: In Sect. 11.2.1, we look at the MAC protocols of BiPS, surveying protocols for wireless sensor networks (WSNs) with multi-MAC support. In

Sect. 11.2.2, we compare the runtime engine of BiPS to other operating systems for WSNs. We are not aware of related work covering both aspects together.

11.2.1 Multi-MAC Support

Communication protocols for wireless networks with more than one MAC scheme can be found in both industrial standards and research, and usually comprise shared and exclusive medium access schemes. A well-known standardized MAC protocol is IEEE 802.15.4 [14], which is also part of ZigBee [23]. In beaconed mode, IEEE 802.15.4 establishes super frames that are composed of contention access period (shared access), contention-free period with guaranteed time slots (exclusive access), and idle period. In WirelessHART [22], a well-known industrial standard, time is also subdivided into superframes, consisting of time slots of fixed length (10 ms), which either are assigned exclusively or are shared by several nodes running CSMA/CA. The same applies to the ISA 100.11a [16] standard, which, however, allows time slots of configurable length.

In research, several proposals combining CSMA-based MAC protocols with TDMA can be found and are often called hybrid MACs [17]. They all try to combine the benefits of TDMA (e.g., more efficient duty cycling and prevention of collisions) and CSMA/CA (e.g., shorter average access delays). Examples are [10], which proposes an extension of IEEE 802.15.4 to mitigate the drawback of the rigid super frame structure, Z-MAC [19], and EE-MAC [18]. ER-MAC [21] addresses typical scenarios of WSNs, where it is assumed that there is a single sink node. In [13], a hybrid MAC protocol is presented that switches dynamically from TDMA to CSMA when exchanging routing information in order to shorten the duration of reactive route discovery.

Comparing these protocols with BiPS, the virtual time regions of BiPS (see Chap. 3) are much more flexible than the (partially) stringent time framings of the industrial standards and proposals from research, in particular w.r.t. the number, placement, and duration of active and inactive slots. Furthermore, all outlined protocols are restricted to one TDMA-based and one CSMA-based protocol. Further protocols such as a binary countdown protocol (e.g., ACTP, see Chap. 6) are not supported and cannot be integrated due to missing network-wide synchronization with bounded offset. Compared to BiPS, a further drawback of most outlined protocols is their reliance on a single sink node, which limits their applicability in ad hoc networks and, in many cases, also implies a topology-dependent slot assignment for TDMA. To assign time slots for TDMA, all protocols either run slot assignment algorithms based on detected communication links or require manual effort to find interference-free schedules. A distinction between communication and interference topology as performed by BiPS (ATDP, see Chap. 4) is not made.

Several papers (e.g., in [4, 12, 20]) point out that the behavior and performance of a protocol highly depends on its implementation. Often, this aspect is neglected, and many protocols are not even implemented, but are assessed analytically or in simulation experiments only. Existing protocol implementations usually rely on general operating systems such as Contiki [6] or TinyOS [8] to execute the protocol in an appropriate and timely manner. Particularly, the handling of concurrency of tasks is often not scrutinized but can affect the behavior adversely. Thus, it is often unclear—in particular, if a protocol is to support real-time guarantees, e.g., due to the incorporation of TDMA—how timing constraints can be satisfied and if provided prioritization measures are actually sufficient to preserve a protocol's conceptual real-time capability.

11.2.2 Runtime Engines of Operating Systems

A widespread event-driven operating system for WSNs is TinyOS [8], which schedules tasks cooperatively by applying first come, first served (FCFS) and earliest deadline first (EDF). Tasks can be preempted by interrupts, commands, and events, but not by other tasks. Since Version 2, there is also support for full multi-threading and synchronization primitives, yet w.r.t. the Imote2 platform, some features like MMU and data caches are no longer used. As TinyOS does not offer a mechanism to run the communication stack in a privileged context, protocols with stringent timing constraints are difficult to implement. Additionally, the modular structure of TinyOS, yet enabling extensions very flexibly, increases the amount and variance of execution delays. Applications for TinyOS are written in a dialect of C called nesC, rendering the learning curve steeper for C/C++ programmers.

Another well-known operating system for WSNs is Contiki [6], with a special focus on strongly resource-constrained devices. Though scheduling in Contiki is basically cooperative, there are extensions for light-weight multi-threading (so-called Protothreads) and full multi-threading support. To execute time-critical code, Contiki provides real-time tasks, which can preempt any low-priority process. Though Contiki subdivides system execution into tasks, memory protection has never been an objective of Contiki. Programs for Contiki are written in constrained C with a strong use of C macros.

Different from TinyOS and Contiki, Nano-RK [7] was developed with a special focus on real-time communication. It is designed for classical multi-hop networks of resource-constrained nodes that send sensor values to a sink periodically. Scheduling with Nano-RK is priority-based and preemptive, with additional support for task synchronization. Nano-RK is considered a static design-time framework, thereby enabling offline real-time schedulability analyses, which reduces runtime flexibility. It follows a reservation paradigm, i.e., tasks can reserve any kind of resource such as CPU, network bandwidth, and sensors/actuators. To incorporate MAC protocols and routing protocols, a basic CSMA/CA template and some infrastructure are provided. Programming language for Nano-RK applications is C.

RIOT [11] is another operating system with support of real-time communication. It comes with full multi-threading support and uses inter-process communication to pass messages between modules, thereby enabling a better decoupling of tasks. Memory protection or advanced measures for fault detection are not provided, and the implementation of time-critical protocols still demands some programming tricks. As in BiPS, applications in RIOT are written in C/C++. Though there is no native porting of RIOT to the Imote2, it can run on this platform in the context of the BAS of BiPS. Thus, higher-level applications running on the Imote2 with BiPS can benefit from both worlds, flexible execution of tasks in BAS/RIOT and real-time communication in BRTS.

Compared to the outlined state-of-the-practice operating systems, BiPS is protocol-centric, i.e., the communication protocols are an inherent part of the system architecture (see Fig. 11.1). Thereby, BiPS can guarantee real-time capabilities of the implemented protocols by design. W.r.t. typical operating system functionalities such as multi-threading, preemption, and inter-process communication, BiPS provides less support compared to the outlined operating systems. However, this drawback can be eliminated by nesting FreeRTOS or RIOT into BiPS.

11.3 Conclusions

In this chapter, we have outlined the real-time-capable implementation framework Black burst integrated Protocol Stack (BiPS), and have surveyed related work. BiPS is the basis for the implementation of the ProNet 4.0 protocol stack on the Imote2 hardware platform. In particular, we have explained structure and operation of the BiPS multiplexer providing uniform access to MAC protocols, the BiPS schedulers BRTS and BAS, and the possibility of nesting state-of-the-practice operating systems into BiPS. In our evaluation, we have assessed memory usage, timer expiry processing delays, and detection delays of message receptions.

A key design objective of BiPS is the provision of a flexible and extendable framework for real-time communication. This is achieved by the incorporation of a protocol for deterministic tick and time synchronization (see Chap. 2), a highly flexible structuring of time into virtual time regions (see Chap. 3), a protocol for automatic topology detection (see Chap. 4), and protocols supporting different medium access schemes (see Chaps. 5 and 6). Another key concern is an implementation of BiPS that complies with the real-time constraints imposed by the structuring of time. This is achieved by implementing BiPS on bare hardware, and by providing real-time protocol-centric operating system support, in particular, with the BiPS real-time scheduler.

Literature

Chair for Networked Systems

1. Gotzhein R (2014) ProNet 4.0—a wireless real-time communication system for industry 4.0. White Paper, Networked Systems Group, Department of Computer Science, University of Kaiserslautern. http://vs.informatik.uni-kl.de/publications/2014/Go14/whitePaperEN-ProNet4.0.pdf. Last accessed 27 Aug 2019
2. Christmann D, Braun T, Engel M, Gotzhein R (2016) BiPS—a real-time-capable protocol framework for wireless sensor networks. In: Proceedings of the 6th international conference on pervasive and embedded computing (PEC 2016), Lisbon, Portugal, 25–27 Jul 2016, pp 17–27
3. Engel M, Kramer C, Braun T, Christmann D, Gotzhein R (2019) BiPS—a real-time-capable protocol framework for wireless networked control systems and its application. In: Obaidat M, Cabello E (eds) E-business and telecommunications (ICETE 2017), communications in computer and information science, vol 990. Springer, pp 313–336

Further References

4. Basmer T, Schomann H, Peter S (2011) Implementation analysis of the IEEE 802.15.4 MAC for wireless sensor networks. In: International conference on selected topics in mobile and wireless networking. Shanghai, China
5. Chipcon AS (2019) CC2420 ZigBee-ready RF transceiver. http://www-inst.eecs.berkeley.edu/~cs150/Documents/CC2420.pdf. Last accessed 27 Aug 2019
6. Dunkels A, Gronvall B, Voigt T (2004) Contiki—a lightweight and flexible operating system for tiny networked sensors. In: 29th annual IEEE international conference on local computer networks, pp 455–462
7. Eswaran A, Rowe A, Rajkumar R (2005) Nano-rk: an energy-aware resource-centric RTOS for sensor networks. In: 26th IEEE international real-time systems symposium (RTSS 2005), pp 10–265
8. Farooq MO, Kunz T (2011) Operating systems for wireless sensor networks: a survey. Sensors 11(6):5900–5930
9. FreeRTOS—Free real-time operating system. https://www.freertos.org/. Last accessed 27 Aug 2019
10. Gilani MHS, Sarrafi I, Abbaspour M (2013) An adaptive CSMA/TDMA hybrid MAC for energy and throughput improvement of wireless sensor networks. Ad Hoc Netw 11(4):1297–1304
11. Hahm O, Baccelli R, Günes M, Wählisch M, Schmidt TC (2013) RIOT OS—towards an OS for the internet of things. In: The 32nd IEEE international conference on computer communications (INFOCOM 2013), Turin, Italy, 14–19 Apr 2013
12. Harvan M, Schönwälder J (2008) TinyOS motes on the internet: IPv6 over 802.15.4 (6lowpan), Praxis der Informationsverarbeitung und Kommunikation 31(4):244–251
13. Hsieh TH, Lin KY, Wang PC (2015) A hybrid MAC protocol for wireless sensor networks. In: 12th IEEE international conference on networking, sensing and control, pp 93–98
14. [IEEE 802.15.4] Institute of electrical and electronics engineers. IEEE standard 802 part 15.4: low-rate wireless personal area networks (LR-WPANs). IEEE Computer Society, New York, USA, Sep 2011
15. MEMSIC Inc.: Imote 2 datasheet. https://vs.cs.uni-kl.de/downloads/Imote2NET_ED_Datasheet.pdf. Last accessed 27 Aug 2019

16. International electrotechnical commission: industrial communication networks—wireless communication network and communication profiles—ISA 100.11a (IEC 62734 ed 1.0), Geneva, Switzerland, Mar 2012
17. Olempia KJ, Pandeeswaran C, Natarajan P (2016) A survey on energy efficient contention based and hybrid MAC protocols for wireless sensor networks. Indian J Sci Technol 9(12)
18. Priya B, Manohar SS (2013) EE-MAC: energy efficient hybrid MAC for WSN. In: Int J Distrib Sensor Netw 9(12)
19. Rhee I, Warrier A, Aia M, Min J, Sichitiu ML (2008) Z-MAC: a hybrid MAC for wireless sensor networks. In: IEEE/ACM Trans Netw 16(3):511–524
20. Silva R, Silva JS, Boavida F (2009) Evaluating 6LowPAN implementations in WSNs. In: 9th Conferencia sobre Redes de Computadores Oeiras, Portugal
21. Sitanayah L, Sreenan CJ, Brown KN (2010) Emergency response MAC protocol (ER-MAC) for wireless sensor networks. In: Abdelzaher TF, Voigt T, Wolisz A (eds) 9th International conference on information processing in sensor networks (IPSN 2010), Stockholm, Sweden
22. International Electrotechnical Commission (IEC). Industrial Communication Networks—Wireless Communication Network and Communication Profiles - WirelessHART (IEC 62591 ed 1.0), Geneva, Switzerland, April 2010
23. ZigBee™ Alliance: ZigBee specification, version 1.0, June 2005. www.zigbee.org, 378 p. Last accessed 27 Aug 2019

Chapter 12
ProNet 4.0 Case Studies

In this chapter, we present three case studies highlighting different functionalities of ProNet 4.0. The first case study [2] is about the operation of different types of control systems over a wireless TDMA network, using exclusive and mode medium access. In the second case study [3, 4], a control application based on Try-Once-Discard (TOD), a protocol from the control systems domain, is conceived, using deterministic network-wide value-based arbitration and data transfer. In the third case study [5], ProNet 4.0 is part of a setting for remote maintenance in a production plant. This industrial case study ranges from field level to management level and comprises a production facility, a wireless sensor network, an autonomous robot, a multimedia system, and an enterprise resource planning (ERP) system. To realize application requirements on field level, communication functionalities on all layers of ProNet 4.0 are exploited.

12.1 Inverted Pendulum and Flow Rate System

The first case study [2] addresses the operation of several independent control systems over a shared wireless communication network. It has been devised during a research project at the University of Kaiserslautern, involving two research groups from the Department of Electrical and Computer Engineering and the Department of Computer Science, as part of the Priority Programme 1305 "Control Theory of Digitally Networked Dynamic Systems" of the German Research Foundation (DFG) [13]. In this chapter, we outline control aspects and address some aspects of the communication solution.

R. Gotzhein, *Real-time Communication Protocols for Multi-hop Ad-hoc Networks*, Computer Communications and Networks,
https://doi.org/10.1007/978-3-030-33319-5_12

12.1.1 Control Aspects

In the case study, several independent instances of different types of control systems are operated over a shared wireless communication system. The inverted pendulum (see Sect. 1.2) is a time-based control system, where sampling and exchange of values occur synchronously at specified points in time. The flow rate system uses an event-driven mode of operation, where exchange of values is triggered by the occurrence of specified events.

Figure 12.1 shows the two types of controlled systems in a schematic way. The *inverted pendulum* consists of a cart that is mounted to a belt. The belt can be accelerated by a motor in both directions such that the cart moves back and forth, with changing velocity. Attached to the cart is a rod (pendulum) with a weight at the end that can swing around one axis. Control objective is to stabilize the rod in an upright position, by moving the cart. Disturbances are caused, for instance, by frictions of the rail, local reaction delays of the motor, and pushes to the rod. The state of the inverted pendulum is sampled periodically and characterized by position x and velocity v of the cart, and by the angle φ of the rod. From the history of samplings, a voltage u is calculated by the controller and applied to the motor.

To achieve the control objective, the following real-time requirements are to be satisfied. Sampling of measured values x, v, and φ has to occur synchronously and periodically, with a maximum sampling period $d_{\text{IPsampling}} = 72$ ms. The maximum end-to-end reaction delay, i.e., the largest time span between value sampling and application of the corresponding steering value, is $d_{\text{IPend2endReactionMax}} = 24$ ms. Finally, loss of measured and steering values must be rare.

The *flow rate system* consists of a hose with several devices attached. The hose is filled with liquid via a tap. A pump then creates a pressure in the hose to establish a flow. Two sensors measure the flow rate, which can be controlled by an adjustment valve. The control objective is to achieve a specified flow rate trajectory, e.g., a ramp signal or a step signal. Disturbances are caused, for instance, by reaction delays of the adjustment valve and by nonlinear overrun. The state of the flow rate system is sampled periodically and characterized by flow rates f_1 and f_2. From the history of samplings, a value v is calculated by the controller and applied to the adjustment valve.

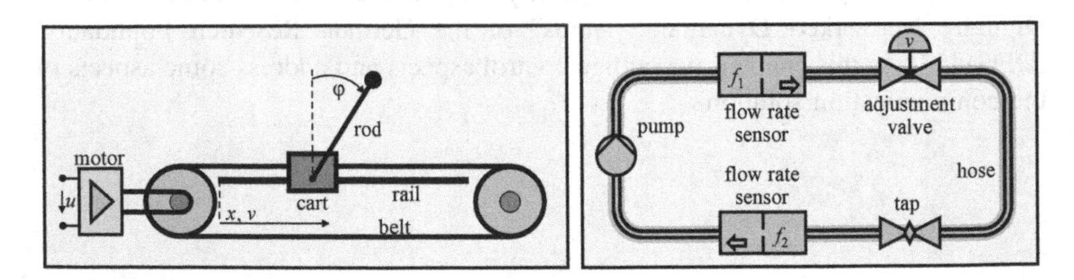

Fig. 12.1 Controlled systems: inverted pendulum and flow rate system

Sampling of measured values f_1 and f_2 has to occur periodically, with a maximum sampling period $d_{\mathrm{FRsampling}} = 160$ ms. Maximum end-to-end reaction delay is $d_{\mathrm{FRend2endReactionMax}} = 60$ ms. Loss of measured and steering values must be rare.

12.1.2 Communication Solution

In the case study, the task is to control two inverted pendulums and two flow rate systems over a shared wireless network, applying ProNet 4.0. Figure 12.2 shows the architecture consisting of controlled systems, sensors, actuators, and controllers. For instance, the inverted pendulum control system IP_1 comprises sensors $s_{1,x}$, $s_{1,v}$, and $s_{1,\phi}$, an actuator $a_{1,u}$, and a controller c_1. The figure also indicates measured and steering values and their exchange over the wireless network. In this case study, all nodes are in single-hop range.

To satisfy the real-time requirements of the control applications, we have selected appropriate medium access schemes, have devised a time structure of virtual time regions, and have defined a communication schedule.

An analysis of the application requirements of the inverted pendulum has shown that for safe operation, it is sufficient to transfer measured and steering values with a communication period of $d_{\mathrm{IPcommBase}} = 72$ ms. In case of disturbances, a shorter

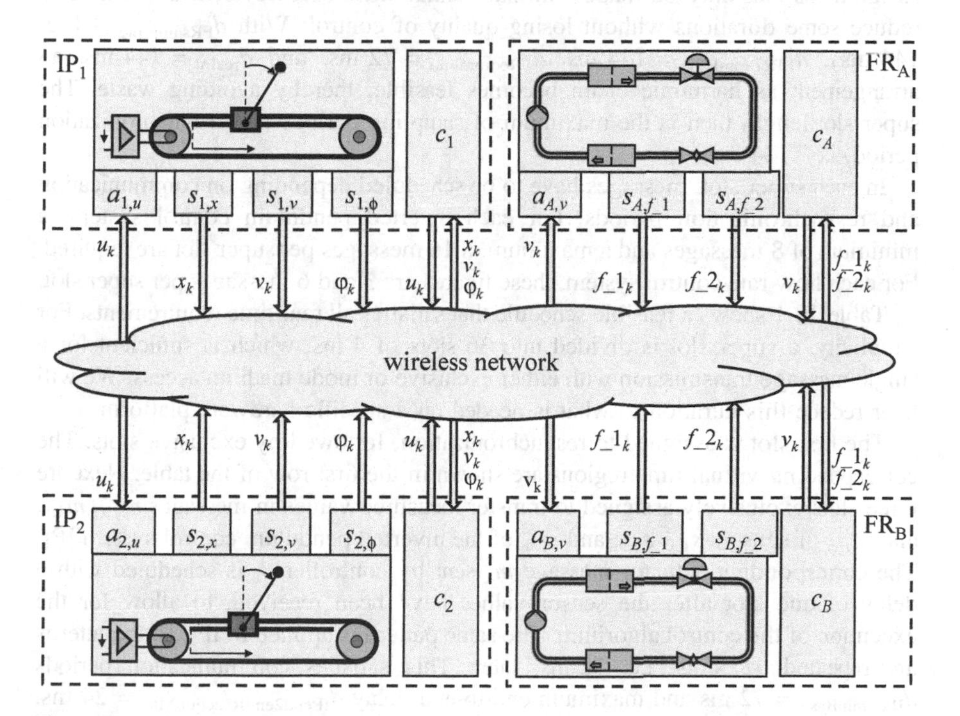

Fig. 12.2 Architecture of the case study

period $d_{\text{IPcommPref}} = 36$ ms yields faster stabilization and could, for instance, be supported if there are free communication resources. This would imply that the sampling period be reduced to $d_{\text{IPsampling}} = 36$ ms. To satisfy these requirements, we use exclusive medium access every 72 ms and mode medium access in-between (see Chap. 5).

Similar considerations apply to the flow control system. Here, it is sufficient to transfer measured and steering values with a communication period of $d_{\text{FRcommBase}} = 160$ ms, and with a period $d_{\text{FRcommPref}} = 80$ ms (based on $d_{\text{FRsampling}} = 80$ ms) to improve quality of control. Different from the inverted pendulum, which is a time-based control system, the control algorithm of the flow rate system uses an event-driven mode of operation. Therefore, reserved slots are only used when events occur. It is not feasible to use shared medium access, as this may cause unpredictable medium access delays and message loss due to collisions.

To structure time, we first consider sampling, communication, and resynchronization periods. For the inverted pendulums, we have sampling periods $d_{\text{IPsampling}} \in \{36, 72 \text{ ms}\}$ and communication periods $d_{\text{IPcommBase}} = 72$ ms and $d_{\text{IPcommPref}} = 36$ ms. For the flow rate system, these periods are $d_{\text{FRsampling}} \in \{80, 160 \text{ ms}\}$, $d_{\text{FRcommBase}} = 160$ ms, and $d_{\text{FRcommPref}} = 80$ ms. For resynchronization, we define the resynchronization interval duration $d_{\text{resInt}} = 200$ ms. Obviously, these periods, in increasing order, cannot be arranged as a harmonic chain, i.e., $d_i = k_i \cdot d_{i-1}$, where $1 < i \leq 9$ and $k_i \in \mathbb{N}$ (see Chap. 3). The drawback is that slot assignments can only be made with substantial waste. However, it is possible to reduce some durations without losing quality of control: With $d_{\text{FRsampling}} \in \{72, 144 \text{ ms}\}$, $d_{\text{FRcommBase}} = 144$ ms, $d_{\text{FRcommPref}} = 72$ ms, and $d_{\text{resInt}} = 144$ ms, an arrangement as harmonic chain becomes feasible, thereby avoiding waste. The super slot length then is the maximum of sampling periods and resynchronization period, i.e., 144 ms.

In each super slot, messages have to be scheduled depending on communication and resynchronization periods. For each inverted pendulum control system, a minimum of 8 messages and a maximum of 16 messages per super slot are required. For each flow rate control system, these figures are 3 and 6 messages per super slot.

Table 12.1 shows a feasible schedule that satisfies all real-time requirements. For simplicity, a super slot is divided into 36 slots of 4 ms, which is sufficient for a single message transmission with either exclusive or mode medium access. We will later reduce this duration to what is needed on a specific hardware platform.

The first slot is assigned to resynchronization, followed by exclusive slots. The corresponding virtual time regions are shown in the first row of the table. Next are three slots exclusively assigned to transfer measured values in messages $m_{1,x}$, $m_{1,v}$, and $m_{1,\phi}$ of sensors $s_{1,x}$, $s_{1,v}$, and $s_{1,\phi}$ of the inverted pendulum control system IP_1. The corresponding actuator message m_1 sent by controller c_1 is scheduled with a delay of one slot after the sensor values have been received, to allow for the execution of the control algorithm. The same pattern is applied to IP_2. Both patterns are repeated 18 slots, i.e. 72 ms, later. This satisfies communication periods $d_{\text{IPcommBase}} = 72$ ms and maximum end-to-end delay $d_{\text{IPend2endReactionMax}} = 24$ ms. Furthermore, 3 slots per flow rate control system are exclusively assigned,

Table 12.1 Virtual time regions and communication schedule. For brevity, modes (see Sect. 5.2.3) are omitted

Virtual region	sync region	exclusive region	exclusive region	exclusive region	exclusive region	exclusive region	exclusive region	exclusive region	exclusive region	exclusive region	exclusive region	mode region	mode region	mode region	exclusive region	mode region	exclusive region	exclusive region
prio = 0		$s_{1,x}$	$s_{1,v}$	$s_{1,f}$	s_{A,f_1}	c_1	$s_{2,x}$	$s_{2,v}$	$s_{2,f}$	s_{A,f_2}	c_2	$s_{1,x}$	$s_{1,v}$	$s_{1,f}$	c_A	c_1	s_{B,f_1}	s_{B,f_2}
prio = 1												$s_{2,x}$	$s_{2,v}$	$s_{2,f}$		c_2		

Virtual region (cont'd)	idle region	exclusive region	exclusive region	exclusive region	exclusive region	exclusive region	exclusive region	exclusive region	exclusive region	idle region	exclusive region	mode region	mode region	mode region	idle region	mode region	idle region	idle region
prio = 0		$s_{1,x}$	$s_{1,v}$	$s_{1,f}$	c_B	c_1	$s_{2,x}$	$s_{2,v}$	$s_{2,f}$		c_2	$s_{2,x}$	$s_{2,v}$	$s_{2,f}$		c_2		
prio = 1												$s_{1,x}$	$s_{1,v}$	$s_{1,f}$		c_1		
prio = 2												s_{A,f_1}	s_{A,f_2}			c_A		
prio = 3												s_{B,f_1}	s_{B,f_2}			c_B		

Legend: sync region, exclusive region, mode region, idle region; IP_1, IP_2, FR_A, FR_B

satisfying $d_{\text{FRcommBase}} = 144$ ms and $d_{\text{FRend2endReactionMax}} = 60$ ms. All exclusive slots are assembled into exclusive regions, as shown in the first row of the table.

To support shorter periods $d_{\text{IPcommPref}}$ and $d_{\text{FRcommPref}}$ to overcome disturbances faster, we configure mode slots, which are assembled into mode regions. In the first part of Table 12.1, there are four slots assigned to two messages each. For instance, the first of these slots is assigned to messages $m_{1,x}$ (prio = 0) and $m_{2,x}$ (prio = 1) of $s_{1,x}$ and $s_{2,x}$, respectively. At runtime, these slots are used only in case of disturbances, with priority given to IP_1. In the second part of Table 12.1, further slots are assigned to messages of IP_1 and IP_2, this time with priority given to IP_2. It follows that in case of disturbances to only one inverted pendulum, the communication rate is doubled, and increased proportionally if both inverted pendulums need higher communication rates. Similar considerations apply to the flow rate control systems.

From the slot assignments, it follows that messages of the communication periods $d_{\text{IPcommBase}}$, $d_{\text{FRcommBase}}$, and $d_{\text{IPcommPref}}$ are scheduled strictly periodically. For periods $d_{\text{FRcommPref}}$, there are waiting times different from 72 ms after the previous transfer of measured and steering values. For instance, message m_{A,f_1} of sensor s_{A,f_1} is scheduled in exclusive slot 5 and mode slot 30, i.e., with an interval of 100 ms instead of 72 ms. In the scenario, it is also possible to assign free slots such that $d_{\text{FRcommPref}}$ is met. However, as the shift is in accordance with the requirements of the control application, we have optimized the schedule in order to save time slots.

By assigning exclusive slots, there is no message loss due to collisions, if the single-network property (see Sect. 1.3) holds. However, in wireless communication, message loss cannot be ruled out entirely. To compensate for message loss, predictive control algorithms can be used. Thus, if a set of measured values is not complete, steering values can be estimated and applied until new values have been received.

As hardware platform, we consider Imote2 motes [10] with the Chipcon CC2420 radio transceiver [8]. Tick synchronization is performed by Black Burst Synchronization (BBS, see Chap. 2), implemented in ProSync of the ProNet 4.0 protocol stack [1]. From the data sheet and the analysis in Chap. 5, it follows that for a frame size of 16 bytes, the length of exclusive slots can be reduced to 0.9 ms; for mode slots, we get 1.5 ms (mode slots with two priorities) and 2.4 ms (mode slots with four priorities). This reduces medium usage from 86.1 to 25.6%, which leaves room to support further control systems over the same wireless communication network or to save energy.

At the time the case study has been defined, the ProNet 4.0 protocol stack was not yet completed. For instance, mode medium access, originally devised for wired bus systems, and the BiPS framework (see Chap. 11) were not available on Imote2. Therefore, only a subset of the functionality described in this chapter has been realized.

12.2 Batch Reactor

The second case study [3, 4] features the use of Black Burst Synchronization (BBS, see Chap. 2) and Arbitration and Cooperative Transfer Protocol (ACTP, see Chap. 6), which are implemented in protocol modules ProSync and ProArb of the ProNet 4.0 protocol stack [1]. In the case study, BBS and ACTP are applied to realize Try-Once-Discard (TOD), a protocol from the control systems domain, in order to control an unstable batch reactor via a wireless network [19]. In this chapter, we outline control aspects and address some aspects of the communication solution.

12.2.1 Control Aspects

A reactor is a container in which controlled reactions such as chemical processes or nuclear fusion take place. A specific reactor type is the batch reactor, which consists of a container for liquid or compressed gas, an agitator for shaking or stirring, and a heating or cooling system. Batch reactors are used, for instance, in chemical and pharmaceutical industry.

In the example in [19], an unstable batch reactor is described as a coupled two-input-two-output networked control system (NCS). For the exchange of sensor values, a protocol called Try-Once-Discard (TOD) is devised. TOD is of special interest in the control systems domain, because its properties can be characterized by Lyapunov functions [11]. This makes it straightforward to incorporate TOD into

Lyapunov-based design of linear and nonlinear control systems, yielding a self-contained theory for system stabilization.

In [9], a Lyapunov-based theory is presented to derive bounds on the maximal allowable transfer interval (MATI) that guarantees asymptotic stability. Furthermore, the maximal allowable delay (MAD) is derived. Stabilization is achieved with TOD, if MATI and MAD can be satisfied. TOD realizes the global scheduling policy maximum-error-first (MEF). Since this policy requires network-wide and value-based arbitration and end-to-end transmission within each sampling interval (bounded by MATI), TOD is difficult to implement, especially in wireless networks.

Walsh et al. [19] characterize TOD by the following requirements:

- Time is divided into time slots of fixed duration, to which all nodes have to synchronize with deterministic accuracy.
- In each time slot, a TOD round consisting of two phases is performed. In the arbitration phase, sensor nodes contend, with the node reporting the greatest weighted error w.r.t. the last reported value winning. In the data phase, the winning node transmits its current value to the controller node.
- For the duration of a time slot, a deterministic upper bound satisfying MATI must be guaranteed, to ensure system stability.

In the original work [19], the Controller Area Network (CAN) protocol [7] used in the automotive and automation domains is identified as a suitable communication technology to implement TOD. When using weighted error values augmented with a unique node identifier as CAN identifiers, contention yields deterministic medium arbitration. We note, however, that by design, CAN is limited to wireline, single-hop networks, and that CAN does not support synchronization, which violates the requirement of synchronized TOD rounds.

12.2.2 Communication Solution

TOD is difficult to implement, especially in wireless networks. In [16], Pereira, Andersson, and Tovar have presented a binary countdown protocol (see Sect. 6.1.2) for deterministic arbitration in wireless single-hop networks called Wireless Dominance (WiDom) protocol. WiDom was later extended to wireless multi-hop networks [17]. However, it cannot avoid concurrent destructive arbitrations. In summary, WiDom and its extensions do not solve the problems of network-wide synchronization, medium arbitration, and value transfer. Therefore, it does not satisfy the communication requirements of TOD.

It has long been an open question whether TOD can be realized in wireless multi-hop networks. In [15], it is stated explicitly that "TOD cannot be implemented directly in wireless networks." Moreover, [18] concludes that "no such arbitration (of TOD) is possible for wireless channels." In [3], we have given a

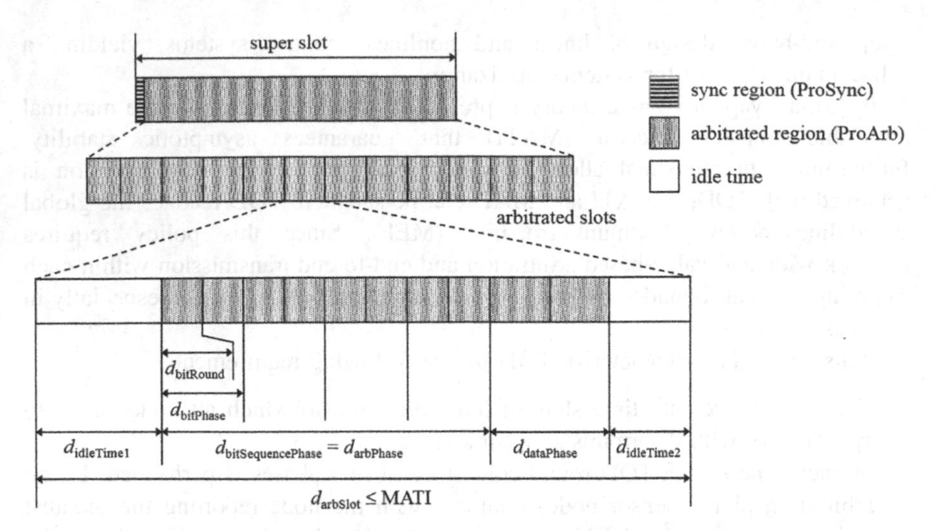

Fig. 12.3 Virtual time slotting for TOD (see also Sect. 6.2.2)

positive answer to the open question, by presenting a functionally complete realization of TOD in wireless multi-hop networks.

To realize TOD, we apply Black Burst Synchronization (BBS, see Chap. 2) and Arbitrating and Cooperative Transfer Protocol (ACTP, see Chap. 6). Figure 12.3 shows the hierarchical global time structure (see Chap. 3). On top level, time is decomposed into super slots consisting of sync region and arbitrated region. The arbitrated region is then divided into arbitrated slots of equal length $d_{arbSlot}$, except for the first slot, which has a duration of $d_{arbSlot} - d_{conv}$. Here, d_{conv} is the convergence delay for resynchronization, i.e., the length of the sync region at the beginning of super slots. The reason for this is that to satisfy the requirements of TOD, all time slots must have fixed duration. For the first time slot, this is achieved by conceptually decomposing it into resync slot and TOD slot, to leave time for resynchronization. Furthermore, arbitrated slot duration has to satisfy MATI, i.e., $d_{arbSlot} \leq$ MATI.

Arbitrated slots are further decomposed into arbitration and data phase—the TOD round—preceded and followed by idle times (see Fig. 12.3). This slightly deviates from the time structure of arbitrated slots in Chap. 6, which does not contain idle times. An exception is again the first arbitrated slot, which does not contain the first idle time. The first idle time of subsequent slots aligns the starting point of the TOD round with that of the first slot, to compensate for the preceding resynchronization, i.e., $d_{idleTime1} = d_{conv}$. The second idle time fills up slots to save energy in case $d_{arbSlot} > d_{arbPhase} + d_{dataPhase} + d_{idleTime1}$.

In the arbitration phase, nodes contend by reporting their weighted error. ACTP ensures that all nodes reporting the greatest weighted error win. To have exactly one winning node, nodes extend contention by sending their unique identifiers. Figure 12.3 shows a scenario where the arbitration phase consists of 4 bit phases,

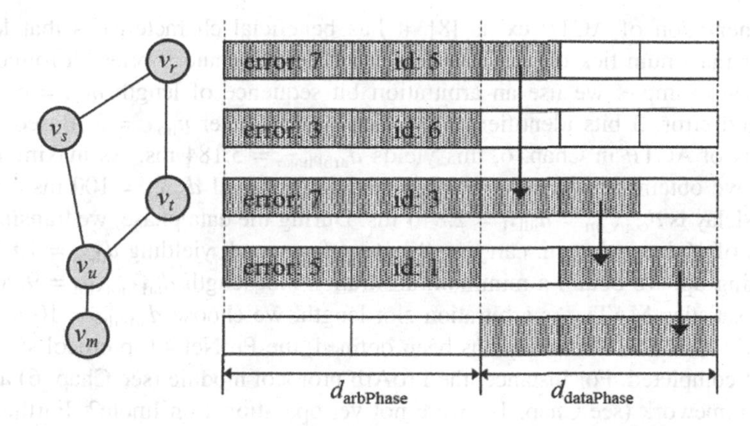

Fig. 12.4 Unstable batch reactor control system—three-hop scenario: sample TOD round

which in turn are decomposed into 2 bit rounds. This indicates that an arbitration bit sequence of length $n_{\text{bits}} = 4$ is used, and that the network diameter is $n_{\text{hops}} = 2$.

The winning node then transmits its data frame in the data phase. In a multi-hop network, each data frame has to be forwarded over one or more hops until it reaches the destination node, in the same data phase. This places a timing constraint concerning the minimum duration $d_{\text{dataPhase}}$. In the two-hop scenario in Fig. 12.3, two slots for data frame transfer are configured.

Figure 12.4 shows a concrete scenario of a batch reactor control system. The topology on the left side consists of a master node v_m acting as controller and sensor nodes reporting errors. In the arbitration phase, sensor nodes contend with their weighted error value followed by their node identifier, with $n_{\text{bits}} = 6$ and $n_{\text{hops}} = 3$. In the example, error values and identifiers are shown in decimal notation. Because nodes v_r and v_t report the same greatest weighted error value 7, they continue and transmit their node identifiers to resolve contention, yielding node v_r as winner. In the data phase, v_r transmits its data frame containing the actual error value to v_m, via nodes v_t and v_u, which completes the TOD round.

Figure 12.4 also shows idle times during the TOD round, indicating potential for energy saving. For instance, the master node v_m does not participate in the arbitration, and therefore can stay idle in the arbitration phase. The drawback is that v_m does not learn about the winner identifier, and therefore, it has to listen for an incoming data frame during the entire data phase. Since the remaining nodes are aware of the winner identifier, and assuming that topology information is readily available, they can be switched to active mode when participating in a frame exchange in the data phase, and to sleep mode otherwise.

In [12], the bound MATI = 10.65 ms for guaranteeing L^2 stability in the presence of bounded perturbations has been derived. We now analyze whether the constraint $d_{\text{arbSlot}} \leq$ MATI can be satisfied for a given hardware platform in the three-hop scenario in Fig. 12.4. For the analysis, we use Atmel's AT86RF230 transceiver [6]. Compared to the Chipcon CC2420 transceiver for which an implementation of ACTP exists [8], it has beneficial characteristics that lead to smaller maximum tick offsets, less convergence delay, and shorter bit rounds.

In the example, we use an arbitration bit sequence of length $n_{\text{bits}} = 6$ (3 bits weighted error, 3 bits identifier) and a network diameter $n_{\text{hops}} = 3$. Based on the analysis of ACTP in Chap. 6, this yields $d_{\text{arbPhase}} = 5.184$ ms. As maximum tick offset, we obtain 56 μs, with resynchronization interval $d_{\text{resInt}} = 100$ ms; convergence delay is $d_{\text{conv\_m}} = d_{\text{idle1}} = 2.846$ ms. During the data phase, we transmit data frames of 16 bytes length, carrying 8 bytes of payload, yielding $d_{\text{data}} = 1.872$ ms. Summing up, we obtain a minimum arbitration slot length $d_{\text{arbSlotMin}} = 9.902$ ms, which satisfies MATI. As arbitration slot length, we choose $d_{\text{arbSlot}} = 10$ ms.

At the time the case study has been defined, the ProNet 4.0 protocol stack was not yet completed. For instance, the ProArb protocol module (see Chap. 6) and the BiPS framework (see Chap. 11) were not yet operational on Imote2. Furthermore, the transceiver AT86RF230 was not available. Therefore, the case study has been of analytical nature. Nevertheless, feasibility and reliability experiments with the transceiver CC2420 have been conducted (see Chap. 6).

12.3 Vertical Integration of Production Processes (VI-P)

The third case study [5] features the entire protocol stack ProNet 4.0 and its prototypical implementation on Imote2 nodes. The case study is part of the demonstrator vertical integration of production processes (VI-P) developed in Task 6.1.4 "Adaptive Services in Networked Production Environments" of the project SINNODIUM (Software Innovations for the Digital Enterprise), funded by the German Federal Ministry of Education and Research. In this chapter, we survey the broader application context and address some aspects of the communication solution that involve ProNet 4.0.

12.3.1 Application Context

Industry 4.0 is said to be the next evolution step in production and automation. By digitization and by making factories intelligent, production processes become more efficient regarding cost, resources, administration, and maintenance. This requires that production processes and products are continuously monitored and controlled, that structures in production plants are flexible, and that the layers of automation (AT) and information technology (IT) are vertically integrated.

With several partners from industry, research institutes, and universities, we have developed technological components and a demonstrator featuring solutions for Industry 4.0 [5]. In the core of the technological foundations are QoS-capable middleware concepts for production plants. Though in the demonstrator, several heterogeneous middleware systems are composed, their interfaces are generic and

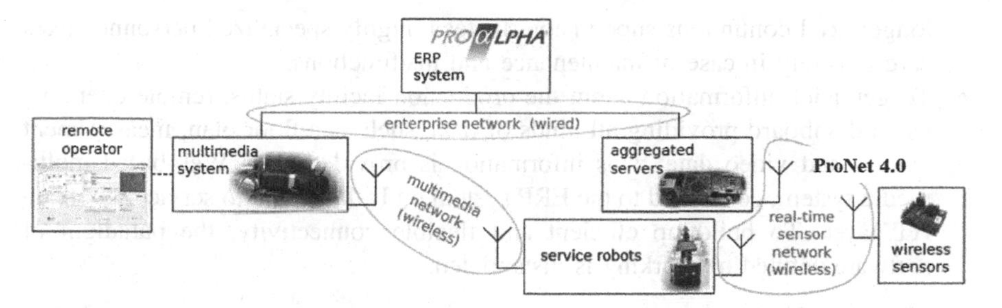

Fig. 12.5 Components of the infrastructure of an enterprise with vertical integration (see [5], Fig. 1)

guided by the idea of emergent software to enable a seamless and systematical vertical integration of the AT and IT worlds. This way, high-level management and enterprise resource planning (ERP) systems are coupled with field and control layers.

Figure 12.5 shows the technological components developed in the research project, and their integration:

- On IT layer, there is an ERP system developed and maintained by proALPHA Software GmbH, a software provider for mid-sized enterprises. This system has been enhanced by interfacing components to the AT layer. In particular, abstract interfacing with aggregated servers by definition of key performance indicators (KPIs) and exchange of KPI values is supported.
- Aggregated servers are responsible for the orchestration of field services and the collection of data from field devices. Services and data are merged into a uniform semantical information model based on Open Platform Communications Unified Architecture (OPC UA) [14], where KPIs are generated on the basis of the model and standardized KPI templates. These KPIs are homogeneous and have a uniform semantical description to make them accessible and processible by business applications like ERP systems.
- Aggregated servers collect raw data provided by sensors on field level. Communication is wireless, via ProNet 4.0, using the real-time services of Production Middleware (ProMid, see Chap. 10), which in turn uses the real-time communication services of the ProNet 4.0 protocol stack.
- Service robots such as automated guided vehicles contribute to a more flexible and safer operation of production facilities. Robot services may include safe and autonomous navigation to target positions, teleoperation, and the provision of robot sensor data and system states. To realize these services, other services available in the production plant, for instance, services of stationary sensors to improve navigation, may be subscribed to via ProMid.
- Remote operators are responsible for running and maintaining production facilities. With a high level of automation and monitoring, these facilities no

longer need continuous supervision. Instead, highly specialized personnel takes care remotely in case of maintenance and malfunctions.
- To get quick information about the production facility status, remote operators use a dashboard providing all kinds of data, such as a floor plan, measurement values, and video data. This information is provided by a Web-based multimedia system, connected to the ERP system on IT layer and to service robots on AT layer. To build up efficient and flexible connectivity, the paradigm of software defined networking is instantiated.

All technological components have been prototypically implemented and deployed in the demonstrator VI-P.

12.3.2 Communication Solution

The technological components developed in the research project form part of the VI-P demonstrator, located in the Living Lab SmartFactory^{KL} of the German Institute for Artificial Intelligence (DFKI) in Kaiserslautern. The laboratory hosts a production plant bottling liquid colored soap. For demonstration purposes, we consider the following scenario. During the production process, a pump fails. The automation system has to detect the failure in real time and to store failure data in a failure variable monitored by the ERP system. To reduce downtime, this automatically starts a workflow to identify the fault causing the failure and to start repairing.

In the scenario, we assume that it is not economical or feasible to have a specialist on standby in the production plant. Therefore, as part of the workflow, a remote specialist is consulted. Via the dashboard as remote interface, the specialist starts narrowing down the problem. He orders a mobile service robot equipped with a camera and further sensors to navigate to the faulty pump and evaluates the video streams. After identifying the origin of the problem, a local mechanic is consulted to repair the pump while being guided by the remote specialist.

Communication between technological components is via enterprise network, multimedia network, and real-time sensor network (see Fig. 12.5). To establish the sensor network, ProNet 4.0 is deployed, connecting stationary sensors of the production plant, mobile sensors of the service robot, and an aggregated server (see Fig. 12.6). Each sensor provides a service, which is published via ProMid and registered in the ProMid service registry (see Chap 10). For instance, sensors monitor the status of the pump and measure temperatures and gas concentration. Aggregated server and service robot can subscribe to these services dynamically. When navigating toward the faulty pump, the service robot subscribes to services along planned routes. If critical sensor values are received, for instance, due to a gas leak, the robot reduces its set of planned routes and finally chooses a safe route or stops if no such route exists.

Fig. 12.6 Production plant bottling liquid soap with Imote2 nodes running ProNet 4.0 (see [20], Fig. 8)

To realize this high-level functionality, communication functionality on all layers of ProNet 4.0 is exploited:

- Production middleware (ProMid, see Chap. 10) provides the abstract interface for service users (aggregated server, service robot) and service providers (sensors) for service publication and subscription.
- Services are dynamically registered and looked up in the ProMid service registry (see Chap. 10).
- To establish the service registry, heterogeneous network clustering with HNC (see Chap. 9) is performed.
- QoS multicast routing (QMR, see Chap. 8) detects routes in this partially mobile TDMA sensor network.
- QMR constructs routing trees and reserves exclusive slots to achieve reliable value transfers, based on topology information collected by Automatic Topology Discovery Protocol (ATDP, see Chap. 4) and global time slotting (see Chap. 3).
- Exclusive medium access is provided by the medium access control protocol ProRes (see Chap. 5).

- Global time slotting, medium access, and application services require network-wide tick and time synchronization, which is provided by the synchronization protocol Black Burst Synchronization (BBS, see Chap. 2).
- Finally, the Black burst integrated Protocol Stack (BiPS, see Chap. 11) framework provides communication and operating system functionality, and integrates further communication protocols.

12.4 Conclusions

In this chapter, we have presented three case studies featuring different functionalities of ProNet 4.0. The first case study addresses the operation of several independent control systems over a shared wireless communication network. Part of the communication solution is global time slotting, synchronization, and exclusive and mode medium access. The second case study realizes the protocol Try-Once-Discard in a wireless multi-hop network. This requires global time slotting, synchronization, and deterministic multi-hop arbitration and data transfer. The third case features the entire ProNet 4.0 protocol stack, exploiting various real-time communication functionalities and the real-time production middleware.

Taken together, the case studies provide evidence that ProNet 4.0 combines rich functionalities that cover a variety of requirements. Most importantly, ProNet 4.0 incorporates deterministic protocols that can satisfy real-time requirements—key is a deterministic protocol for tick and time synchronization. Furthermore, ProNet 4.0 offers novel functionalities, in particular, mode medium access and deterministic multi-hop arbitration and value transfer. Automatic topology detection and a QoS multicast routing protocol for partially mobile networks support operation of multi-hop networks. Finally, a production middleware provides an abstract interface for production and control applications, incorporating a distributed service registry with replicated service entries.

Literature

Chair for Networked Systems

1. Gotzhein R (2014) ProNet 4.0—A wireless real-time communication system for Industry 4.0. White Paper, Networked Systems Group, Department of Computer Science, University of Kaiserslautern. http://vs.informatik.uni-kl.de/publications/2014/Go14/whitePaperEN-ProNet 4.0.pdf. Last accessed 27 Aug 2019
2. Haupt A, Gotzhein R et al (2014) Control and communication co-design. In: Lunze J (ed) Control theory of digitally networked dynamic systems. Springer, pp 328–348

3. Christmann D, Gotzhein R, Siegmund S, Wirth F (2014) Realization of try-once-discard in wireless multi-hop networks. IEEE Trans Ind Inf 10(1):17–26
4. Duc LH, Christmann D, Gotzhein R, Siegmund S, Wirth F (2015) The stability of try-once-discard for stochastic communication channels—theory and validation. In: 54th conference on decision and control, Osaka, Japan, 15–18 Dec 2015
5. Berg M, Gotzhein R et al (2015) Vertical integration and adaptive services in networked production environments. In: Proceedings of ERP Future 2015, Munich, Germany, 16 Nov 2015. Springer LNBIP 245, pp 1–16

Further References

6. Atmel Corporation: AVR Low Power Transceiver AT86RF230, 5131E-MCU Wireless-02/09, http://ww1.microchip.com/downloads/en/DeviceDoc/doc5131.pdf. Last accessed 27 Aug 2019
7. Etschberger K (2001) Controller area network—basics, protocols, chips, and applications. IXXAT Press
8. Chipcon AS (2019) CC2420 ZigBee-ready RF transceiver. http://www-inst.eecs.berkeley.edu/~cs150/Documents/CC2420.pdf. Last accessed 27 Aug 2019
9. Heemels W, Teel A, van de Wouw N, Nesic D (2010) Networked control systems with communication constraints—tradeoffs between transmission intervals, delays and performance. IEEE Trans Autom Control 55(8):1781–1796
10. MEMSIC Inc. Imote 2 datasheet. https://vs.cs.uni-kl.de/downloads/Imote2NET_ED_Datasheet.pdf. Last accessed 27 Aug 2019
11. Isle APB, Kozin F (1972) On the almost-sure sample stability of systems with randomly time-varying delays. Automatica 8:755–763
12. Jentzen A, Leber F, Schneisgen D, Berger A, Siegmund S (2010) An improved maximum allowable transfer interval for L^p-stability of networked control systems. IEEE Trans Autom Control 55(1):179–184
13. Lunze J (ed) (2014) Control theory of digitally networked dynamic systems. Springer
14. Mahnke W, Leitner SH, Damm M (2009) OPC unified architecture. Springer
15. Nesic D, Liberzon D (2009) A unified framework for design and analysis of networked and quantized control systems. IEEE Trans Autom Control 54(4):732–747
16. Pereira N, Andersson B, Tovar E (2007) WiDom: a dominance protocol for wireless medium access. IEEE Trans. Ind Inf 3(2):120–130
17. Pereira N, Andersson B, Tovar E, Rowe A (2007) Static-priority scheduling over wireless networks with multiple broadcast domains. In: 28th IEEE international real-time systems symposium (RTSS 2007), Tucson, Arizona, USA, 3–6 Dec 2007
18. Tabbara M, Nesic D, Teel AR (2007) Stability of wireless and wireline networked control systems. IEEE Trans Autom Control 52(9):1615–1630
19. Walsh G, Ye H, Bushnell L (2002) Stability analysis of networked control systems. IEEE Trans Control Syst Technol 10(3):4438–446
20. Engel M, Kramer C, Braun T, Christmann D, Gotzhein R (2019) BiPS—a real-time-capable protocol framework for wireless networked control systems and its application. In: Obaidat M, Cabello E (eds) E-Business and Telecommunications (ICETE 2017), Communications in Computer and Information Science, vol 990. Springer, pp 313–336

Chapter 13
Conclusions and Future Research

This chapter briefly summarizes the results reported in this book, presents conclusions, and reports on future research.

13.1 Results and Conclusions

In this book, we have presented core functionalities and protocols for wireless real-time multi-hop networking with time-division multiple access (TDMA), their integration into a flexible, versatile, fully operational, self-contained communication system called *Production Network (ProNet) 4.0* [1], and their operation in a number of case studies. Altogether, ProNet 4.0 provides rich functionalities that cover a large variety of communication requirements.

Real-time behavior is supported by deterministic protocols, providing, for instance, tick and time synchronization, mode medium access, and multi-hop medium arbitration. With protocol modules for automatic topology detection and quality of service (QoS) multicast routing for partially mobile networks, ProNet 4.0 supports the operation of wireless multi-hop networks. A customized production middleware provides an abstract interface for real-time applications in production and control systems, incorporating a distributed service registry with replication of service entries.

- In Chap. 2, we have presented *Black Burst Synchronization (BBS)*, a deterministic protocol with low upper bounds for tick offset, time offset, and convergence delay. Further features of BBS are low complexity of computation, storage, time and structure, and robustness against topology changes. During the development of BBS, we have gained three key insights. First, nodes can synchronize by agreeing on network-wide reference ticks, without synchronizing their local clocks. This reduces the amount of synchronization data to be exchanged among nodes. Second, time synchronization can be achieved on top

R. Gotzhein, *Real-time Communication Protocols for Multi-hop Ad-hoc Networks*, Computer Communications and Networks,
https://doi.org/10.1007/978-3-030-33319-5_13

of tick synchronization, without much additional exchange of synchronization data. Third, by encoding synchronization messages with black bursts, potential collisions are rendered non-destructive. This reduces convergence delay and enables deterministic operation.

- In Chap. 3, we have presented a novel approach for *global time slotting*. Here, fixed, strictly periodic physical time slotting and variable, weakly periodic virtual time slotting provide a high degree of flexibility. This enables customized sync, exclusive, shared, mode, arbitrated, and idle regions and thereby reduces waiting delays and energy consumption.
- In Chap. 4, we have presented *Automatic Topology Discovery Protocol (ATDP)*, which detects communication, interference, and sensing topologies in wireless multi-hop TDMA networks. Solid information about these topologies is vital for other network functionalities, in particular, for routing and clustering. Furthermore, knowledge about the interference topology enables space-division multiple access (SDMA) to increase network throughput.
- In Chap. 5, we have presented *medium access schemes*. Exclusive access and shared medium access are well known and supported by ProNet 4.0. Furthermore, we have presented a novel scheme called *mode medium access* with mode-triggered scheduling and fast mode-signaling. Mode medium access restricts shared access to defined subsets of nodes. By assigning and signaling unique mode priorities, contention in single-hop range is rendered predictable, which avoids collisions.
- In Chap. 6, we have presented *Arbitrating and Cooperative Transfer Protocol (ACTP)*. ACTP is a binary countdown protocol providing deterministic arbitration and value transfer within a configurable hop radius in wireless networks. This supports applications such as leader election and distributed consensus. ACTP has low time and space complexity and is robust against node movements and node failures.
- In Chap. 7, we have presented a flexible and energy-efficient *duty cycling* approach for wireless networks supporting weakly periodic idle periods. Flexibility of placing idle regions is due to very flexible global time slotting (see Chap. 3), which enables customized idle periods. In addition, given the type of virtual time region, further potential for duty cycling can be exploited, with exclusive regions performing best.
- In Chap. 8, we have presented *QoS multicast routing (QMR)*, a specialized routing protocol for partially mobile wireless TDMA networks. QMR incrementally constructs and prunes routing trees, and assigns and releases exclusively reserved slots at runtime. Taking routing and slot scheduling decisions in a centralized way prevents mutual blocking of resources due to concurrent route searches, substantially reduces management traffic, and avoids the problem of inconsistent reservation status due to lost or delayed management messages.
- In Chap. 9, we have presented *heterogeneous network clustering (HNC)* for wireless TDMA networks consisting of stationary nodes with different capabilities and resources. HNC clusters a heterogeneous network into a 3-hop

connected 1-hop dominating set, if such a clustering is feasible. Clustering objectives are a small number of clusters and efficient connectivity.

- In Chap. 10, we have presented *production middleware (ProMid)*, a customized middleware for production and control applications. ProMid adopts a client–server architecture, with service providers publishing services, and service users subscribing to services. ProMid operates a distributed service registry, where service entries can be replicated by setting a replication radius.
- In Chap. 11, we have presented *Black burst integrated Protocol Stack (BiPS)*, a framework comprising operating system and communication functionalities. In particular, BiPS comprises a real-time scheduler controlling the timely execution of processes and protocols. Furthermore, BiPS implements core protocol functionalities, in particular, tick and time synchronization, global time slotting, topology detection, and several medium access schemes.
- In Chap. 12, we have presented three case studies featuring different functionalities of ProNet 4.0. Taken together, the case studies provide evidence that ProNet 4.0 combines rich functionalities that cover a large variety of communication requirements, and, most importantly, real-time requirements.

ProNet 4.0 has been fully implemented as prototype and has been successfully deployed and operated in an industrial setting as part of the demonstrator vertical integration of production processes (VI-P, see Chap. 12). Target platform are Imote2 nodes [9] equipped with Chipcon CC2420 transceivers [8]. Building on the real-time-capable implementation framework BiPS, ProNet 4.0 classifies as bare-metal solution.

The successful implementation and operation of ProNet 4.0 on the chosen target platform provide evidence that this is feasible on customary hardware. We observe, however, that efficiency of operation could be improved on a hardware platform that is specifically devised to support the functionality of ProNet 4.0, in particular black bursts and tick synchronization. For instance, the CC2420 transceiver produces substantial delays when switching between rx and tx modes, and when performing clear channel assessment (CCA). Here, Atmel's AT86RF230 transceiver [7] is much faster. Furthermore, the low data rate of 250 Kbps is a bottleneck. Finally, a better (and more expensive) processor quartz could reduce maximum clock skew.

In Chap. 2, we have analyzed BBS and its performance on the CC2420 and AT86RF230 transceivers. With a specialized transceiver that offers, for instance, d_{maxCCA} = 4 μs, $r_{maxClockSkew}$ = 5 ppm, a data rate of 1 Mbps (d_{sym} = 4 μs), and switching delays $d_{rxtx} = d_{txrx}$ = 4 μs, synchronization accuracy and convergence delay could be improved as shown in Table 13.1. Compared to the CC2420, the improvements are dramatic and still substantial w.r.t. to the AT86RF230.

In Sect. 12.2, we have analyzed the minimum arbitration slot length to be $d_{arbSlotMin}$ = 9.90 ms, based on the values of the AT86RF230 transceiver. This satisfies maximal allowable transfer interval (MATI) of 10.65 ms, but leaves no bandwidth for other usage. For the CC2420, we even get $d_{arbSlotMin}$ = 29.37 ms, which violates MATI. With the specialized transceiver, we obtain

Table 13.1 Analysis of BBS: CC2420, AT86FR230, and specialized transceiver

	CC2420		AT86RF230		Specialized transceiver	
$n_{maxHops}$	1	4	1	4	1	4
d_{resInt}	1 s		1 s		1 s	
d_{sym}	16 μs		16 μs		4 μs	
d_{maxCCA}	128 μs		16 μs		4 μs	
d_{rxtx}, d_{txrx}	192 μs, 192 μs		17 μs, 33 μs		4 μs, 4 μs	
$r_{maxClockSkew}$	40 ppm		40 ppm		5 ppm	
$d_{maxTickOffset\_m}$	208 μs	592 μs	96 μs	144 μs	14 μs	26 μs
$d_{maxTickOffset\_d}$	400 μs	1360 μs	113 μs	212 μs	18 μs	42 μs
d_{conv_m}	1.60 ms	8.32 ms	0.82 ms	3.86 ms	0.21 ms	1.00 ms
d_{conv_d}	1.56 ms	13.94 ms	0.66 ms	3.42 ms	0.17 ms	0.89 ms

$d_{arbSlotMin}$ = 2.47 ms, resulting in a duty cycle of only 24.7%. Altogether, this shows that a specialized transceiver is crucial for the efficient implementation of ProNet 4.0.

13.2 Future Research

At the end of this book, the question of future research in the area of wireless real-time communication systems remains to be addressed. How is the work reported here to be continued? Following the style of this monograph, we will provide a concise and focused outlook of our research instead of a broad treatment of developments and trends in the entire research community and industry.

The development of the ProNet 4.0 real-time protocol stack has greatly benefitted from the decisions to use the Imote2 hardware platform with the CC2420 transceiver, to develop a bare-metal solution and to replace the medium access control (MAC) layer of the IEEE 802.15.4 standard by a customized MAC layer supporting black bursts and a variety of medium access schemes. This of course has created substantial hardware dependencies, which makes it cumbersome to change platform components. In fact, production of Imote2 motes has been discontinued several years ago.

Given this situation, the question arose whether it is feasible to build wireless real-time communication systems on off-the-shelf widespread communication technologies such as IEEE 802.11 (Wi-Fi), using general-purpose boards running customary operating systems such as Linux. Such an approach would greatly simplify changing hardware platforms and transceivers, as there would be less hardware dependencies. Furthermore, it would reduce development effort, as a common well-maintained operating system and an existing MAC layer would be used. On the other hand, real-time capabilities of bare-metal solutions would no longer be within reach. In particular, it would be very difficult to achieve accurate tick and time synchronization, as concepts such as black bursts cannot be implemented on top of the IEEE 802.11 MAC layer. Furthermore, Linux does not support a protocol-centric real-time approach as in BiPS.

Starting point for answering the question has been our decision to explore ways of devising real-time capable wireless communication systems on Linux-driven low-priced boards, equipped with widely available Wi-Fi cards employing, for instance, the Atheros chipset. Different from ProNet 4.0 and IEEE 802.15.4, additional communication functionality is to be built on top of the MAC layer, not on top of the PHY (PHYsical) layer, which fosters portability of implementations. Furthermore, data rates of IEEE 802.11 are substantially higher than data rates of IEEE 802.15.4.

In the following, we report on our activities to build up communication functionality that contributes to the real-time behavior of IEEE 802.11 networks. As network-wide tick synchronization that is sufficiently accurate to establish short time slots for single message transfer is out of reach, exclusive slot reservation is not feasible. Instead, we strive for statistical approaches for link quality assessment, bandwidth reservation, traffic monitoring, fair bandwidth usage, routing, and channel quality assessment.

Automatic Topology Detection

Solid information about communication topologies is essential for other network functionalities, in particular, for routing and clustering. In Chap. 4, we have presented Automatic Topology Detection Protocol (ATDP), which detects communication, interference, and sensing topologies of wireless TDMA networks. For this, it observes the reliability of links until deterministic stability criteria are met.

As accurate time slotting for single message exchange is not available in IEEE 802.11 networks, automatic topology detection is restricted to communication topology. Furthermore, medium access is contention-based, therefore, deterministic stability criteria for the reliability of links are out of reach. To support reliability, we have therefore decided to use a network model $G = (V, r)$ with nodes $v \in V$ and a *statistical reliability function* $r\colon V \times V \rightarrow \mathbb{R}_{0,1}$ for single-hop communication [3]. Based on r, the set of links is then defined as $E =_{\mathrm{df}} \{(v, v') \in V \times V \mid r(v, v') > 0\}$. This network model is suitable for statistical reliability routing, as explained below. It is the task of automatic topology detection to establish and maintain G, via observation of statistical link reliabilities during network operation.

Time Token Bucket Traffic Monitoring and Traffic Shaping

In IEEE 802.11 networks, nodes follow a greedy strategy to satisfy communication requests of applications as quickly as possible. With increasing network density, this leads to overload situations resulting in an increase of collisions, frame loss, retransmissions, and frame delays. This problem can, for instance, be tackled if each node is prepared to shape its traffic locally such that network overload is avoided. Traffic shaping in turn requires traffic monitoring to assess the current network load, in order to make local decisions about medium usage.

For traffic shaping in IEEE 802.11 networks, we have devised and implemented a variant of the token bucket algorithm called *time token bucket* [2]. With this algorithm, it is possible to control medium access such that local bandwidth usage conforms to the reserved relative bandwidth, frames exceeding it are delayed or dropped, and traffic is distributed uniformly to avoid high contention situations. For traffic monitoring, we have conceived the novel metric *unusable wasted bandwidth ratio (UWBR)*, which is computed locally from the number of unusable wasted tokens. Other than the commonly used metric channel busy time (CBT), UWBR is hardware-independent.

Fair Bandwidth Scaling

In IEEE 802.11 networks, applications with quality of service (QoS) requirements such as Voice over IP (VoIP) apply scaling based on the perceived data rate. However, this happens in a greedy way, where each node tries to maximize its own throughput. This problem can be solved, for instance, if nodes cooperate to get fair shares of the available bandwidth.

For fair bandwidth scaling in IEEE 802.11 networks, we have devised and implemented a novel algorithm [6] based on reserved minimum and preferred relative bandwidth $bw_{min,v}$ and $bw_{pref,v}$ of nodes $v \in V$ using our time token bucket algorithm for traffic monitoring and traffic shaping. The idea is to divide the difference $bw_{pref,v} - bw_{min,v}$ into an equal number of levels and to balance the levels of all nodes such that each node v gets (almost) the same relative share of bandwidth on top of $bw_{min,v}$. Using the local metric UWBR and information about the levels of its neighbors, each node decides locally when to scale up or down. First experiments in a single-hop test bed of nodes equipped with off-the-shelf Wi-Fi adapters have shown that after a short stabilization phase, the QoS fairness index is close to 1, i.e., fair bandwidth distribution is reached.

r_{min}-Routing

In multi-hop IEEE 802.11 networks, routes are required to provide end-to-end connectivity. To assess the quality of routes, route metrics are applied. The most prevalent route metric is probably shortest path (SP), which minimizes the number of hops between source and destination. However, this is not a good idea in wireless networks, as it gives preference to links that cover more distance and are therefore less reliable, due to signal attenuation and fading.

For routing in IEEE 802.11 networks, we have conceived a novel algorithm called *r_{min}-routing* [3]. The idea of r_{min}-routing is to discover routes matching a specified statistical reliability target r_{min}. Route discovery is based on a network model $G = (V, r)$, where V is the set of nodes and r is a statistical reliability function for single-hop communication. G is determined and maintained by automatic topology detection, as already explained. To increase the number of route candidates, r_{min}-routing uses well-directed retransmissions. To select among a set of route candidates, it applies several quality criteria. Key is a new routing metric called *smallest maximum number of transmissions (SMTX)*, which determines the minimum number of transmissions to match a specified statistical reliability r_{min} on a path p and distributes them among the links of p such that r_{min} is satisfied.

Channel Hopping Schedules

IEEE 802.11 networks can operate on different channels. For instance, when using the 2.4 GHz range, up to 14 channels are available. When setting up a Wi-Fi network, one of these channels is selected, possibly based on a channel quality assessment to avoid or reduce overlap with other networks. For improved performance, it is also possible to dynamically measure channel qualities, select the best channels, and perform channel hopping based on a common schedule.

In [4], we have presented an approach for the dynamic computation and adjustment of channel hopping schedules for, but not limited to, cognitive radio networks. Based on channel quality, metrics for the number of channel utilizations in a hopping schedule and for the distribution of channel usage in the schedule are defined. Using these metrics, optimal channel hopping schedules are computed and dynamically adjusted. The approach gives preference to channels of higher quality, by using them more frequently. Furthermore, dynamic schedule adjustments due to changing channel quality usually lead to small changes, keeping network operation more stable.

Operation of Wireless Networks with Channel Hopping

When using channel hopping, network nodes have to agree on a common hopping schedule. In [5], we have presented a three-dimensional stabilization protocol for, but not limited to, multi-hop cognitive radio networks with channel hopping. For stabilization in the dimensions time and space, nodes agree on a common time structure of super slots and macroslots. For stabilization in the dimension channel, nodes determine and share global channel hopping schedules.

These are examples of communication functionalities that contribute to a more reliable and more predictable operation of multi-hop IEEE 802.11 networks. In our future research work, we will integrate, tune, and augment these functionalities.

Literature

Chair for Networked Systems

1. Gotzhein R (2014) ProNet 4.0—a wireless real-time communication system for Industry 4.0. White Paper, Networked Systems Group, Department of Computer Science, University of Kaiserslautern. http://vs.informatik.uni-kl.de/publications/2014/Go14/whitePaperEN-ProNet4.0.pdf
2. Mathews K, Kramer C, Gotzhein R (2017) Token bucket based traffic shaping and monitoring for WLAN-based control systems. In: 28th IEEE annual international symposium on personal, indoor, and mobile radio communications (PIMRC 2017), Montreal, Canada, 8–13 Oct 2017

3. Kohlstruck C, Mathews K, Gotzhein R (2019) $r_{\min}$-routing—discovery and operation of routes in wireless ad-hoc networks with specified statistical minimum reliabilities. In: 44th IEEE conference on local computer networks (LCN 2019), Osnabrück, Germany, 14–17 Oct 2019
4. Engel M, Gotzhein R (2018) Dynamic computation and adjustment of channel hopping sequences for cognitive radio networks based on quality metrics. In: International conference on embedded wireless systems and networks (EWSN 2018), Madrid, Spain, 14–16 Feb 2018
5. Aragao P, Engel M, Gotzhein R (2018) A three-dimensional stabilization protocol for time-slotted multi-hop cognitive radio networks with channel hopping. In: The 32nd IEEE international conference on advanced information networking and applications (AINA 2018), Cracow, Poland, 16–18 May 2018, pp 32–39
6. Kramer C, Mathews K, Gotzhein R (2019) Cooperative fair bandwidth scaling in contention-based wireless networks using time token bucket. In: 38th IEEE international performance computing and communications conference (IPCCC 2019), London, UK, 29–31 Oct 2019

Further References

7. Atmel Corporation (2019) AVR Low Power Transceiver AT86RF230, 5131E-MCU Wireless-02/09, http://ww1.microchip.com/downloads/en/DeviceDoc/doc5131.pdf. Last accessed 27 Aug 2019
8. MEMSIC Inc. (2019) Imote 2 datasheet. https://vs.cs.uni-kl.de/downloads/Imote2NET_ED_Datasheet.pdf. Last accessed 27 Aug 2019
9. Chipcon AS (2019) CC2420 ZigBee-ready RF Transceiver. http://www-inst.eecs.berkeley.edu/~cs150/Documents/CC2420.pdf. Last accessed 27 Aug 2019

Correction to: Real-time Communication Protocols for Multi-hop Ad-hoc Networks

Correction to: R. Gotzhein, *Real-time Communication Protocols for Multi-hop Ad-hoc Networks*, Computer Communications and Networks, https://doi.org/10.1007/978-3-030-33319-5

In the original version of the book, the following belated corrections have been incorporated:

In chapter 2, the term "$d_{\text{round m}}$" has been replaced with "$d_{\text{round\_}m}$" and "$d_{\text{maxTickOffset m}}$" has been replaced with "$d_{\text{maxTickOffset\_}m}$."

In chapter 4, Figures 4.4 and 4.5 have been replaced.

In chapter 8, the equation "$\text{FSF}(p) =_{\text{df}} \{n_{\text{Free}}(v, v') > 0 \wedge$" has been replaced with "$\text{FSF}(p) =_{\text{df}} \{(v, v') \in \text{p} \mid (v, v') \in \text{NoDec}(p) \wedge n_{\text{Free}}(v, v') > 0 \wedge$."

The erratum book has been updated with the changes.

The updated versions of these chapters can be found at
https://doi.org/10.1007/978-3-030-33319-5_2
https://doi.org/10.1007/978-3-030-33319-5_4
https://doi.org/10.1007/978-3-030-33319-5_8

R. Gotzhein, *Real-time Communication Protocols for Multi-hop Ad-hoc Networks*, Computer Communications and Networks,
https://doi.org/10.1007/978-3-030-33319-5_14

Index

R. Gotzhein, *Real-time Communication Protocols for Multi-hop Ad-hoc Networks*, Computer Communications and Networks,
https://doi.org/10.1007/978-3-030-33319-5

Printed in the United States
By Bookmasters

Printed in the United States
By Bookmasters